Indian Medicinal Plants and Therapeutic Uses

Srinath Rao and Akula Ramakrishna

CRC Press
Taylor & Francis Group
Boca Raton London New York

CRC Press is an imprint of the
Taylor & Francis Group, an **informa** business

Designed cover image: Professor Chandralekha CT

First edition published 2026
by CRC Press
2385 NW Executive Center Drive, Suite 320, Boca Raton FL 33431

and by CRC Press
4 Park Square, Milton Park, Abingdon, Oxon, OX14 4RN

CRC Press is an imprint of Taylor & Francis Group, LLC

ISBN: 978-1-032-20424-6 (hbk)
ISBN: 978-1-032-20425-3 (pbk)
ISBN: 978-1-003-26350-0 (ebk)

DOI 10.1201/9781003263500

Typeset in Times
by Apex CoVantage, LLC

Contents

A

B

C

F

G

H

U

V

W

X

About the Authors

Prof. Srinath Rao is retired professor from the Department of Botany, Gulbarga University, Kalburgi, Karnataka, India. He has 34 years of teaching postgraduate students of botany and biotechnology in the field of in vitro propagation of medicinal plants and in vitro enhancement of secondary metabolites. He has successfully guided 25 MPhil and 27 PhD students. He has made significant contributions on in vitro propagation and enhancement of secondary metabolites in some important medicinal plants. He has presented papers in several international conferences held in the United States, the United Kingdom, Germany, France, and Sri Lanka. Apart from this, he has given 50 invited lectures at national conferences and international conferences. He published more than 100 research papers in national and international journals. He has completed 7 major research projects funded by the University Grants Commission, the Council of Scientific & Industrial Research (CSIR), and the Department of Science and Technology.

Dr. Akula Ramakrishna is currently a scientist at Bayer Science and Innovation Private Limited, Bangalore, India. Dr. Ramakrishna holds a master's degree from Sri Krishna Devaraya University, Anantapur, Andhra Pradesh, India. He obtained his PhD in biochemistry from University of Mysore, Council of Scientific & Industrial Research CSIR-Central Food Technological Research Institute, Mysuru. He is a senior research Fellow of CSIR, New Delhi. He is involved in various research fields, such as plant biotechnology, biochemistry, plant secondary metabolites, food science and technology, and vegetable quality analytics. He has published several research papers in the leading international journals, published books, reviews, and book chapters. He is also serving as editorial board member and reviewer for reputed international journals. He has presented more than 20 research papers in symposia in India and abroad, as well as few invited lectures in India. He attended the Fifth International Symposium on Plant Neurobiology held in 2009 in Florence, Italy. He also attended the Technical Community of Monsanto held in 2016 in St. Louis, Missouri. He is a member of the Indian science congress association and society for biotechnologists (India). He is a fellow of the Society for Applied Biotechnology, India (2012), and recipient of Global Vegetable Research Excellence Award (2017), Global Technology Recognition, Rapid Recognition Award, and Special Recognition from the Monsanto company.

Preface

India is a major center of origin and diversity of medicinal plants. It is one of the twelve mega diversity nations in the world. This volume offers profiles of approximately 200 medicinal plant species. The book also lists the detailed description and popular nomenclature (Bengali, English, Gujarati, Hindi, Malayalam, Marathi, Sanskrit, Tamil, and Telugu), chemical constituents, properties and actions, pharmacological principles, parts used, and therapatic uses. The World Health Organization (WHO) has focused much on rapid activation of research and development activities in herbal and modern systems of medicine. The WHO estimated that 80% of the people in developing countries of the world rely on traditional medicine for their primary health care needs and that about 85% of traditional medicine involves the use of plants. The medicinal plants are known to be rich in antioxidants, vitamins, proteins, carbohydrates, dietary fibers, amino acids and minerals, and antiviral and antibacterial properties. The medicinal plant extracts contain several phytoconstituents such as flavonoids, alkaloids, polyphenols, and terpenoids. Interestingly, they have been demonstrated to exert their pharmacological activities by interacting simultaneously with numerous biological targets, thereby increasing their therapeutic potential. To boost the immune system, without causing any side effects, AYUSH scientists have identified the potential phytomedicines and reduced the risk of various diseases. The National Medicinal Plants Board and other government agencies are actively involved in promoting the conservation, cultivation, and trade of medicinal plants.

In this book, we have also included some of the important medicinal plants that can boost the immune system. This volume provides special interest to practitioners of alternative medicine, students of Ayurveda, researchers and those in industries associated with medical botany, pharmacologists, sociologists, and medical herbalists.

Prof. Srinath Rao

Dr. Akula Ramakrishna

Acknowledgments

We are very much thankful to Renu Upadhyay, Assistant Commissioning Editor; Jyotsna Jangra and Tisha Sharma, Editorial Assistants and Aruna Rajendran, Project Manager, Apex CoVantage; Danielle Zarfati, Project Manager; and the production team from the Taylor and Francis Group for their help in bringing out the volume in a beautiful and appealing manner, and all the other staff members of CRC, Taylor and Francis who were directly or indirectly associated with the project for their unconditional support and help, valuable suggestions to make this publication in time. The authors are thankful to Dr Chandralekha CT, Dr R.S. Modi, Dr Kotresh for providing photographs of medicinal plants for publication in the book. The authors are thankful to our families who have encouraged us to take up this task, permitting us to take the time off from the family. We are also thankful to our friends and academic colleagues for their support in successful accomplishment of the book.

Prof. Srinath Rao

Dr. Akula Ramakrishna

A

Abroma augusta (L.) L.f.

Synonyms: *Abroma alatum* Blanco; *Abroma sinuosa* Nichols.; *Abroma wheleri* Retz.; *Ambroma fastuosa* Jacq.; *Theobroma augusta* L.

Family: Sterculiaceae

Common names: Bengali: Ulatkambal; **English:** Perennial Indian hemp, Devil's cotton; **Hindi:** Ulatkambal; **Kannada**: Devva hatthi; **Marathi:** Ola katambol; **Sanskrit:** Pishaacha Kaarpaasa, Pivari; **Tamil:** Sivapputtuti; **Telugu:** Konda gogu

Distribution: Throughout the hotter and moister parts of India, from Punjab and Uttar Pradesh and eastward to Arunachal Pradesh, Assam, Meghalaya, and Tripura, ascending to 1200 m, southward in Peninsular India.

Description: Devil's cotton is a large spreading shrub, or a small tree, with fibrous bark and irritant hairs. It grows up to 2.5 m tall with hairy branches. Leaves are ovate-oblong long-pointed, with a heart-shaped base, 10–21 cm long and 5.5–13 cm wide. The leaf blade is 3–7 nerved, with unevenly toothed margins. The flowers are maroon, up to 5 cm across, looking down, in few-flowered clusters in leaf axils. The sepals are lance-shaped, fused at base. The petals are 5, which soon fall off, concave below, prolonged above into a spoon-shaped blade. The capsule is papery, 5-winged, and cutoff at the tip.

Flowering and fruiting: June–September

Chemical constituents: The root contains abromine (betaine), friedelin, abromasterol, abromasterol A, choline, beta-sitosterol, stigmasterol, and octacosanol. The leaves contain taraxerol, itsacetate, and lupeol (Khare, 2007).

Used in: Ayurveda, homeopathy, traditional.

Parts used: Leaves, root bark

Uses: Rootbark—emmenagogue (used for dysmenorrhea, amenorrhea), abortifacient, and galactotrophic. Dosage leaf juice: 10–20 ml. Root bark powder: 3–6 g (Khare, 2007). Leaves are used as a remedy for diabetes, inflammation, rheumatic pain of joints, uterine disorders, and headache. The leaf is traditionally used to treat diabetes in India and Southern Asia (Islam et al., 2013; Khanra et al., 2015). In India, the mother tincture of *A. augusta* is widely used in homeopathic medicine to treat uterine disorders and diabetes mellitus. Its root bark is highly useful as uterine tonic and is used to treat amenorrhea and dysmenorrhea, with abortifacient and antifertility activities, whereas leaves are used for diabetes, rheumatic pains, gonorrhea, headache, and sinusitis (Khanra et al., 2015; Miah et al., 2020).

Methods of propagation: Seeds

DOI: 10.1201/9781003263500-2

Acacia leucophloea Willd.

Synonyms: *Acacia alba* Willd.; *Acacia arcuata* Decne.; *Mimosa leucophloea* Roxb.

Family: Mimosaceae

Common names: Bengali: Safed babul; **English:** White babul; **Hindi:** Nimbar, Safed babul, Safed kikar; **Kannada:** Bili Jali; **Malayalam:** Chenkaringali; **Marathi:** Himvar; **Sanskrit:** Arimedha, Swetha Barhura, Arimeda; **Tamil:** Valval, Velvayalam

Distribution: Dry regions of the country, especially in Punjab, Rajasthan, and Madhya Pradesh.

Description: Trees growing about 20 m high; bark yellow to yellowish-brown, rough, exfoliating in irregular scales; branchlets yellowish, velvety; Leaves bipinnate, alternate, stipulate; stipular spines in pairs, up to 1 cm, gray, straight, and puberulous; rachis 9–14 cm wide, pulvinate, grooved above, and pubescent; pinnae 5–15 pairs, opposite, even pinnate, 3.5–6 cm, with a gland in between terminal 3–4 pinnae and between basal pairs on upper side, leaflets 30–54, opposite, sessile; lamina 5–10 × 2–3 mm, linear, base oblique, truncate, apex subacute, margin ciliate, glabrous above, pubescent below, chartaceous; nerves obscure. Flowers yellow or yellowish-white, 1 mm across, sessile, in heads arranged in terminal panicles; peduncle zigzag; bracts ovate, minute; bracteoles approximately 5 in an involucel, in the middle of the peduncle; calyx 1–1.3 × 1 mm, campanulate, villous; teeth triangular; corolla 2–2.5 mm long, puberulous without, lobes ovate-oblong; stamens many, 2.5–5 mm long with glandular appendages; ovary stipitate, 0.3–0.5 mm long; style 1–1.5 mm long. Fruit a pod 6–15 × 0.8–1 cm, flat, strap-shaped, straight, or sometimes curved, somewhat bullate along seeds, thin walled, round, or often mucronate at apex, minutely brown tomentose, sessile to subsessile, tardily dehiscent; seeds 5–12, ovate, 5–6 × 3–4 mm.

Flowering and fruiting: June–January

Chemical constituents: The rootbark contains leucophleol, leucophleoxol, and leucoxol (Khare, 2007).

Used in: Traditional

Parts used: Leaves, stem, and bark

Uses: Bark: bitter, demulcent, and cooling; used in biliousness and bronchitis. Seeds: hemagglutinating activity has been reported. Leaves: anti-syphilitic and antibacterial. Gum: demulcent (Khare, 2007). The barks of *Acacia leucophloea* (Fabaceae) are used in traditional medicine as an astringent, a bitter, a thermogenic, a styptic, a preventive of infections, an anthelmintic, a vulnerary, a demulcent, an expectorant, an antipyretic, an antidote for snakebites and in the treatment of bronchitis, cough, vomiting, wounds, ulcers, diarrhea, dysentery, internal and external hemorrhages, dental caries, stomatitis, and intermittent fevers and skin diseases (Imran et al., 2011).

Methods of propagation: Seeds

DOI: 10.1201/9781003263500-3

Photo: Dr. Srinath Rao

Acacia mellifera (M. Vahal) Benth

Synnonyms: *Acacia detinens* Burch; *Mimosa mellifera* M. Vahl; *Senegalia mellifera* (M. Vahl) Seigler & Ebinger

Family: Leguminosae (Fabaceae)

Common names: English: Blackthorn

Distribution: Native of Sudan, occurs widely as a shrub on rangelands throughout the savanna in western, eastern and southern Africa, as well as on the Arabian Peninsula. In India found in Kerala State

Description: Shrubs or trees, 2–5 m high; on stem covered with black bark that becomes ash-gray to light brown on the branches, bearing small, short, sharply hooked spines in pairs. branchlets flattened, 1.5–5 mm thick, glabrous; stipules deciduous. Leaves are characterized by 2 pairs of pinnulae, each with a single pair of leaflets. Leaflets are elliptic 0.6–2 cm long and 0.6–1.2 cm wide, glabrous, and highly colored beneath, circular, sessile, about 0.5-mm diameter unequal at base, rounded to acute-mucronate at apex, glabrous, green when dry; main vein sub-central with 2 prominent accessory nerves ascending about 2/3 way up the lamina; lateral nerves prominent rachis 2–5 cm long, slender, sparsely puberulous to glabrous; glands more or less midway on petioles. Inflorescence: axillary spikes, up to 8 cm long. Flowers: Flowers sweetly scented, especially at night, in elongated spikes, cream to white in spiciform racemes, up to 3.5 cm long; pedicels 0.5–1.5 mm long; calyx up to 1 mm long; corolla 2.5–3.5 mm long, lobes oblong, 1.5–2.5 mm long, glabrous; stamens ca 5 mm long; ovary shortly stipitate, glabrous: Pods: papery with 2–3 seeds are reticulate, flat, elongated, 2.5–5.5 cm long, 6 cm wide, hemmed, sometimes more or less narrowed between the seeds. obliquely oblong, 4.5–6.5 × 2–2.5 cm, flat, thin, unlobed, rounded or obtuse-mucronate at apex, greenish brown when dry, glabrous, prominently and transversely reticulate-veined; seeds 4–5.

Flowering: May–July **Fruiting:** August

Chemical constituents: Three triterpenoids (20*S*)-oxolupane-30-al, (20*R*)-oxolupane-30-al, and betulinic acid were reported from the bark.

Used in: Ayurveda

Parts used: Bark

Use: The liquid of boiled bark is used to treat stomach problems, sterility, pneumonia, malaria, and syphilis. A decoction of the roots is a medicine for stomach pain (https://apps.worldagroforestry.org/treedb/AFTPDFS/Acacia_mellifera.PDF). The bark is astringent. It is used in the treatment of diarrhea and dysentery and can also be helpful in cases of internal bleeding. Applied externally, often as a wash, it is used to treat wounds and other skin problems, hemorrhoids, perspiring feet, some eye problems, and as a mouthwash, among others (https://tropical.theferns.info/viewtropical.php?id=Senegalia+mellifera&redir=Acacia+mellifera).

Methods of propagation: Direct sowing of seeds is the common method of propagation. Seed are soaked in concentrated sulfuric acid for 5–15 min or in hot water, left overnight and planted the next morning. Seeds usually germinate from the 5th day onward. Due to its high rate of seeding, the long viability of the seeds, and the ability to overcome arid climates, profuse natural regeneration is generally common.

Reference: http://www.worldagroforestry.org/treedb2/speciesprofile.php?Spid=72

DOI: 10.1201/9781003263500-4

Acacia senegal Willd.

Synonyms: Acacia verek Guillem and Perr; *Acacia oxyosprion* Chiov

Family: Fabacea (Mimosaceae)

Common names: English: Gum arabic tree; **Gujrati:** Kagar, goraduja baval; **Hindi:** Kumttha, khor, goraduja baval, kumat, kumta, humath; **Kannada:** Mullimara; **Sanskrit:** Swetha babbuula, Sneta khadira

Distribution: Native to Sudan. Throughout India, except Jammu and Kashmir, Himachal Pradesh, Sikkim, Arunachal Pradesh, Assam, Nagaland, Meghalaya, Manipur, Tripura, Mizoram cultivated in Western parts of India.

Description: It is a deciduous shrub, growing to 5–10 m tall and usually branched from the ground, branches fork repeatedly forming a round canopy. The bark is grayish white to dark and scaly. Thorns: 3–5 mm long, with enlarged bases, powerful hooked appear at the nodes of the branches. Leaves: bipinnate, 3–8 pinnae glands present between uppermost and lowermost pinnae; up to 2.5 cm long. Flowers: yellowish white and fragrant, in 5–10-cm-long axillary pedunculate spikes, like bottlebrush-like clusters. Calyx: 5 deeply lobed petals; 5 stamens; many pistils inconspicuous. The pods are straight about 7–8 cm long and 2 cm broad, thin, flat, and oblong green and pubescent when young, turn to shiny bronze when old often with dark patches. seeds 3–6, flat, small, smooth, dark brown, and shiny.

Flowering: June–July **Fruiting:** August–December

Chemical constituents: The gum consists mainly of arabin. It is the salt of an organic acid, arabic acid, with metals such as calcium, magnesium, and potassium. The stem bark contains beta-amyrin, beta-sitosterol, its glucoside, erthrodiol octacosanol, and uvaol. From the leaves, an alkaloid, dimethyltryptamine, has been isolated.

Used in: Ayurveda, Siddha, and Unani

Parts used: Gum, stem bark, and root

Uses: Gum demulcent used as an ingredient in compounds for treatment of diarrhea, catarrh, bechic, antihemorrhagic, anti-inflammatory. Stem bark: anti-inflammatory, spasmolytic. Root: used for dysentery and urinary discharges.

Methods of cultivation: Direct sowing of seeds

DOI: 10.1201/9781003263500-5

Acalypha ciliata Forsk

Synonyms: *Ricinocarpus ciliatus* (Forssk.) Kuntze; *Acalypha rubra* Wight ex Benth., nom. nud.

Family: Euphorbiaceae

Common names: English: Fingered acalypha; **Gujrati:** Daadari, Ruchalo; **Hindi:** Dadri; **Kannada:** Kuppigida; **Malayalam:** Kuppameni; **Marathi:** Kupi, Lahan khokhali; **Sanskrit:** Haritamanjari; **Telugu:** Nugukuppinta, Tella muripinda

Distribution: Tropical Africa, Sri Lanka, and India. Generally found on open and wooded grassland and in deciduous and coastal bushland. Often in rocky or damp localities, near lakes, and on floodplains at elevations from sea level up to 1650 m. In India, common throughout South India, common in plains as a weed in gardens; also in wastelands.

Description: An erect herb growing up to about 85 cm with velvet-hairy branches. Leaves: simple 7 cm long and 4 cm broad, ovate, wedge shaped, membranous, round at base and tapering at tip, fringed margins with hairs, leaf stalk up to 5 cm. Flowers: borne in spike in leaf axils, unisexual on the same inflorescence, male flowers on the upper half, minute very small, greenish with yellow anthers, female flowers crowded below, up to 2 cm, bracts nearly 5mm, bell-shaped, sepals 4, velvet-hairy, about 12–15 toothed. Ovary: 1–1.5 mm. Fruit: a capsule about 3 mm and 3 lobed, smooth (Mapura and Timberlake, 2004).

Flowering and fruiting: November–August

Chemical constituents: The plant contains kaempferol; leaves and twigs contain acalyphamide and other amides, quinone, sterols, cyanogenic glycoside.

Used in: Ayurveda and Folk

Parts used: Leaves and entire plant

Uses: Used in arterial hemorrhage, arthrosis, cancer, congestion, constipation, Plant: emetic, expectorant (used in bronchitis, asthma, pneumonia). A tincture of fresh plant is used in homoeopathy for incipient phthisis with bloody expectorations, emaciation. A decoction of the leaves is drunk as a remedy for female sterility.

Methods of cultivation: Seeds

DOI: 10.1201/9781003263500-6

Achillea millefolium L.

Synonyms: *Achillea albida* Wild; *Achillea anethifolia* Fisch. Ex Herder; *Achillea lamulosa* Nutt.

Family: Asteraceae

Common names: English: Milfoil, Yarrow, Thousand leaf; **Hindi:** Bhut kesi, Gandrain, Puthkanda; **Kannada:** Rosemarri; **Marathi:** Rosemarri; **Tamil:** Achchilliya

Distribution: In India, common yarrow is found in the Himalayan region of Jammu and Kashmir, Himachal Pradesh, and Uttarakhand at an altitude range of 1050–3600 m.

Description: Erect perennial herb 1–2 ft tall. Leaves: alternate, linear to lanceolate, 2–3-pinnatifid, to 12 × 2.5 cm at the base. Corymbs: terminal, to 10 cm wide. Capitula: numerous, radiate; peduncle 4 mm. Involucre: oblong. Phyllaries: several-seriate, lanceolate. Receptacle: convex. Outer: ray florets pistillate, about 6 in number rose; ligule 3-toothed. Inner: disk florets bisexual. Yellow- to light cream–colored. Achenes compressed. Pappus 0.

Flowering and fruiting: March–October

Chemical constituents: The plant contains flavonoids, alkaloids (achilleine), polyacetylenes, triterpenes, coumarins, tannins, salicylic acid, a volatile oil containing linalool, camphor, sabinene, chamazulene, and other azulenes (Khare, 2007).

Used in: Traditional and Folk

Parts used: Whole plant and leaves

Uses: Anti-inflammatory, antispasmodic (used in cold, flatulent colic, heartburn), emmenagogue, cicatrizant, antidysenteric, antihemorrhagic, antipyretic, diaphoretic, diuretic, urinary, and antiseptic. Key application: In dyspeptic ailments, such as mild, spastic discomforts of the gastrointestinal tract. As an astringent, antispasmodic, choleretic, and antibacterial (German Commission E, 1998). As diaphoretic (*The British Herbal Pharmacopoeia*, 1996). Internally for feverish conditions, common cold, and digestive complaints; topically for slow-healing wounds and skin inflammations (*The British Herbal Compendium*, 1993; Khare, 2007).

Methods of propagation: Cuttings taken in early spring/grafting

DOI: 10.1201/9781003263500-7

Achyranthes bidentata Blume

Synonyms: *Acyranthes hispida, Acyranthes japonica, Acyranthes chinensis*
Family: Amaranthaceae (Acanthaceae)
Common names: English: Ox knee; **Hindi:** Putkanda; **Kannada**: Kadu uttharani; **Malayalam:** Perumkadalaad, Cheruk adaladii; **Marathi:** Aghsda; **Sanskrit:** Apamarga; **Tamil:** Nayuruvi
Distribution: It occurs in Nepal, China, and Japan. Within India it is found throughout the hilly regions between an altitude of 1200–3200 m and is common in waste places and in shady oak forests (https://www.flowersofindia.net/catalog/slides/OX/20knee.html).
Description: It is an erect or straggling herb, 0.6–2 m in height, and much-branched. Stem and branches: indistinctly quadrangular, striate or sulcate, sub glabrous to moderately hairy. Leaves: elliptic-oblong to broadly oval, short or long acuminate, gradually or more abruptly narrowed below, about 9–22 cm long and 3–8 cm broad, usually thinly hairy rarely densely appressed-hairy on the lower surface, petioles of main stem leaves 3–20 (–30) mm, shortening above and below. Inflorescences at first dense, finally lax and elongating to as much as 10–15 cm, peduncles 1–4 cm. Bracts: narrowly lanceolate, brownish-membranous, 3–5 mm, glabrous. Bracteoles: 3.5–5.5 mm, perianth segments 5, 4–7 mm, the outer longest, all narrowly lanceolate, very acute, with a distinct midrib. Stamens: 5, the filaments 2–3 mm, alternating with sub-quadrate pseudo-staminodes which are typically simple and shortly dentate or sub-entire, more rarely. Style: slender, 1–2 mm. Capsule: 2–3 mm. Seed: filling the capsule, cylindrical smooth. *Achyranthes bidentata* is distinguishable from *Achyranthes aspera* by its simple pseudo-staminodes and much less hairy leaves and stem.
Flowering and fruiting: September–December
Chemical constituents: *A. bidentata* root extract is rich in active phytochemicals compounds such as saponins (Li et al., 2005), ketosteroids (Zheng et al., 2008), and flavonoids (Stefan et al., 1996).
Used in: Ayurveda and Folk
Uses: *A. bidentata* is a medicinal herb with the property of strengthening bones and muscles and ensuring proper downward flow of blood (Suh et al., 2014). Its root juice is used for toothache. The plant is used externally in the treatment of leech bites. Decoction prepared from plant is used as a diuretic (Sharma et al., 2001). It is also used for osteoporosis and arthritis and known for its anti-inflammatory effects. It is reported that the polysaccharides from this medicinal plant have oxidative stress protective effect (Xue et al., 2009), anti-HIV activity (Peng et al., 2008).
Methods of cultivation: Seeds

DOI: 10.1201/9781003263500-8

Aconitum chasmanthum Stapf ex Holmes.

Synonyms: *Aconitum hians* Watt; *Aconitum violaceum* var.robustrum:Spatf
Family: Ranunculaceae.
Common names: English: Gaping monkshood, Wolfsbane, Indian napellus; **Hindi:** Mohari, Meethaa Zahar; **Sanskrit:** Sthavaravisa VishaShringika, Vajranaga
Distribution: Subalpine and alpine zones of western Himalayas at an altitude of 7000–12000 ft
Description: *A. chasmanthum* is an erect, herbaceous perennial plant producing an unbranched stem 60–120 cm tall. The stem grows from a l tuberous root that produces a new tuber each year; the old tuber dies after the plant flowers. Leaves: numerous, usually more distant in the lower part and crowded in the upper, the lowest leaves petiolate, petioles up to 7.5 cm. long, the upper shortly petioled or subsessile, passing into the floral leaves, glabrous, somewhat fleshy, lower and intermediate blades orbicular, reniform in outline, 4–6 cm high, 5–9 cm across. Inflorescence: a long, narrow, stiff, dense, or loose raceme, often over 30 cm long, often leafy below, and sometimes with slender, erect additional branches from the base, crispo-pubescent; rachis stout; floral leaves like the preceding leaves, but still less divided or entire, passing into the linear to filiform bracts; bracteoles, if any, small; pedicels slender, the lowest at length 2.5–3.5 cm long and adpressed to the rachis when mature. Flowers: blue or whitish and variegated with blue, crispo-pubescent or almost glabrous; uppermost helmet-shaped, 12–18 mm long from the tip to base, 5–7 mm wide, lateral margin conspicuously concave. Sepals: blue or whitish and variegated with blue. These sepals are often described as crispo-pubescent or almost glabrous; boat-shaped or helmet-shaped. The lateral sepals are 12–15 mm long and broad, 5–10 mm high, and 12–18 mm long from the tip to the base and 5–7 mm wide (seen from the side); lateral margin conspicuously concave, and lateral sepals not contiguous with the helmet, except near the base, obliquely suborbicular or almost square, shortly or obscurely clawed, 12–15 mm long and broad, lower sepals oblong, obtuse, 9–12 mm long. Nectaries: extinguisher-shaped glabrous, claw 5–6 mm long, leaning forward in the upper part; hood short, wide, very obtuse, honey-gland occupying the whole top. Filaments: glabrous or very sparingly hairy, winged; wings gradually or abruptly attenuated. Carpels: 5, glabrous, rarely hairy on the back, conniving abruptly, contracted into the short style, back convex. Follicles: oblong, truncate, 10–16 mm. long, contiguous or with slightly divergent tips, glabrous. Seeds: brown obovoid to obpyramidal, 3.5 mm. long, equally 3-winged, wings thin, faces smooth. Roots: paired, tuberous 2.5–3.5 cm, 12–18 mm. thick, bearing numerous root-fibres, dark brown to blackish brown, smooth or wrinkled when dry,
Chemical constituents: Indaconitine, A & B; chasmaconitine and chasmanthinine, chasmanine
Used in: Ayurveda, Folk, and Unani
Parts used: Roots
Uses: Sedative, antirheumatic, analgesic, antitussive, antidiarrheal. The dried root is analgesic, anodyne, diaphoretic, diuretic, irritant and sedative. This is a very poisonous plant and should only be used with extreme caution and under the supervision of a qualified

DOI: 10.1201/9781003263500-9

practitioner. Diabetes, gout, paralysis, cephalgia, asthma, colic, and hypotension (Chiej, 1984; Launert, 1981; Singh and Kachroo, 1976).

Methods of cultivation: Through tubers, thrives in most soils and in the light shade of trees. Grows well in heavy clay soils. Prefers a moist soil in sun or semi-shade. Prefers a calcareous soil.

Aconitum ferox Wall. ex Ser.

Family: Ranunculaceae

Common names: English: Indian aconite, Monkshood, Wolfsbane; **Gujrati**: Bachang, Bachang; **Hindi:** Bachang, Mithaa vish, vatsanabha; **Kannada:** Vatsanabhi; **Malayalam:** Vatsanabhi; **Marathi:** Bachang, Vachnag; **Sanskrit:** Vatsanaabha (because the root resembles the navel of a child) Mahavisha; **Tamil:** Vasanaavi, Karunaab; **Telugu:** Telia Visha. Nabhi

Distribution: *A. ferox* grows at altitudes of 2100–3600 m. The alpine Himalayas from Sikkim to Garhwal and Assam in India, Bhutan, and Nepal

Description: Herbaceous perennial reaching to about 3 ft in height.

Chemical constituents: It contains large quantities of the extremely toxic alkaloid pseudaconitine also known as nepaline, after Nepal (Jyothi et al., 2016). Other alkaloids include aconitine, picro aconitine, benzylo aconitine, and homonopelin.

Flowering and fruting: August–October

Used in: Folk, Ayurveda, Siddha, and Unani

Parts used: Fruits and Root

Uses: Extremely poisonous. Due to its poisonous nature it is used in 60–125 mg of purified substance. The plant is used with due caution in regard to its extreme potency and toxicity and the consequent danger of overdose; the plant has long been used in the folk, Ayurvedic, and sometimes in homeopathic medicine of the Indian state of Sikkim on the southeastern edge of the Tibetan plateau (Rai and Sharma, 1994). The plant is used after purification by immersing it in cow's urine for 3 days or by boiling in cow's milk for 3–6 hours. It is used as narcotic, sedative, antileprotic, and anti-inflammatory. Roots possess depressant activity, but after mitigation in cow's milk for 2–3 days, they exhibit stimulant activity (Khare, 2007). Its other uses include being a cure for diabetes mellitus, asthma, ear infections, nasal infections, leprosy, paralysis, and arthritis. The root is also used as an antidote to *hartal*, believed to be a lethal poison (Rai and Sharma, 1994). The fruits of the plant are broncho-dilatory, anti-asthmatic, and anti-inflammatory. The herb is used as a cardiac tonic and aphrodisiac for treating seminal ejaculations. It is also used in treating enlarged spleen (Bhavaprakash Nighunta, 2010).

Plant material or extracts derived from *A. ferox* and various *Aconitum spp.* has, on occasion, been added, in small quantities, to Indian alcoholic drinks to heighten their intoxicating effects, a dangerous practice that has not only caused drinkers to exhibit symptoms of poisoning, which led some times to fatalities (Jyothi et al., 2016).

Methods of cultivation: Seeds

DOI: 10.1201/9781003263500-10

Actaea acuminata Wall. ex Royle

Synonyms: *Actaea spicata* var. acuminata (Wall. ex Royle) H. Hara

Family: Ranunculaceae

Common names: English: Himalayan baneberry

Distribution: Afghanistan, Pakistan, Bhutan, East and West Himalayas, Kashmir, Himachal Pradesh in India

Description: Erect perennial rhizomatous herbs. Stems: 45–100 cm, sheathed by ovate scales at base. Leaves: alternate, 3–4 times ternately compressed, 18–35 cm, leaflets ovate, 4–8 × 1.5–3 cm, acuminate, base rounded, margins irregularly serrate, pubescent at first; petioles 2–10 mm. Flowers: actinomorphic in short terminal racemes. Sepals: 3–5, elliptic petaloid, caducous, 5 × 2mm. Petals: 3–5(–10), spathulate. 3 × 1 mm, whitish, weakly 3-nerved. Stamens: many. Carpels: solitary, stigma rounded, sessile. Fruit a berry, ellipsoid, 7–8 × 5mm, blackish purple (https://biodiversity.bt/species/show/6152).

Flowering and fruiting: May–August

Chemical constituents: Preliminary phytochemical screening of methanol extract showed the presence of alkaloids and polyphenols.

Used in: Traditional and Homeopathy

Parts used: Stem, bark, and whole plant

Uses: Possesses antianxiety, anticonvulsant, antidepressant and antistress activities and mild sedative activity. The plant has been traditionally used in the treatment of rheumatism, inflammation, rheumatic fever, lumbago, scrofula, nervous disorders, chorea, and as emetic, expectorant, laxative, stomachic, and purgative (Chopra et al., 1956; Kirtikar and Basu, 1975; Khare, 2007; Biswas et al., 2014). In traditional health care system of Uttarakhand, decoction of *A. acuminata* roots is used in the treatment of bronchial asthma (Gairola and Biswas, 2008), root powder of plant is given to children for treatment of cough (Tiwari et al., 2010), and leaves of plant are used in the treatment of sciatica (Phondani, 2011).

Methods of propagation: Seeds

DOI: 10.1201/9781003263500-11

Actaea spicata Linn.

Synonym: *Actaea acuminata* Wall. ex Royle; *Actae christophoriana* Goun; *Actaea corymbosa* Stokes; *Actaea racemosa* Gilib; *Christophoriana spicata* Moench.

Family: Ranunculaceae

Common names: English: Baneberry grapewort; **Hindi:** Mamira; **Kannada:** Visha-phale; **Sanskrit:** Vishaphala

Distribution: Native to most of Europe, including Britain, through temperate and arctic Asia to China. Generally found on Ash woods on limestone and in damp stony woods to elevation about 1500 m. In Asia it is found in deciduous woodland, or clearings in coniferous woods, at elevations up to 500 m. In Pakistan and India at an altitude of 2000 m. Himachal Pradesh, Uttar Pradesh, Arunachal Pradesh, Sikkim.

Description: It is an herbaceous perennial plant growing up to 2 ft tall. The basal leaves are large, bipinnate. The leaflets are more or less toothed. Inflorescence is an erect raceme. The flowers are white, with 3–6 petaloid sepals. The fruit is a berry, black when ripe, and 10–13 mm across.

Flowering and fruiting: May–August

Chemical constituents: It is reported to contain isoquinoline alkaloids magnoflorine, corytubrine, and triterpene glycosides, including actein and trans-aconitic acid.

Used in: Ayurveda, Folk, and Homeopathy

Parts used: Roots

Uses: Root: antirheumatic, anti-inflammatory, nerve sedative, purgative; used in the treatment of chorea, lumbago, nervous disorders, rheumatic fever, scrofula, Its active against Ehrlich's ascites tumors. In folk medicine, roots are used in cases of ovarian neuralgia, uterine tenderness, and sub-involution. Berries are poisonous; used topically for skin diseases (Khare, 2007; Chopra et al., 1956; Duke et al., 2008; Madaan et al., 2011). The plant has also been used in traditional systems of medicines of various countries for the treatment of snakebites, asthma, and externally for skin complaints. It used in for treatment of arthritis and joint pain (Kirtikar and Basu, 1975, Madaan et al., 2011)

Methods of cultivation: Seed—best sown as soon as it is ripe in the autumn outdoors in a moist, shaded seedbed. Germination is poor.

DOI: 10.1201/9781003263500-12

Actinodaphne aungustifolia Hook. fil. & Thomsex Mesin

Synonym: *Actinodaphne hookeri* Meissn.; *Actinodaphne gullavara* (Buch.-Ham. ex Nees) M. R. Almeida; *Actinodaphne heterophylla* Blume *Jozoste hookeri* (Mesin) Kuntze. *Tetranthera angustifolia* (Blume) Nees; *Litsea angustifolia* Blume; *Tetranthera saligna*; *Tetranthera angustifolia* (Blume) Nees

Family: Lauraceae

Common names: English: Pisa; **Hindi** Pisa; **Kannada:** Galavaar, Haggodimara, Todgenasu; **Malayalam:** Malavirimvi, Neyaram, Puvaca; **Marathi:** Pichli, Pisaa; **Tamil:** Thali, Paratathali

Distribution: Found in the evergreen forests of Eastern and Western Ghats, Karnataka, Orissa, Sikkim, and North-East India.

Description: It is a moderate evergreen tree about 8–10 ft. Young branchlets: terete, rusty tomentose. Bark: smooth. Leaves: simple, alternate, spiral, subverticilate; petiolate, petiole about 2.5 cm long, terete, rusty tomentose; lamina 12–27 cm long 4–6.5 cm broad, narrow-elliptic to oblanceolate, apex acuminate, base cuneate to acute, margin undulate, chartaceous, glaucous, fulvous tomentose when young, later glabrous except midrib beneath; midrib raised; secondary nerves 7–10 pairs; tertiary nerves nearly closely and horizontally percurrent, not prominent. Flowers: unisexual, dioecious; female flowers in umbels on short peduncle. Fruit: a berry, is globose, about 1 cm across seated on a thickened bell-shaped perianth tube; seed 1 (http://www.flowersofindia.net/catalog/slides/Pisa.html).

Flowering and fruiting: August–May

Chemical constituents: The leaves contain a very small amount of an amorphous alkaloid. They also contain beta-sitosterol, hentriacontanone, hentriacontanol, and quercetin-3-rhamnoside and β-sitosterol (Anonymous, 2003; Chopra et al., 1956). The bark gives an alkaloid, actinodaphnine (Khare 2007). The roots contain a flavanone glycoside.

Used in: Siddha and Folk

Parts used: Leaves, stem, and bark

Uses: Infusion of leaves as urinary tract disinfectant, antidiabetic, spasmolytic (Anonymous, 2003; Kirtikar and Basu, 1995)

Methods of cultivation: Seeds

DOI: 10.1201/9781003263500-13

Adansonia digitata L.

Synonyms: *Adansonia baobab* Gaertn.; *Adansonia sphaerocarpa* A. Chev.;
Family: Bombacaceae (Malvaceae)
Common names: English: Baobab, Monkey bread tree, African calabash, Upside-down tree; **Gujarati:** Sumpura; **Hindi:** Gorakh-imli, Hathi-khatiyan; **Sanskrit:** Sheet-phala, Ravanaamlikaa, Gorakshi, Panchparni; **Tamil:** Papparapuli, Anaipuliya-marum; **Telugu:** Brahma-mlinka, Seemasinta
Distribution: Native to tropical Africa; common along the west coast of India.
Description: *A. digitata* is a massive deciduous tree, up to 20–30 m tall with swollen trunk with a diameter up to 2–10 m at adult age. The bark is smooth, reddish brown to gray and soft. Leaves alternate, digitately 3- to 9-foliate; leaflets oblong to ovate, 5–15 × 3–7 cm, lower leaflets being the smallest and terminal leaflet the largest; leaflets dark green, with short, soft hairs; lateral veins looping; apex and base tapering; margin entire; petiole up to 12 cm long. The flowers are a waxy white, up to 20 cm in diameter, axillary, solitary, pendulous, bisexual; all floral parts in 5s; calyx deeply lobed, with white, silky hairs inside; large, crinkly, spreading petals; many stamens on a large central column that is shed with the petals; ovary superior, 5–10 chambered; petals bruise easily and become brown; flowers have an unpleasant scent. Fruit ovoid, 12 cm or more in length, with a hard, woody shell, covered with yellowish-gray velvety hairs, indehiscent; seeds smooth, embedded in a whitish powdery pulp, have little or no endosperm.
Flowering and fruiting: September–November
Chemical constituents: The roots contain aflavanone glycoside. The fruit pulp is a source of vitamin C (175.0–445.4 mg/100 g); dried pulp contains calcium and vitamin B1 Furfural (9.6%) is obtained after distillation of the fruit (Khare, 2007). The fruit pulp has very high vitamin C, calcium, phosphorus, carbohydrates, fibers, potassium, proteins, and lipids content (Rahul et al., 2015).
Used in: Traditional
Parts used: Leaves, bark, and seeds
Uses: Cooling, refrigerant (allays burning sensation). Leaves: diaphoretic (used as a prophylactic against fevers). Fruit: antidysenteric, antiseptic, antihistaminic. In Africa, dried leaves provide much of the dietary calcium. Aqueous extract of the bark is used for treating sickle cell anemia. An infusion of the leaves and flowers is given in respiratory disorders. (Powdered leaves prevented crisis in asthma induced by histamine in guinea pigs.) Dried fruit pulp also gives relief in bronchial asthma, allergic dermatitis and urticaria (Khare, 2007). The various parts of the plant (leaves, bark, and seeds) are used as a panacea, that is, to treat almost any disease and specific documented uses include the treatment of malaria, tuberculosis, fever, microbial infections, diarrhea, anemia, dysentery, toothache, and more. The leaves and fruit pulp are used as febrifuge as well as an immune stimulant. In India, it is reported that baobab pulp is used externally with buttermilk for the relief of diarrhea and dysentery, while the young leaves are crushed and used to treat painful swellings (Rahul et al., 2015).
Methods of propagation: Seeds

DOI: 10.1201/9781003263500-14

Adenanthera pavonina Linn

Synonyms: *Adenanthera gerseni* Scheff, *Adenanthera polita* Miq. *Corallaria parvifolia* Rumph

Family: Leguminosae (Mimosaceae)

Common names: Bengali: Ranjana; **English:** Coral wood, Redwood; **Gugrati:** Badigumchi; **Hindi:** Badi gumchi; **Kannada:** Aane gulgunj; **Malayalam:** Manjadi, Sem; **Marathi:** Thorla goonj; **Sanskriṭ:** Ksharaka, Rakta Kanchana, Rakta Kambala; **Tamil:** Aanai Kundumani; **Telugu:** Gurivenda

Distribution: Bangladesh, Cambodia, China, Malaysia; in India, Andamans and Nicobar Islands, Assam, Kerala, Uttar Pradesh, the Western Ghats, and sub-Himalayan tract; also cultivated

Description: A large deciduous tree, with pale pinkish-gray bark. Leaves: Bipinnate, up to 24 cm long, rachis about 10–40 cm long, petiole 5–10 cm long, pinnae opposite, 2–6 pairs, 7–12 cm long, leaflets 7–15 pairs, sessile, alternate, obovate, 1.5–4.0 cm long, 5–25 mm broad, margins slightly curved backward. Inflorescence: An axillary and terminal raceme, about 7.5–15 cm long. Flower: Pale yellow, pedicellate, pedicel 3.5–4.0 cm long. Calyx 5, gamosepalous, 5 toothed, 1 mm long. Corolla pale yellow, 2.5–3 mm long, petals united at the base. Stamens: 10, free, anthers with a gland at the tip. Fruit: A pod, curved, 10–22.5 cm long, and 0.7–10 mm broad. Seeds: Many, bright red in color, like beads, 7–8 mm wide.

Flowering and fruiting: March–May

Chemical constituents: The seed contains an anti-inflammatory active principle, O-acetylethanolamine. The leaves contain octacosanol, dulcitol, glucosides of beta-sitosterol and stigmasterol. The bark contains sitgmasterol glucoside (Khare, 2007).

Used in: Ayurveda, Folk, and Siddha

Parts used: Leaves and seeds

Uses: Is an astringent, used in diarrhea, hemorrhage from the stomach, hematuria, anti-inflammatory, used in rheumatic affections, gout. Seeds are anticephalgic; also used for the treatment of paralysis. A decoction is given in pulmonary affections (Khare, 2007). A red powder made from the wood is also used as an antiseptic paste.

The ground seeds are used to treat boils and inflammations. A decoction of the leaves is used to treat gout and rheumatism; the bark is used to wash hair (https://www.flowersofindia.net/catalog/slides/Red%20Bead%20Tree.html). Traditionally, the ground seed is widely used for the treatment of various human ailments such as treatment of arthritis, boils, blood disorders, cholera, convulsion, epilepsy, inflammation, indigestion, paralysis, rheumatism, and spasm (Burkill, 1966; Balogun et al., 2004).

Methods of cultivation: Seeds. Seed has physical dormancy due to hardiness of testa.

DOI: 10.1201/9781003263500-15

Aerva persica (Burm. f.) Merr.

Synonyms: *Aerva javanica* (Burm. f.) Schult; *Aerva wallichii* Moq.; *Aerva tomentosa* Forssk; *Achyranthes javanica* Pers.; *Aerva tomentosa* Forssk; *Celosia lantana* L.; *Iresine javonica* Burm. f.

Family: Amaranthaceae

Common names: English: Desert cotton, Javanese wool plant, Snow bush; **Gugrati:** Bour, Patharphod; **Hindi:** Safed bui; **Kannada:** Dodda hindi gida; **Marathi:** Khul; **Sanskrit:** Astmabayda; **Tamil:** Periya pulai; **Telugu:** Magabira, Pedda pindi

Distribution: India, Myanmar, and Pakistan. Wildly distributed in Gujarat, Saurashtra, and Kachchh

Description: Much branched, erect, perennial under-shrubs, up to 1 m high; stem terete, covered with thick, easily detachable tomentum. Leaves alternate, 5–12 × 1.5–2.5 cm, linear oblong or oblong-spathulate, obtuse or acute at apex, rarely retuse, tomentose on both surfaces, sessile. Flowers unisexual, dioecious, dull white, sessile, in linear, 2.5–15 cm long, spikes, the spikes arranged in terminal, leafless panicles; bracteoles broadly ovate, acute, white, hyaline. Male flowers: perianth lobes 5, elliptic, oblong, subobtuse, wooly on dorsal surface; pistillode ovoid, acute, usually shortly stipulate with short style; stigma minutely bifid. Female flowers: perianth 2.5 mm long, lobes oblong, subacute, apiculate; stigma 2 as long as the style. Fruit an orbicular-ovoid utricle.

Flowering and fruiting: January–October

Chemical constituents: The essential oil of *Aerva javanica* leaves was found to be rich in hentriacontane (21.48%), heptacosane (19.78%), hexacosane (2.04%), nonacosane (20.59%), octacosane (3.47%), pentacosane (5.58%), and triacontane (2.81%). The essential oil of stems is rich in hentriacontane (18.32%), heptacosane (22.48%), nonacosane (23.26%), octacosane (3.42%), squalene (2.07%), and triacontane (2.24%). It also contains kaempferol, sterol, triterpenes, flavonoids, ί-sitosterol, α-amyrin, palmitic acid, stearic acid, linoleic acid, myristic, oleic acid, palmitic acid, aervanone, alkaloids, Chrysin-7-Ogalactosidem (Garg et al., 1979).

Used in: Ayurveda and Siddha

Parts used: Flowers, seeds, and root

Uses: Cutaneous affections, dysentery, gonorrhea headache, and kidney disorders. The plant is used as a stuffing material and tooth cleaner. It is used as a purgative, antidiarrheal, and anthelmintic medicine in cattle (Khan, 2009; Qureshi et al., 2009). In India, boiled seeds are fed to animals suffering from foot-and-mouth disease (Mistry et al., 2003), possessing nephroprotective activity (Movaliya and Zaveri, 2014). Roots and flowers are reported to possess medicinal properties against rheumatism and kidney troubles (Gupta et al., 1966). It is reported as anthelmintic, diuretic, and demulcent. It is used for the treatment of headache (Chopra et al., 1956). The decoction of the plant is administered to remove swellings (Kirtikar and Basu, 1996). The plant seeds are utilized to alleviate migraine and furthermore stiffness. *The plant* possess hypoglycemic, antioxidant, antimalarial, anthelmintic, analgesic, antivenin activities, and medicinal applications against kidney troubles and rheumatism (Chawla et al., 2012). It also showed different activities, including antiviral (Baltina et al., 2003), anti-plasmodial (Ahmed et al., 2010), and antidiabetic (Reddy and Reddy, 2009).

Methods of cultivation: Seeds

DOI: 10.1201/9781003263500-16

Aganosma dichotoma (Roth) K. Schum

Synonyms: *Aganosma caryophyllata* G. Don; *Agaosma herein* (Spreng.) I.M. Turner Chonemorpha dichotoma

Family: Apocynaceae

Common names: Bengali: Haparmali; **English:** Malati, Clove-scented echites; **Hindi:** Malati; **Kannada:** Maalatilata; **Malayalam:** Cherupaval; **Marathi:** Malati; **Sanskrit:** Asanamallika, Asphota, Bhadramuja, Madhumaalati; **Telugu:** Paalamalle, Mogari

Distribution: Native to India found in Assam, West Bengal, Bihar, Orissa, Andhra Pradesh, and Tamil Nadu; often cultivated in Indian gardens.

Description: Large climbers, woody, pubescent, with milky sap, bark brown, fissured. Leaves: ovate or elliptic, 7–12 × 4–7 cm long, acute, obtuse or shortly acuminate at apex, glabrous or tomentose beneath, with 3 pairs of nerves, leathery in texture. Flowers: in lax white woolly cymes, tomentose. Flowers possess spicy aroma, shorter than calyx segments. Calyx: 5 sepals, segments linear, spreading. Corolla: 5 petals white, about 3.5 cm across; throat and tube cylindrical. Follicles: 7.5–12.5 cm long, stout. Seeds: ovoid or oblong approximately 1.25 cm long (Singh et al., 2001).

Flowering and fruiting: Throughout the year

Chemical constituents: The leaves contain quercetin, kaempferol, and phenolic acids. Shoot tips and flower buds contain tannin as well as some other phytochemical constituents are tannin, flavonoid, flavanol, alkaloid, and saponin (Khare, 2007).

Used in: Ayurveda

Parts used: Leaves flowers and seeds.

Uses: Antiseptic; anodyne (an ingredient in massage oils for paraplegia, neuralgia, sciatica); also anthelmintic and emetic (Khare, 2007). It has been traditionally used in the treatment in various ailments, including emesis, anthelmintic, bronchitis, leprosy, skin disease, ulcers, inflammation, and disease of mouth while flowers are used in disease of eye and leaves are used in biliousness. It is also used as an ingredient in massage oils for paraplegia, neuralgia, and sciatica (Maurya et al., 2021). It is also used in treating snakebites (Payyappallimana, 2010; Jain 1991). Traditionally, it is used for skin conditions, ulcers, fever, arthritis, purulent discharge of the ear, and oral disease (Kirtikar and Basu, 1994). Flowers are good for eye diseases and leaves cure biliousness. Plants are mentioned in Agni Purana for the treatment of semen. Root decoction is treated orally with a touch of pepper powder twice every day in fever for 3 days. Also used in treating snakebite and diabetes (Jain, 1991).

Methods of cultivation: Layering and seeds

DOI: 10.1201/9781003263500-17

Agave angustifolia Haw.

Synonyms: *Agave vivipara* L.
Family: Agavaceae
Common names: English: Caribbean agave, Dwarf aloe; **Hindi:** Khetki; **Marathi:** Chota Ghayal; **Sanskrit:** Kantala
Distribution: This plant was introduced to India as an ornamental plant in the earlier period. These are naturalized in the sub-Himalayan hills, outer Himalayas and throughout most of Peninsular India. However, the origin of this plant is seated in tropical America and Mexico. This species is one of the widest ranging with a natural distribution and it occurs in tropical savanna, thorn forest, and drought deciduous tropical forests from sea level to about 5,000 ft in elevation.
Description: These plants are short-stemmed perennials. The leaves are grayish green in color and dagger- or sword-shaped. These are 40–60 cm long with flat spines and slender-cusped prickles. The leaves are fairly stiff, narrow (2–4 in.), and concave toward the middle, with a broad central band colored pale green and strong creamy white margins that often have a flush of pink. The flowers are 4–5 cm long, greenish in color with an unpleasant odor. The fruits of this plant are strongly stipulate and beaked with large seeds. When in flower, which does not occur until this plant is at least 10 years old and often much later, it sends up an 8–10-ft spike with a well-branched panicle bearing greenish-white flowers.
Flowering and fruiting: Flowers once in a lifetime
Chemical constituents: It contains phenolic compounds, saponins, and fatty acids. *Agave* leaf powder has a high amount of total dietary fiber (38.40%), total sugars (45.83%), and protein (35.33%), with a relatively low content in ash (5.94%) and lipid (2.03%).
Used in: Traditional
Parts used: Leaves and roots
Uses: These plants are important for their medicinal purposes. The roots are considered diuretic and diaphoretic. The fresh juice of the leaves is applied externally to relieve bruises and as a remedy for psoriasis (https://www.indianetzone.com/38/khetki.htm). The uses of *A. angustifolia* in traditional medicine are diverse. The fiber is used to treat urticaria; the sap, cooked leaves, and root infusions have been used for "internal injuries"; a concoction of the root has been used against dysentery; fresh leaves are useful to stop bleeding and alleviate wound pain, to treat skin pimples, and for coughs; roasted leaves are used to relieve rheumatic pain and as a remedy for sprains or broken bones. Interestingly, the plant has also been used to treat psoriasis, an autoimmune disease that causes joint pain and inflammation (Cano, 1998; Monroy-Ortiz and Castillo-Espana, 2007).
Methods of cultivation: Through suckers

DOI: 10.1201/9781003263500-18

Photo: Dr. Srinath Rao

Aglaia roxburghiana Miq. Hiern

Synonyms: *Aglaia elaegnoidea* (A. Juss.) Benth.; *Milnea roxburghiana* (Miq.) Wight

Family: Meliaceae

Common names: Bengali: Priyangu; **English:** Droopy leaf; **Hindi:** Priyangu; **Kannada:** Gadagayya, Kempu nola, Thottilu, Priyangu; **Malayalam:** Nyalei, Punniyava, Cheeralam cempuli; **Sanskrit:** Gundraphala, Phalini, Priangu; **Tamil:** Gnaazhal, Kokkalai, Kanni-k-kompu; **Telugu:** Yerra anduga, Konda anduga

Distribution: Western Ghats, tropical forests in the hills of Andhra Pradesh, Karnataka, Kerala, and Andaman and Nicobar islands

Description: An evergreen tree growing up to 10 m tall. It has grayish-brown bark. The leaves' compound is alternate to sub-opposite, 10–20 cm long, with 3–7 leaflets. Leaflets are opposite to sub opposite, elliptic, or oblong-elliptic, 6–12 cm long, 2.5–5.5 cm wide. Inflorescence axillary. Flowers are around 2.5 mm diameter, are roundish, yellow are borne in panicles. The leaf axis is shorter than leaves, pedicel covered with small scale like stellate rust-colored trichomes. Calyx is around 0.6 mm, sepals 5, rounded, outside rust-colored or yellowish scaly. Petals are 5, oblong, 1–1.5 mm, smooth or yellowish scaly, tip rounded. Stamen tube is nearly spherical, slightly shorter than petals, giving the flowers ball-like appearance anthers 5 included. Ovary: ovoid densely covered with scale-like stellate trichomes, stigma sessile. Fruits: brown, squamate, yellowish brown when mature, ellipsoid or obovoid 1–1.5 cm in diameter calyx persistent and are indehiscent. Seeds: 1–2 per fruit.

Flowering: June–October **Fruiting:** July–December

Chemical constituents: Bisamide alkaloids are present in the leaves. The triterpenes roxburghiadiol A and B isolated from this plant (Janaki et al., 1999); 14 alpha-methylsterols, roxburghiadiol A and roxburghiadiol B, were isolated from the leaves and fruits of *Aglaia roxburghiana* (Balakrishna et al., 1990).

Used in: Ayurveda and Folk

Parts used: Leaves, fruits, and seeds

Uses: Fruit—cooling (in febrile complaints), antipyretic, astringent, antidiarrheal, antidysenteric, anti-inflammatory. Seeds are used for painful micturition. Fruits are also used for treating obstinate skin diseases and tumors. Leaves exhibit anticancer activity (by inhibiting the growth of vinblastine-resistant KB cells; Khare, 2007). The plant is considered as a remedy for dysentery, skin diseases, leprosy, inflammation, leukoderma, and abdominal pain. Fruits have cooling effects and is useful in burning sensation of the body and painful micturition (Khare, 2007).

Methods of cultivation: Seeds

DOI: 10.1201/9781003263500-19

Agrimonia eupatoria L.

Synonyms: *Agrimonia pilosa* Hook.f. non Ledeb. *Agrimonia pilosa* Ledeb. var. nepalensis (D. Don) Nakai

Family: Rosacae

Common names: English: Church steeples, Cocklebur, Sticklewort, Stickwort

Distribution: The Himalayas from Kashmir to West Bengal at 900–1300 m and in Arunachal Pradesh, Nagaland, and Meghalaya

Description: The plant is a slim, erect leafy deciduous perennial herb. The stem is about 60–90 cm long, scarcely branched. The plant is deep green in color covered with soft silky hairs, and when slightly bruised, it exhales a peculiar but pleasant aromatic smell. Leaves: Leaves are pinnately compound, lower leaves 10–18 cm; leaflets 6–21 in number toothed, hairy on both surfaces very unequal, elliptic, ovate or obovate, 1.3–3.8 cm. Upper leaves gradually smaller and with few leaflets. Stipules: adnate to the base of leafstalk. Inflorescence is a raceme. Flowers are very numerous, 6.0mm in diameter, yellow in color, with terminal spike-like racemes. Each flower in the axial of a small, 3-cleft bract with two smaller, 3-toothed bracteoles. Calyx: tube top–shaped, grooved, bearing outside its mouth a ring of small, hooked bristles; limb 5-lobed. Petal: 5, oblong. Stamen: 15, carpels 2, free, enclosed within the calyx tube; styles thread like protruding; stigmas terminal, dilated; ovule solitary. Fruit is an achenes, 1 or 2, in closed in the hardened bristly calyx crowned with a ring of hooked bristles. The calyx of fruits encircled with a thick whorl of hooked prickles. Each flower contains 2 seeds (Al-Snafi, 2015a).

Flowering: June–August **Fruiting:** August–September

Chemical constituents: The herb contains condensed tannins up to 8%, coumarins, flavonoids (glucosides of luteolin, apigenin, and quercetin), polysaccharides, volatile oil. Luteolin 7-glucoside shows a cholagogue action. Essential oil is antibacterial, active against *Bacillus subtilis* (Bunney, 1984; Khare, 2007). The phytochemical analysis of the aerial parts of the plant revealed the presence of carbohydrates, tannins, terpenoids, phenolic compounds (flavonoids), agrimony, agrimony lactone, glycosides, and oils (Ghaima, 2013; Pan et al., 2008). Aerial parts also contained polysaccharides, triterpenoid (α-amyrin, ursolic acid, euscapic acid), silicic acid, salicylic acid, traces of essential oil, flavonoids, organic acids, ascorbic acid, nicotinamide complex (about 100–300 pg/g leaf), thiamine (about 2μg/g leaf), and vitamin K. The fresh herb contained agrimoniolide, palmitic and stearic acids, ceryl alcohol, phytosterols, and volatile oil 0.2%. Seeds contained oil consisted of oleic, linoleic, and linolenic acids (Bisset and Wichtl, 1994; Bilia et al., 1993; Carnat et al., 1991).

Used in: Ayurveda, Unani, and Folk

Parts used: The whole plant is generally used; however, dried leaves are mainly used in medicine.

Uses: Astringent, anti-inflammatory, hepatic, cholagogue, diuretic, mild hemostatic, antibacterial. Used for irritations and infections of the intestinal tract, gallbladder diseases, hyperacidity, colic, urinary disorders (bed-wetting, incontinence), sluggish liver, mucus membrane inflammations; externally for ulcerated and discharging skin, psoriasis, and seborrheic eczemas. Key application: in mild, non-specific, acute diarrhea and in inflammation of oral and pharyngeal mucosa; as astringent. The ethanolic extracts of the herb are used for their antiviral properties. It is an antitumor agent, astringent, cardiotonic, coagulant, diuretic, sedative, and anti-asthmatic and used for corns or warts (Bunney, 1984; Khare, 2007).

Methods of cultivation: Through seeds. Easily grown in most soils, preferring a calcareous soil. Thrives in a dry, lightly shaded position.

DOI: 10.1201/9781003263500-20

Agropyron repens Beauv.

Synonyms: *Agropyron firmum* J. Presl; *Agropyron repens* (L.) P. Beauv.; *Elymus repes* (L.) Gould; *Elytrigia repens* (L.) Desv. ex Nevski; *Triticum firmum* (J. Presl) Link; *Triticum repens* L

Family: Gramineae, Poaceae

Common names: English: Couch grass, Dog grass, Wheat grass

Distribution: The western Himalayas and Kashmir at altitudes between 2700 and 3600 m.

Description: Quackgrass is a strongly rhizomatous perennial plant. The rhizomes are long, highly branched, yellowish white, sharp-pointed, and somewhat fleshy. Stems are erect and usually 1–3 ft tall. Leaf blades are 0.25–0.5-in. wide, flat, and pointed and have small auricles at the junction of blade and sheath. Leaf blades often have a diagnostic slight constriction near the tip and are sparsely hairy above and hairless below. The spikelets are arranged in two long rows and borne flatwise to the stem. The florets are awnless to short-awned. Seeds are elliptical and pale yellow to brown (Al-Snafi, 2015b).

Chemical constituents: The rhizome contains triticin, a carbohydrate allied to starch, a fructosan polysaccharide, inositol, mannitol; volatile oil up to about 0.05%, consisting mainly of agropyrene; vanillin glucoside; mucilage, gum, large quantities of silica; iron, minerals, vitamins, K salt. Quackgrass seed contained oil 12%, triticin, mucilage 10%, saponins, sugar alcohols (mannitol, inositol, 2%–3%), essential oil with polyacetylenes or carvone (0.01%–0.05%), small amounts of vanilloside (vanillin monoglucoside), vanillin, and phenolcarboxylic acids, silicic acid, and silicates (Deane, 2014).

Used in: Ayurveda and Folk

Parts used: Rhizome, seeds, and root

Uses: Demulcent (used in cystitis, nephritis), aperient, diuretic and urinary antiseptic, and anticholesterolemic. Key application: In irrigation therapy for inflammatory diseases of the urinary tract and for the prevention of kidney gravel. It is contraindicated in edema due to cardiac or renal insufficiency. The juice of rhizomes is used for cystitis, nephritis, scirrhous liver; decoction for tonsils and as an adjuvant for cancer; also used for gout, rheumatism, and chronic skin disorders. Extracts show uric acid solvent properties (Khare, 2007; Al-Snaf, 2015b).

Methods of cultivation: Seeds

DOI: 10.1201/9781003263500-21

Ailanthus excelsa Roxb.

Synonyms: *Alianthus wightii* Tiegh.

Family: Simaroubaceae

Common names: Bengali: Mahaneem; **English:** Tree of Heaven; **Gugrati:** Aduso, Ardusi, Bhutrakho, Moto adusa; **Hindi:** Maharuk, Ardu, Ardusi, Arua, Horanim; **Kannada:** Doddamaru, Hemaraheera mara, Dodumani; **Malayalam:** Perumaram, Mattipongilyam; **Marathi:** Mahanimb, Maharukh; **Sanskrit:** Dirghavranta, Puutivrksha, Madala; **Tamil:** Perumaruttu, Periamaram, Peru, Perumaran, Pimaram, Pinar; **Telugu:** Peddamamanu, Putta Pee vepachettu

Distribution: Indigenous to Central and Southern India. Also distributed in Australia, China, Japan, and Malasia

Description: *A. excelsa* is a large deciduous tree, 18–25 m tall; trunk straight, 60–80 cm in diameter; bark light gray and smooth, becoming gray-brown and rough on large trees, aromatic, slightly bitter. Leaves: alternate, pinnately compound, large, 30–60 cm in length; leaflets 8–14 pairs, long stalked, ovate or broadly lance shaped from very unequal base, 6–10 cm long, 3–5 cm wide, often curved, long pointed, with hairy gland; edges coarsely toothed and often lobed. Flower: clusters droop at leaf bases, shorter than leaves, much branched; flowers many, mostly male and female on different trees, short stalked, greenish yellow; calyx 5-lobed; 5 narrow petals about 6 mm across; stamens 10; on male flowers, female flowers with 2–5 separate pistils, each with elliptical ovary, 1 ovule, and slender style. Fruit: a 1-seeded samara, lance-shaped, flat, pointed at ends, 5 cm long, 1 cm wide, copper red, and strongly veined (http://apps.worldagroforestry.org/treedb/AFTPDFS/Ailanthus_excelsa, PDF).

Flowering: Cold season **Fruiting:** Onset of monsoon

Chemical constituents: The bark contains several quassinoids and ailanthone derivatives. It also contains sitosterol and lactic acid.

Used in: Ayurveda and Siddha

Parts used: Leaves, stem, and bark

Uses: Bark, bitter, astringent, anthelmintic, antispasmodic, expectorant (used in asthma, bronchitis), and febrifuge. Also used for dysentery as a substitute for *Holarrhena antidysenterica*. Bark and leaves, used as tonic in debility, especially after childbirth. Leaves used as adulterant for *Adhatoda zeylanica* leaves. Along with other therapeutic applications, the *Ayurvedic Pharmacopoeia of India* indicates the use of stem bark in high fevers and giddiness. Dosage: stem bark 1–3 g. A decoction of 50–100 ml exhibits antitumor activity against P-388 lymphocytic leukemia and are cytotoxic against Kleihauer-Betke (KB) test system (Khare, 2007; Kumar et al., 2010). The root bark is used to cure epilepsy and heart troubles. The root bark is used to cure epilepsy and heart troubles. In Africa, the plant is used to treat cramps, gonorrhea epilepsy, tape warm infestation, and high blood pressure (Sharma, 1996)

Methods of cultivation: The tree can be raised from both seeds and stumps.

DOI: 10.1201/9781003263500-22

Ajuga bracteosa Wall. ex Benth.

Synonyms: *Ajuga integrifolia* Buch-Ham; *Ajuga remota* Benth

Family: Labiatae; Lamiaceae

Common names: English: Bugleweed, Carpet bugle; **Hindi:** Neelkanthi; **Sanskrit:** Neelkanthi

Distribution: From Kashmir to Nepal, sub-Himalayan tract, the plains of Punjab, and the upper Gangetic plains of India

Description: A low, diffuse, much-branched hispid herb, usually less than 20 cm tall, with rhizome and erect to ascending branches. Leaves: opposite, simple, oblanceolate to narrowly obovate or sub spatulate, 4–8 cm long 2–3 cm wide margin undulate; petiole short in lower leaves, absent in upper leaves; stipules absent. Inflorescence: an axillary verticillaster. Its yellowish flowers are congested in axillary spirals. Flowers are bisexual, 0.6–1.3 cm long, 5-merous; calyx campanulate, oblique, with triangular teeth; corolla gamopetalous, 2-lipped with upper lip very short and lower lip long and 3-lobed, white or pale blue; stamens 4, exserted; ovary superior, shortly 4-lobed, style bifid at the apex. Fruit: consist of 4 dry 1-seeded schizocarpous nutlets enclosed in the persistent calyx; nutlets obovoid, 1.5–2 mm long, 1 mm broad, shallowly rugose, reticulate, yellowish, with large scar (https://uses.plantnet-project.org/en/Ajuga_bracteosa_(PROSEA).

Flowering and fruiting: March–December

Chemical constituents: From the plant, ceryl alcohol, β-sitosterol, γ-sitosterol, daucosterol, hexacosan-1-ol, triacontan-1-ol-docosanoate, and cerotic and palmitic acids have been isolated. *A. bracteosa* contains various compounds such as neo-clerodane diterpenoids, flavanol glycosides, iridoid glycosides, ergosterol-5,8-endoperoxide and phytoecdysones (Kayani et al., 2014; Castro et al., 2011).

Used in: Ayurveda and Folk

Parts used: Powder of the whole plant and leaves

Uses: The herb is recommended in Ayurveda for the treatment of rheumatism, gout, palsy and amenorrhea. The herb has also been credited for its astringent, febrifugal, stimulant, tonic, and diuretic properties (Anonymous, 1985). Juice of the leaves: blood purifier. The powder is used for burns and boils. The leaves are used in fever as a substitute for cinchona. An aqueous extract of the leaves showed diuretic activity. The plant exhibited anticancer activity (Khare, 2007). Some investigations on *A. bracteosa* have reported the cancer chemopreventive and anti-plasmodial activities (Chandel and Bagai, 2010, 2011). *A. bracteosa* is regarded an alternative to cinchona (Pai and Pawar, 2011). Mothan et al. (2009) reported in vitro anti-proliferative activities of the crude methanolic extracts against three human cancer cell lines.

Methods of cultivation: Seeds

DOI: 10.1201/9781003263500-23

Albizia lebbeck (Linn.) Willd.

Synonyms: *Acacia lebbeck* (L.) Willd.; *Acacia macrophylla* Bunge; *Acacia speciosa* (Jacq.) Willd.; *Albizia latifolia* B. Boivin; *Mimosa lebbeck* Linn.; *Mimosa sirissa* Roxb.

Family: Mimosaceae

Common names: English: Siris tree, East Indian walnut; **Kannada:** Baage; **Malayalam:** Karimathakara, Karivaka, vaga; **Sanskrit:** Shirisha, Bhandi, Bhandila, Shitapushpa, Mridu pushpa, Kapitana; **Tamil:** Siridam, Vaagai

Distribution: All over India, from the plains up to 900 m in the Himalayas; also in the Andamans

Description: It is a tree growing to a height of 18–30 m tall with a trunk 50 cm to 1 m in diameter. Trunk: cylindric with extensively spreading branches. Bark: pale brown with irregular flakes. Leaves compound, bipinnate, 7.5–15 cm long, with 1–4 pairs of pinnae, each pinna with 6–18 leaflets. Inflorescence an axillary corymbose. Flowers: bisexual, greenish white, in subglobose heads, 3.5–3.7 cm long, solitary or 2–4 together in racemes; peduncle 6–8 cm; bracts linear, caducous; pedicels to 3 mm; calyx tube funnel-shaped pubescent, 3–4.5 mm in length, 1.5–3 mm broad, teeth triangular; corolla infundibuliform 7–10 mm long, greenish, lobes 5, lanceolate, pubescent, stamens many, filament tube shorter than corolla tube, filaments long exserted, green or pink; anthers very small; ovary subsessile, glabrous, style 2.5 cm, filiform; stigma minute. Fruit: a pod 20–30 cm long, 4–5 cm broad, flat, oblong, compressed, straw-colored, base and apex obtuse, turgid above the seeded region; seeds 8–12, 6–10 × 5–8 mm, ovate, dull dark brown, flattened (Kavitha et al., 2022).

Flowering: April–May **Fruiting:** December–January

Chemical constituents: The bark contains condensed tannin; also d-catechin and d-leucocyanidin. It showed anti-dermatophytic activity. Anthraquinone glycosides and its aglycone exhibited antibacterial activity.

Used in: Unani and Siddha

Parts used: Bark, seeds, flowers, and fruits

Uses: Antiseptic, antibacterial, antiallergic, anti-dermatosis, antidysenteric. Bark: used in bronchitis; bark and seeds in piles; root in hemicrania; flowers in cough, bronchitis, tropical pulmonary eosinophilia, and asthma. Pod: antiprotozoal. Along with other therapeutic applications, the *Ayurvedic Pharmacopoeia of India* indicates the use of stembark in rhinitis, sinusitis, and migraine (Khare, 2007).

Methods of cultivation: Seeds

DOI: 10.1201/9781003263500-24

Albizzia saman (Jacq.) F. Muell.

Synonyms: *Acacia propinqua* A. Rich; *Calliandra saman* (Jacq) Grisebn; *Enterolobium saman* (Jacq.) Prain; *Feuilleea saman* (Jacq.) Kuntze; *Inga cinerea* Willd; *Inga salutaris* Kunth; *Inga saman* (Jacq.) Wild; *Mimosa pubifera* Poir; *Mimosa saman* Jacq; *Pithecellobium cinereum* Benth; *Pithecellobium saman* (Jacq.) Benth; *Samanea saman* (Jacq.) Meer; *Zygia saman* (Jacq.) A. Lyons

Family: Fabaceae (Leguminosae)

Common names: English: Rain tree, Coco tamarind, French tamarind, Saman, French tamarind, Saman, monkeypod; **Bengali:** Biliti siris; **Gugrati:** Shirish; **Hindi:** Gulabi siris, Vilaiti siris; **Kannada:** Bhagaya mara; **Malayalam:** Chakkarakkay maram; **Marathi:** Vilaiti siris; **Sanskrit:** Shiriisha; **Tamil:** Amaivagai, Thoongumooni maram; **Telugu:** Nidra ganneru

Distribution: Native to Central and South America but has been widely introduced across the tropics, especially South and Southeast Asia. Found throughout India and Bangladesh

Description: Saman is a wide-canopied tree with a large symmetrical umbrella-shaped crown. It usually reaches a height of 50–80 ft and a diameter of 90 ft. Its branches have velvety and hairy bark. Stem is erect, cylindrical, branched, solid and green. Leaves, bipinnately compound, 15–40 cm long, velvety, with a circular gland at the base and usually between each of the pinnae; pinnae 4–6 opposite, 7–15 cm long, velvety, with small glands between most of the leaflets, leaflets 4–8 pairs, opposite, progressively larger upward, the end pair 4–5 cm long, 18–32 mm broad, unsymmetrical with the midrib curved inward and the outer margin more curved than the inner. The leaves fold in rainy weather and in the evenings, hence the names rain tree and five o'clock tree. Flowers: Rose pink in color, Calyx funnel-shaped, corolla tubular, Stamens 20, rose-colored, much exserted (https://apps.worldagroforestry.org/treedb/AFTPDFS/Albizia_saman.PDF). Fruit is a pod, large, black thick; 12–20 cm long, 1–2 cm long, 1.2 cm thick, indehiscent, curved and leathery; containing numerous seeds embedded in the pulp (https://en.wikipedia.org/wiki/Samanea_saman).

Flowering and fruiting: March–October

Chemical constituents: Chemical constituents of *S. saman* revealed the presence of octacosanol, α-spinasterol, β-Dglucose of α-spinasterol, kaempferol, and pithecolobine from the different parts of the tree. The volatile constituents of *S. saman* have been reported with palmitic acid (55.5%), 1,8-cineole (15.9%), and oleic acid (7.4%) as the major constituents (Isika et al., 2006). Another study reported the isolation of lupeol and epilupeol from the whole plant of *S. saman* (Ferdous et al., 2010).

Ragasa et al. (2014) reported from the dichloromethane extract of *S. saman* epilupeol and lupenone from the leaves and lupeol from the peduncles; α-spinasterol and α-spinasterone from the twigs. The alkaloids, pithecolobin, samarin, and steroidal compounds are also present in the plant (Anonymous, 1976).

Used in: Ayurveda and Folk

Parts used: Leaves, fruit, seeds, and stem bark

Uses: Medicinally, the plant is used in the treatment of diarrhea, stomach pain, and sore throat. It is also used as a laxative. A decoction of the inner bark and fresh leaves is used as a treatment for diarrhea, acute bacillary dysentery, enteritis, colds, sore throat and headache.

DOI: 10.1201/9781003263500-25

A brew of small sections of the bark is taken to treat stomachache. A crude aqueous or alcoholic extract of the leaves is observed to have an inhibiting effect on *Mycobacterium tuberculosis*. The alkaloid fraction of the leaves is effective on the central nervous system and pheripheral nervous system. An infusion of the leaves is used as a laxative. The fruit decoction is used as a central nervous system sedative. The seeds are chewed to treat sore throat (https://pfaf.org/user/Plant.aspx?LatinName=Samanea+saman). A decoction of fresh material is applied as external wash for anaphylactic dermatitis, eczema, and skin pruritus.

Methods of cultivation: Seeds

Alhagi pseudalhagi (Bieb.) Desv

Synonyms: *Alhagi camelorum* Fisch. ex DC. *Alhagi maurorum* Medic; *Alhagi pseudalhag* M. Biebi; *Hedysarum alhagi* L.; *Hedysarum pseudalhgi* (M.Bieb) Desv. ex B. keller & Shap.

Family: Papilionaceae (Fabaceae).

Common names: English: Camel thorn, Caspian manna, Persian manna plant; **Gugrati:** Javasau; **Hindi:** Javasa, Oont jhadi; **Kannada:** Nelingala; **Sanskrit:** Dhanvayasa, Duhsparshaa, Duraalabhaa, Kunaashak; **Tamil:** Punaikanjuri, Thurnjbin; **Telugu:** Tella giniyachettu

Distribution: The plant is found in Afghanistan, Iran, Iraq, North Africa, Pakistan, Russia, Syria, and Turkey. In India it is found in the drier parts of Rajasthan, Gujarat, Punjab, and Uttar Pradesh.

Description: It is an under shrub 2–3 ft tall, hairless. Leaves simple, entire 1–2.5 cm long 3–8 mm broad, elliptic-oblong or obovate, leaf stalk about 2 mm, stipulate stipules minute. Flowers are born in lateral leaf axis in racemes. Flowers: pink to reddish violate 6–9 mm long. Calyx: 2–2.5 mm teeth triangular. Fruit: 2–3.5 mm long broad hairless constricted between the seeds (https://www.flowersofindia.net/catalog/slides/Camel%20Thorn.html).

Flowering: April–September **Fruiting:** October

Chemical constituents: The aerial parts contain flavonoids, tannins, sterols, triterpenes, saponins and anthroquinones. The proanthocyanidins derived from the compounds prevented an increase in rat serum cholesterol and triglycerides, and they decreased the manifestation of atherosclerosis. A polymeric proanthocyanidin, extracted from. Ethanolic extract of the aerial parts.

Used in: Ayurveda, Folk, Siddha, and Unani

Parts used: Whole plant and flowers

Uses: It is used locally in folk medicine as a treatment for glandular, tumors, nasal polyps, and ailments related to the bile ducts (Srivastava et al 2014). It is an anti-bilious, diaphoretic, diuretic expectorant, and laxative. Leaves: used for fever, headache, rheumatism. Flowers: blood coagulant, used in the treatment of piles. Alhagi manna is antiemetic, expectorant, and laxative. Along with other therapeutic applications, The *Ayurvedic Pharmacopoeia of India* indicates the use of dried whole plant in gout and hemorrhagic disorders. The plant possesses hypolipidemic and anti-atherosclerotic properties. The plant, improved energy metabolism and increased the work capacity (Khare, 2007). Plant powder, about 20–50 g, is used in preparing decoction.

Methods of cultivation: Seeds

DOI: 10.1201/9781003263500-26

Allamanda cathartica Linn.

Synonyms: *Allamanda aubletii* Phol; *Allamanda hendersonii* W. Bull ex Dombrain; *Allamandalatifolia* C.presl; *Allamanda linnae* iPhol; *Allamanda salicifolia* hort.; *Echites verticifolius* Willd. ex Roem. & Schult

Family: Apocynaceae

Common names: English: Golden trumpet; **Hindi:** Pilaghanti; **Kannada:** Arsinhu; **Malayalam:** Kolambi; **Marathi:** Pivlee ghunti, Zahari Sontakkaa

Distribution: Native to Central America and Brazil. Grown in Indian gardens

Description: The plant is a vigorous, rambling woody climber that can get up to 15 ft tall. The leaves are obovate wavy-margined, about 16 cm long, and 6 cm wide and are born in whorls. Stipules are transformed into 4 small interpetiolar glands. Flowers are arranged in an axillary cume. Flowers are large about 5 in. across; trumpet-shaped blossoms are composed of 5 large overlapping petals that flare widely. There are both single and double forms. The clusters of flowers occur at the tip of new shoots. There can be one to several flowers in each cluster. Sepals: 5, lanceolate-ovate, acuminate, glabrous, 7–10 mm long. Corolla: tube about 2.5 cm long, throat infundibuliform, about 1.25 cm in diameter, lobes orbicular-rotundate, 3.5 × 4 cm, glabrous. Stamens inserted in the throat, acute, 3.5 mm long; filament very short; style, 2.5 cm long, filiform; fruit a globose-subglobose prickly capsule, about 2–3 cm long, prickles soft, 7.5 mm long, seeds many, obovate, flat, and winged. All parts of the parts contain white milky sap, which may cause rashes, blisters, and itching when in contact. https://indiabiodiversity.org/species/show/228638

Flowering and fruiting: Summer season

Chemical constituents: The root contains antileukemic iridoid lactone, allamandin, and two other iridoids, allamandicin and allamdin. The stems and leaves contain beta-amyrin, beta-sitosterol, and ursolic acid. Petals gave flavonoids—kaempferol and quercetin (Khare, 2007).

Used in: Folk

Parts used: Leaves, stem bark, and root

Uses: Leaves: cathartic (in moderate doses; emetic in large doses). Bark: hydragogue and in ascites. The purgative property of the aqueous extract of leaves was confirmed pharmacologically in rats. The extract also showed antifungal activity against ringworm-causing fungi. Flower extract inhibits fungal growth. Ethanol extract of roots showed in vivo activity against P-388 leukemia in mice and in vitro against human carcinoma cells of nasopharynx (Khare, 2007).

Methods of cultivation: Propagation by cuttings taken from April to May.

DOI: 10.1201/9781003263500-27

Allium ascalonicum Linn.

Synonyms: *Allium angolense*; *Allium aobanu*
Family: Liliaceae
Common names: English: Shallot, Spring onion; **Hindi:** Ek-daanaa Lehsun
Distribution: Native to Palestine; cultivated in India
Description: The shallot has a tapered shape and fine-textured coppery skin that differentiates it from onion. Inside shallots do not have concentric structures but are sectioned in to cloves like garlic. Shallot is a plant that can reach 20–30 cm in height. A cluster of radical leaves emerges from the underground bulb, cylindrical and fleshy, having sheathy leaf bases with parallel venation. Inflorescence: Scapigerous, that is, the inflorescence axis (peduncle) arising from the ground bearing a cluster of flowers at its apex. Pedicels are of equal length, arising from the apex of the peduncle, which brings all flowers to the same level. Flowers are small, white, bracteate, ebracteolate, pedicellate, complete, trimerous, actinomorphic, and hypogynous. Flowers are protandrous. Perianth: sepals 6, white, arranged in two whorls of three each, syntepalous showing valvate aestivation. Androecium: stamens 6, arranged in two whorls of three each, epitepalous, apostemetous, free and opposite to tepals. Anthers dithecous, basifixed, introrse, and dehiscing longitudinally. Gynoecium: tricarpellary and syncarpous. Ovary superior, trilocular with two ovules in each locule on axile placentation. Style simple, slender with simple stigma. Fruit: a loculicidal capsule. Seeds are minute and black-colored.
Chemical constituents: The aqueous extract of the bulbs contains mainly fructose (50%), besides glucose, arabinose, galactose, mannose, rhamnose, glucosamine, and uronic acid. It also contains lysine, glutamic acid and aspartic acid (Khare, 2007) in addition to vitamins it contains quercetin and kaempferol, two substances that are released when the shallot is crushed and which in turn produce another substance, allicin
Used in: Ayurveda and Unani
Parts used: Bulbs and leaves
Uses: The bulb is anthelmintic, anti-inflammatory, antiseptic, antispasmodic, carminative, diuretic, expectorant, febrifuge, hypoglycemic, hypotensive, lithontriptic, stomachic and tonic (Lust, 1983; Chiej, 1984). Also anticoagulant, fibrinolytic, and hypocholesterolemic. It shows significant activity against P-388 lymphocytic leukemia in mice. In experiments on rabbits, the alcoholic extracts of the bulb showed significant anticoagulant, fibrinolytic, and hypocholesterolemic properties (Khare, 2007).
Allicin released from the bulb also has very important property to lower the levels of bad cholesterol in the blood with beneficial effects in the prevention of arteriosclerosis and heart disease like heart attack and stroke
Methods of cultivation: Usually propagated by means of their bulbs, each one dividing up in the growing season to produce from 2 to more than 12 new bulbs.

DOI: 10.1201/9781003263500-28

Alocasia indica (Lour.) Spach.

Synonyms: *A locasia macrorrhiza* (Linn.) G. Don

Family: Araceae

Common names: English: Giant taro, Elephant ear; **Hindi:** Mankanda; **Kannada:** Balaraaksh, Karimanasangi, Mundi; **Marathi:** Kaasaalu; **Sanskrit:** Chhatraptra, Maankanda, Kasaalu, Hastikarni; **Tamil:** Merukan, kizhangu

Distribution: A genus of tropical Asia, Malayasia, and the Pacific. Found wild and cultivated all over India

Description: It is a massive perennial with huge elephant ear like leaves, 3–6 ft in length and 2–4 ft wide, borne on a leaf stalk about 2–4 ft long. The leaf stalk arise from an upright, stout trunk, which is about 6 ft tall. Flowers grow at the end of a short stalk (http://www.flowersofindia.net/catalog/slides/Giant%20Taro.html).

Chemical constituents: All parts of the plant, except tubers, contain cyanogenic principle, a mixture of triglochinin and isotriglochinin. The tubers contain sterols and high concentration of soluble oxalates (prolonged use may lead to calcium deficiency and oxaluria). The tubers contain a trypsin/chymotrypsin inhibitor. The plant contains HCN 0.0027%. It is found to be mitogenic (Khare, 2007).

Used in: Ayurveda and Siddha

Parts used: Leaf, petiole, and root

Uses: Rootstock, mild laxative, diuretic (in anasarca); used in inflammations and diseases of abdomen and spleen. Leaf, astringent, styptic, antitumor. Root and leaf, rubefacient. Tubers, used as vegetable after eliminating oxalate content. Dosage: Tuber—5–10 g powder (Khare, 2007). Plant used as a laxative, chopped-up roots and leaves as a rubefacient, and juice from the petiole against cough. The plants are applied for stimulating the skin, for example, in cases of fever and to remove blotches. The rhizome is sometimes used as a poultice to treat furuncles. The pounded stems are applied as a paste to snakebites and scorpion stings (https://uses.plantnet-project.org/en/Alocasia_(PROSEA_Medicinal_plants)#:~:text=macrorrhizos).

Methods of cultivation: Through corm and tuber

DOI: 10.1201/9781003263500-29

Althaea officinalis Linn.

Synonyms: *Althaea kragujevecensis* Pancisex Diklic & Stevan; *Althaea micrantha* Wiesb ex Borbas; *Althaea taurinensis* DC.; *Althaea vulgaris* Bubani; *Malva althaea* E.H.L. Krause; *Malva maritima* Salsib.

Family: Malvaceae

Common names: English: Marshmallow, Hollyhock; **Hindi:** Khatmi, Khaira; **Kannada:** Utpanna; **Marathi:** Khatmi; **Sanskrit:** Gulkairo, Khatmi; **Tamil:** Shemaitutti, Khatmii; **Telugu:** Khatmi

Distribution: Native to Eastern Europe found in Kashmir and Himachal Pradesh

Description: This is an herbaceous perennial that has stems that die down in the autumn. They typically grow to a height of 90–120 cm and put out only a few lateral branches. The leaves are shortly petioled, roundish, ovate-cordate, 50–75 mm long, and about 30 mm broad, entire or 3–5 lobed, irregularly toothed at the margin, and thick. They are soft and velvety on both sides, due to a dense covering of stellate hairs. Flowers hermaphrodite, lilac- to pink-colored are either axillary or in panicles, more often the latter, sepals and petals 5; stamens monoadelphous, purple to lilac-colored. Fruit are small, round, and hairy, which are popularly called "cheeses".

Flowering and fruiting: August–October

Chemical constituents: The root contains starch, mucilage, pectin, flavonoids, phenolic acids, sucrose, tannins, and asparagines. Mucilage (18–35%) consists of a number of polysaccharides. Flavonoids include kaempferol, quercetin, and diosmetin glucosides. Polyphenolic acids include syringic, caffeic, salicylic, vanillic, and p-coumaric acids (Khare, 2007). Investigations on seeds led to the isolation of three new phytoconstituents, identified as *n*-hexacos-2-enyl-1,5-olide (altheahexacosanyl lactone), 2β-hydroxycalamene (altheacalamene), and altheacoumarin glucoside, along with the known phytoconstituents lauricacid, β-sitosterol, and lanosterol (Rani et al., 2010).

Used in: Siddha and Unani

Parts used: Leaves, flowers, seeds, and roots

Uses: Demulcent, emollient, antitussive (used for cough, bronchitis, gastritis, enteritis, and cystitis), antilithic, and diuretic; leaf and root used in irritation of the oral and pharyngeal mucosa and associated dry cough; in mild inflammation of the gastric mucosa. As demulcent in gastroenteritis, peptic and duodenal ulceration, and common and ulcerative colitis. Topically for varicose veins, skin ulcers, abscesses, cuts, burns. The mucilages have proven biological activity including stimulation of phagocytosis in vitro. The root counters excess stomach acid, peptic ulceration and gastritis (Khare, 2007). Seeds, expectorant, anti-catarrhal, laxative, divergent, demulcent, concoctive of phlegm, emollient for stomach and organs, diuretic and emmenagogue. Leaves are anti-inflammatory; a poultice of leaves is used to treat abscesses, wounds, ulcers, and boils. Flowers are expectorant, exhilarant, and a tonic to heart and constipative. Roots analgesic, astringent demulcent, resolvent lubricant hemostatic.

Methods of cultivation: Seeds. Germination of seeds is poor; stratification of seeds can improve rate of germination.

DOI: 10.1201/9781003263500-30

Altingia excelsa Noronha

Synonyms: *Liquidambar altingiana* Blume.; *Liquidambar excesla* (Noronha) Oken.; *Liquidambar rasmala* Blume.

Family: Altingiaceae (Hamamelidaceae)

Common names: English: Storax, Oriental sweet gum; **Hindi:** Rasmala; **Malayalam:** Rasamalla; **Sanskrit:** Pindita, Shilaarasa, Silhaka, Turushka; **Tamil:** Neriyurishippal; **Telugu:** Sila rasam

Distribution: Bhutan, China, India, Malaysia, Myanmar, Nepal, Bangladesh, Cambodia, East Himalaya, Laos, Indonesia, Thailand, Tibet, Vietnam, Assam, Arunachal Pradesh

Description: Trees are about 20 m tall, young branches glabrous or sparsely pubescent, older growth lenticellate. Bark: Light gray, smooth with narrow, longitudinal fissures and irregular flakes. Leaves: Alternate, simple, elliptical to oblong or ovate to ovate-lanceolate, 6–12 cm long by 2.5–5.5 cm wide, with finely toothed margins. Stipules 2–6 mm; petiole slender, 2–4 cm, sparsely pubescent; leaf blade ovate or narrowly ovate, 8–14 × 4–7 cm, abaxially glabrescent, base rounded or subcordate, margin crenate-serrate, apex acuminate; lateral veins 6–8 on each side. Male inflorescences globose, many, arranged in raceme. Female inflorescence usually solitary in leaf axils, 14–22 flowers. Monoecious/unisexual. Male and female flowers are on separate inflorescences but on the same tree. Female flowers 4–18 rounded heads are borne together, 1.2–2.5 cm in diameter. Male racemes 6–14 heads per inflorescence, 1–2 cm long, consisting of masses of the stamen. Petals and sepals are absent. Peduncle 2–4 cm, pubescent. Male flowers: stamen filaments about 1 mm, glabrous; anthers somewhat longer than filaments. Female flowers: staminode teeth absent; styles 3–4 mm, pubescent. Fruit a capsules brown in color with four compartments Seeds many flattened, obovate and surrounded by a sweetly aromatic narrow wing, brown. (https://indiabiodiversity.org/species/show/248945; https://www.bimbima.com/health/altingia-excelsa/3839)

Chemical constituents: Aromatic alcohols: phenylpropyl-, cinnamic, benzyl alcohol; cinnamic acid (up to 30%): cinnamic acid esters, including among others cinnamyl cinnamate (styracine) and cinnamic acid ethyl esters (Khare, 2007).

Used in: Ayurveda, Folk, and Siddha Unani

Parts used: Resin, stem, and bark

Uses: Resin is carminative, stomachic, antiscorbutic expectorant, antipyretic, anti-inflammatory, antistress, carmative hepatoprotective. Externally used in scabies and leukoderma (Kapoor, 2001). The leaves are used to cure cough. The resin of the plant is dissolved in hot water or with honey for curing cough, dyspnea (difficulty in breathing), retention of urine, burning micturition, and dysmenorrhea. The resin is aphrodisiac (https://www.easyayurveda.com/2016/07/14/silhaka-altingia-excelsa-uses-research-side-effects/). The resin contained in the stems is anodyne, antiphlogistic, antiscorbutic, antiseptic, carminative, expectorant, stimulant, and stomachic. It is used in the treatment of affections of the throat; skin diseases such as scabies and leukoderma; it is smeared over the abdomen of children to relieve the pains of colic; it is applied to inflamed testicles to treat orchiditis and is useful in the early stages of hydrocele (Kapoor, 2001).

Methods of cultivation: The seed has a short period of viability and is best sown as soon as it is ripe. It is best soaked in cold water for 24 hours prior to sowing. The sowing medium is a mix of sand and soil (1:1). Germination starts about 10 days after sowing. One-month-old seedlings can be transplanted to plastic bags, preferably in a growing medium enriched with organic matter and fertilizer.

DOI: 10.1201/9781003263500-31

Amaranthus tricolor Linn.

Synonyms: *Amaranthus gangeticus* Linn.; *Amaranthus melancholicus* Linn.; *Amaranthus polygamus* Linn. Hook. f.; *Amaranthus tristis* Linn

Family: Amaranthaceae

Common names: English: Chinese spinach, Garden amaranth, Fountain plant; **Hindi:** Laal Saag, Laal Marashaa; **Kannada:** Dantina soppu; **Marathi:** Chavalaayi, Ranmaath, **Sanskrit:** Alpamarisa, Maarisha-rakta; **Tamil:** Arai-keerai, Sirukeerai, Mulakkerai; **Telugu:** Thota kura

Distribution: Cultivated throughout India

Description: It is annual erect stem stout, diffuse herb Attaining 1–2 m high, usually much branched. Roots have a tap-root shape cylindrical, yellow in color, with roots lets. Stem is purplish-pink with ridges and furrows, fracture, short, leaf simple 4–11 cm long, 2–8 cm wide very variable in shape. rhomboid, ovate, lanceolate, obtuse apex, petiolate, membranous, and purplish-pink seed 1.5 mm in diameter, biconvex, smooth, shiny black-colored. Inflorescences: an axillary, globase cluster up to 2.5 cm in diameter, the upper cluster sometimes forming a terminal spike with male and female flower intermixed: brown or shiny black seed faintly reticulate (Mahbubur Rahman and Iffat Ara Gulshana, 2014).

Chemical constituents: Plant possesses large numbers of chemical constituents that show various biological activity. The betalains pigments are betacyanins and betaxanthin. Are the major pigments present in *Amaranthus tricolor l.* and glycosylated flavonoids (kaempferol and quercetin; Das et al., 2013). Leaf possesses pigments, such as chlorophyll a (Chl a), chlorophyll b (Chl b), carotenoid (Ctn), betacynins, betaxanthine, and betalains (B). Total phenolic content: hydroxybenzoic acids—gallic acid (3,4–5trihydroxybenzoic acid), vanillic acid (4-hydroxy-3-methoxybenzoic acid), syringic acid (4-hydroxy-3,5-dimethoxybenzoic acid), p-hydroxybenzoic acid (4-hyroxybenzoic acid), salicylic acid (2-hydroxybenzoic acid), ellagic acid (2,3,7,8-tetrahydroxy-chromeno)—[5,4,3-cde]chromene-5,10-dione], benzoic acids. Hydroxycinnamic acids—caffeic acid (3,4-dihydroxy-trans-cinnamate), chlorogenic acids (3,4-[dihydroxycinnamoyl]quinic acid), P-coumarin acid (4-hydroxycinnamic acid), M-coumaric acid (3-hydroxycinnamic acid). Flavanoids—isoquercitin (quercetin-3-glucoside), rutin (quercetin-3-rutinoside), isorhamnetin, kaempferol, quercitin (Sowjanya et al 2015). Betalains are immonium derivatives of betalamic acid [4–(2-oxoethylidene)-1, 2, 3, 4tetrahydropyridine-2, [6-di carboxylic acid]. The plant is well known for its purple betalain pigments, such as amaranthine and isoamaranthine (Miguel, 2018). *Amaranthus* L. has calcium, iron, magnesium, potassium, zinc, protein, and vitamin C, A (Shukla et al., 2006; Kachiguma et al., 2015). The fatty acids and sterols are present in *Amaranthus* tricolor L. The major unsaturated fatty acid in the seeds and stems was linoleic acid, in leaves it was linolenic. The major saturated fatty acids found in seeds, stems and leaves was palmitic acid. Linolenic, lignoceric, and arachidic present in trace amounts; sterols and spinasterol was present in the highest amount, and 24-methylenecycloartenol was found in the seeds only (Fernando and Bean, 1984).

Used in: Ayurveda and Folk

Parts used: Whole plant

Uses: The whole plant is traditionally used as an astringent (Chopra et al., 1956) *Amaranthus tricolor* is one of the traditional medicines used in many folk claims, and the plant has been extensively used in Ayurveda and Sidda for treating menorrhagia, diarrhea, dysentery, hemorrhages, cough, and bronchitis. It is also used externally as an emollient poultice

DOI: 10.1201/9781003263500-32

or a mouthwash to treat ulcerated conditions of the throat and mouth (Misra, 2004) This plant is also used for variety of treatments for various ailments like coughs, throat infections, toothache, eczema, piles, diarrhea, gonorrhea, leucorrhea, and impotence (Aneja et al., 2011). Whole plants are used by folk medicinal practitioners of Bangladesh for treatment of pain, anemia, skin diseases, diabetes, and as a blood purifiers. The root decoction of *Amaranthus tricolor*, along with *Cucurbita moschata*, is used to control hemorrhage following abortion (Duke and Ayensu, 1985). The plant decoction is taken internally to strengthen the liver and improve vision. Traditionally root paste mixed with warm water, when given internally, induces vomiting and purges toxic matter from bowels. Powdered root is effective in onychia. Root-paste in combination with honey and rice water is useful in leucorrhea (Prajapati and Kumar, 2003). The leaves and young stems may be used as salad and cooked as a leafy vegetable. Scientific study on the plant suggests that it may inhibit calcium retention (Larsen et al., 2007). The plant possesses hepatoprotective activity (Al-Dosari, 2010).

Methods of cultivation: Seeds

Amberboa divaricata Kuntz.

Synonyms: *Amberboa ramosa* (Roxb.) Jafri.; *Oligochaeta divaricata*; *Volutarella divaricata* Benth. and Hook. F.; *Tricholepis procumbens* Wight

Family: Asteraceae (Compositae)

Common names: English: Branched sweet sultan; **Gugrati:** Bhoydandi; **Hindi:** Badaward; **Marathi:** Sakaj; **Sanskrit:** Brahmadandi

Distribution: Found in Afghanistan, Pakistan, and India, found in Uttar Pradesh, Bihar, Orissa, Madhya Pradesh, Rajasthan, Andhra Pradesh, Karnataka, and Tamil Nadu.

Description: It an annual straggling herb, dichotomously branched, spreading; branches 2.5–5 cm long, angled, smooth or rough. Leaves are usually oblong or obovate, entire, toothed or pinnately cut, 2.5–5 × 0.6–1.5 cm; lobes with a short sharp point, often wavy or crisped. Flower heads are about 1.2 cm across, pink. Involucral bracts are ovate, with long spreading or recurved spinescent awns, glabrate, reddish; spines about 6–8 mm long, smooth. Receptacular bristles short. Flowers are about 1.2 cm long, straight, pale purple. Achenes are narrow about 5 mm long, 4–5-angled, grooved and dotted between angles. Pappus is silvery brown, about 1.2 cm long.

Flowering and fruiting: April–December

Chemical constituents: Contains acid resins, alkaloids campesterol, cycloartane-type triterpenoids, cynaropicrin, flavanol glycosides, green essential oils, guaianolides, gums, fructose, jaceosidine, lupeol, organic acids, sigmasterol, sitosterol, sesquiterpenoids, and triacontane (Ibrahim et al., 2010). From the aerial parts lupeol acetate, stigmasterol, β-sitosterol, lupeol, aguerin-B, cynaropicrin, desacyl cynaropicrin, betulinic acid, and β-sitosterol-D-glucoside were reported by Singh and Mewara (2018).

Used in: Ayurveda, Unani

Parts used: Whole plant and seeds

Uses: Deobstruent, aperient, febrifuge, nervine (used in debility), antiseptic (used in leukoderma), antidiarrheal, antiperiodic, antipyretic, anticough, and skin disorders. (Iqbal et al., 2014). Seeds are antidote, astringent, and resolvent (Bhattacharjee and De, 2005) For blood purification, juice of fresh plant is used with black pepper (Qureshi and Bhatti, 2008).

Methods of cultivation: Seeds

DOI: 10.1201/9781003263500-33

Ammania baccfera L.

Synonyms: *Ammannia apiculata* Koehne; *Ammannia attenuata* Hochst. ex A. Rich.; *Ammannia densiflora* Miq. ex C.B. Clarke; *Ammannia indica* Lam.; *Ammannia prostrata* Buch.-Ham. ex

Family Lythraceae

Common names: Bengali: Banmarich; **English:** Blistering ammannia, Monarch red-stem; **Hindi:** Jungli mehendi, Ban mirich, Kuranta; **Kannada:** Kaadugida; **Malayalam:** Kallur vanchi, Nirumelneruppu; **Marathi:** Aginbuti, Dadmari; **Sanskrit:** Agnigarbha, Kshetrabhusha, Mahasyama, Pasanabheda; **Tamil:** Kal-l-uruvi; **Telugu:** Agnivendrapaku, Pasanabheda, Nelachitramoolam

Distribution: Asia, Africa, Australia, India: Assam, Bihar, Gujarat, Kerala, Madhya Pradesh, Maharashtra, Rajasthan, Odisha, Tamil Nadu; found in paddy fields and other wetlands

Description: The plant is erect, branched, smooth, slender, annual, more or less purplish herb 10–50 cm in height. The stems are somewhat 4-angled. The leaves are oblong, oblanceolate, or narrowly elliptic, about 3.5 cm long—those on the branches very numerous, small, and 1–1.5 cm long—with narrowed base and pointed or somewhat rounded tip. The flowers are small, about 1.2 mm long, greenish or purplish, and borne in dense axillary clusters. The capsules are nearly spherical, depressed, about 1.2 mm in diameter, purple, and irregularly circumscribe above the middle. The seeds are black (Wiart, 2006b).

Flowering and fruiting: September–December

Chemical constituents: The plant containβ-sitosterol-3-O-β-glucopyranoside, quercctin-3-rutinoside (rutin), kaempferol-3-O-β-glucopyranoside and quercetin-3-O-α-L-rhamnoside (quercitrin). The plant is also reported to contain hentriacontane, dotriacontanol, betulinic acid, lupeol, ellagic acid, and lawsone (Joshi, 2007; Al-Snafi, 2015a).

Used in: Ayurveda and Siddha

Parts used: Entire plant and leaves

Uses: The plant has antirheumatic, costiveness, febrifuge laxative, and stomachic properties; it also promotes digestion and stimulates venereal desire. Leaves are used externally for ringworm, herpes eruptions, and other skin diseases; they are also used to increase appetite and relieve the bowels of costiveness to promote digestion and stimulate venereal desire. The bitter herb is an appetizer and stomachic and is useful in treating biliousness, stomach pains, constipation, and excessive flatulence; the leaves are beneficial for removing phlegm from the lungs and trachea. According to Ayurvedic pandits, the herbal extract is a good remedy for tuberculosis and typhoid fever. The plant juice mixed with ginger extract is helpful for curing fevers. Tribal believe that the herb is an effective remedy for all blood diseases. In India, the leaves are used to reduce the sexual libido in animals. The leaves are acrid and find application in folk medicine for the treatment of rheumatic pain, as laxative, rubefacient, and external remedy for ringworm (Kirtikar, 1975). This plant is found to possess hypothermic, hypertensive, anti-urolithiasis, antibacterial, and CNS-depressant activities (Dhar et al., 1973; Al-Shamma and Miyscher, 1979). Recent studies have demonstrated that the ethanol extract from *A. baccifera* possess anti-steroidogenic (Dhanapal et al., 2005), antioxidant, and hepato-protective activities (Lavanya et al., 2010).

Methods of propagation: Seeds

DOI: 10.1201/9781003263500-34

Amomum aromaticum Roxb

Synonyms: *Alpinia fasciculata* (Roscoe) Steud.; *Amomum fasciculatum* (Roscoe) Benth. & Hook.f. ex B.D. Jacks; *Cardamomum aromaticum* (Roxb.) Kuntze; *Geocallis fasciculata* (Roscoe) Horan; *Renealmia fasciculata* Roscoe

Family: Zingiberaceae

Common names: Bengali: Morang elachi; **English:** Black cardamom, Bengal cardamom; **Gugrati:** Elcha; **Hindi:** Badi elaachi, Morang-elaaichi; **Sanskrit:** Brihad, Sthula-elaa; **Tamil:** Pera elam; **Telugu:** Tengu elakuku

Distribution: Nepal, Bangladesh, Sikkim, Meghalaya, Arunachal Pradesh, Nagaland, and West Bengal

Description: A rhizomatous perennial herb with leafy stems. Leaves oblong-lanceolate, glabrous on both sides, 15–30.5 × 5–10 cm. Spikes radical, shortly peduncled, globose; corolla-tube 2.5 cm; segments obtuse, nearly as long as the tube, white; lip pale yellow, as long as the corolla-segments, outer half deflexed. Fruit a capsule 1 in., oblong, trigonous ovate, 2.5 cm, somewhat rugose (https://indiabiodiversity.org/species/show/258865).

Flowering and fruiting: Winter

Chemical constituents: The seeds contain a chalcone (cardamonin), a flavonoid (alpinetin), petunidin-3, 5-diglucoside, and leucocyanidin glucocide; also a aurone glycoside subulin. The seeds yield on essential oil (1%–1.2%) containing cineole as the principal constituent (Khare, 2007).

Phytochemical analysis revealed that the major components of the volatile oils included 1,8-cineole (48.22%), geranial (9.24%), neral (6.72%) α-pinene (2.43%), and α-terpineol (2.28%) (Dang et al., 2020).

Used in: Ayurveda, Folk, and Siddha

Parts used: Seeds and pericarp

Uses: Stomachic, antiemetic, anti-bilious, astringent, alexipharmic; used for the treatment of indigestion, eradicate bad odor, biliousness, abdominal pains, vomiting, in congestion of liver. Pericarp: in headache and stomatitis. Dosage: seed—1–3 g powder. The oils of seeds have been used in India for benefiting the digestive system, applied to the eyelids to eliminate the inflammation (Khare, 2007; Sharma et al., 2004; Kala, 2005).

Methods of cultivation: Seeds

DOI: 10.1201/9781003263500-35

Amorphophallus campanulatus (Roxb.) Blume ex Decne

Synonyms: *Amorphophallus paeoniifolius* (Dennst.) Nicolson

Family: Araceae

Common names: English: Elephant-foot yam; **Hindi:** Zamikand; **Malayalam:** Chena; **Sanskrit:** Suurana, Kanduula, Arshoghna, Kand-ayak, Kandala; **Tamil:** Chenaikkizhangu, kaaraa karanai kizangu; **Telugu:** Kandagadda

Distribution: Native to tropical Asia; cultivated throughout India

Description: This is a robust herbaceous plant, with an erect solitary stem usually 1–2.5 m in height and bearing at the top one or two tripartite leaves, each part of which is deeply dissected into numerous segments. Toward the end of the plant's cycle (usually 4–6 years), a large terminal inflorescence is produced, consisting of a short stalk and spathe and a spadix. The pistillate (female) and staminate (male) flowers are on the same plant and are crowded in cylindrical masses as an inflorescence. The top part is responsible for secreting mucus that gives off a putrid, pungent smell that is used to attract pollinating insects, the middle part of the inflorescence contains staminate, and the base of the inflorescence contains pistillate flowers.

Chemical constituents: The corm contains an active diastatic enzyme amylase, betulinic acid, tricontane, lupeol, stigmasterol, beta-sitosterol, and its palmitate and glucose, galactose, rhamnose and xylose (Khare, 2007). It is also high in vitamin B6. It is loaded with potassium, magnesium, and phosphorous, as well as with trace minerals like selenium, zinc, and copper.

Used in: Ayurveda, Siddha, and Unani

Parts used: Corm

Uses: Corm is prescribed in bronchitis, asthma, abdominal pain, emesis, dysentery, enlargement of spleen, piles, elephantiasis, diseases due to vitiated blood and rheumatic swellings. Along with other therapeutic applications, the *Ayurvedic Pharmacopoeia of India* indicates the use of corm in prostatic hyperplasia. (The corm is an irritant due to the presence of calcium oxalate. It can be consumed after it is washed well and boiled in tamarind water or butter milk; Khare, 2007.) The tuber is reported to be useful in treatment of piles (Dey et al., 2016).

Methods of cultivation: Corm

DOI: 10.1201/9781003263500-36

Amorphophallus konjac K. Koch

Synonyms: *Amorphophallus rivieri* Durieu ex Carriere; *Amorphophallusmairei* H. Lev
Family: Araceae
Common names: English: Voodoo lily, Snake palm, Elephant yam; **Malayalam:** Chena
Distribution: It is native to Yunnan in China and cultivated in warm sub- to tropical East and Southeast Asia, from China and Japan south to Indonesia and Vietnam
Description: Plants can grow up to 4–6 ft tall. It is a perennial plant, growing from a large corm up to 25 cm in diameter. The single leaf is up to 1.3 m across, bipinnate, and divided into numerous leaflets. The fleshy leaf stalk (petiole) is mottled pinkish gray and olive green. The single intricate leaf consists of a horizontal blade on the vertical petiole, which is divided into three sections, giving an umbrella-like effect. The flowers are produced on a spathe enclosed by a dark purple spadix up to 55 cm long. The globose tuber can grow up to 50 lb and 1 ft in diameter (https://en.wikipedia.org/wiki/Konjac).
Chemical constituents: The dried corm of the konjac plant contains approximately 40% glucomannan gum. This polysaccharide makes konjac jelly highly viscous and may be responsible for many of its health benefits as used in traditional Chinese medicine (TCM), detoxification, tumor suppression, blood stasis alleviation, and phlegm liquefaction (Chang Zhongfei et al., 2012).
Used in: Traditional medicine and Ayurveda
Parts used: Corms
Uses: Used in reduction of serum cholesterol and in constipation, control of blood glucose. In TCM, a gel prepared from the flour has been used for detoxification, tumor suppression, blood stasis alleviation, and phlegm liquefaction and for the treatment of asthma, cough, hernias, breast pain, burns, and skin disorders (Niwa et al., 2000; Khokar and Menghani, 2015). For more than 2,000 years, it has been consumed by the indigenous people of China for the treatment of asthma, cough, hernia, breast pain, and burns, as well as hematological and skin disorders (Liu et al., 1998). The root is oxytocic and sialagogue. It is used in the treatment of cancer (Duke and Ayensu, 1985; https://tropical.theferns.info/viewtropical.php?id=Amorphophallus+konjac). It also relieves constipation and irritable bowel syndrome, as well as reducing serum cholesterol levels (Brown, 2000).
Methods of propagation: Tubers. Tubers are planted when the pinkish growing tips start to show. Because the roots do not come from the base of the tuber but from the top and grow out horizontally to help support the leaf, the tubers should be placed deeply below the surface.

DOI: 10.1201/9781003263500-37

Anacyclus pyrethrum DC

Synonyms: *Anacyclus officinarum* Hayne; *Anacyclus depressus* Ball.; *Anacyclus freynii* Wilk.; *Anthemis pyrethrum* L.

Family: Compositae; Asteraceae.

Common names: English: Spanish pellitory, Pyrethrum root; **Hindi:** Akkal kara; **Marathi:** Akkal Kadha; **Sanskrit:** Aakaarakarabha; **Tamil:** Akkiraakaaram

Distribution: Native to the Mediterranean region; cultivated in Algeria; can be found in Jammu and Kashmir, and Bengal

Description: This is a perennial herb with numerous stems that lie on the ground for part of their length, before rising erect. Leaves alternate, pinnate, smooth. They are pale green in color and have deeply cut segments. Each of the stem bears one large flower, the disk being yellow and the rays white, tinged with purple beneath.

Chemical constituents: Phytochemicals like alkaloids, coumarins, flavonoids, and tannins are present in the Anacyclus pyrethrum variety. The root extract contains free fatty acids, sterols, and unsaturated amides. Pellitorin, anacyclin, phenylethylamine, inulin, polyacetylenic amides, and sesamin. The root contains anacycline, isobutylamide, inulin and a trace of essential oil (Usmani et al., 2016).

Used in: Ayurveda, Siddha, and Unani

Parts used: Root

Uses: Stimulant, cordial, rubefacient. A gargle of infusion is prescribed for relaxed vulva. Root used for toothache, rheumatic and neuralgic affections, and rhinitis. Roots, along with the root of *Withania somnifera* and *Vitis vinifera*, are used in epilepsy. Along with other therapeutic applications, the *Ayurvedic Pharmacopoeia of India* indicates the use of the root in sciatica, paralysis, hemiplegia, and amenorrhea. Use of the drug in patients with insulin-dependent diabetes mellitus reduces the dose of insulin. It decreased the plasma glucose and serum cholesterol levels after oral administration for 3–6 weeks. (The plant is mixed with *Helleborus niger* in a ratio of 1:3.) Dosage: Root powder 500 mg–1 g (Khare, 2007). It has antidepressant benefit. It is beneficial in enhancing memory It also shows cognitive improvement It has also shown anticonvulsant activity It also reduces oxidative stress in the brain (Pahuja et al., 2012) Extract of *Anacyclus pyrethrum* have shown a potent libido-enhancing ability (Zaidi et al., 2013; Usmani et al., 2016). Akarkara can be used in powdered form. The powder can be taken with honey (Pandey et al., 2018).

Methods of cultivation: Seeds

DOI: 10.1201/9781003263500-38

Anchusa italica Retz

Synonyms: *Anchusa azurea* Mill.
Family: Boraginaceae
Common names: English: Cow's tongue plant, Garden anchusa, Italian bugloss, or Italian alkanet; **Sanskrit:** Gojihvikaa
Distribution: Distributed in the temperate, especially in Mediterranean and tropical regions. Cultivated in Indian gardens, and hills
Description: It is a bristly perennial which may reach 3–5 ft and 60 cm wide. It has straight lance-shaped leaves that are dark green coarse and upright. Flowers are born in panicles on erect stems. Flowers: 10–15 mm across, petite tubular about 15 mm across with 5 sepals and bright violet-blue petals. Stamens: inserted at the top of the tube overlapping the scales. Fruit: nutlets about 7–10 mm long and 3–4 mm broad oblong and straight.
Flowering: June–August **Fruiting:** July–September
Chemical constituents: Oil is a rich source of vitamin E (0.72%). The nutlets show positive tests for alkaloids and tannins. The flowers yield anthocyanins and the leafy stems yield bornesitol (Khare, 2007).
Used in: Ayurveda and Unani
Parts used: Leaves and flowers
Uses: Stimulant, tonic, demulcent; used in bilious complaints, fever, cough, asthma; as diuretic for bladder and kidney stones (Khare, 2007). *A. italica* was used traditionally as stimulant, tonic, demulcent, in bilious complaints, fever, cough, and asthma and as diuretic for bladder and kidney stones (Kebriaee-Zadeh, 2003; Amin, 2005; Khare, 2007). It is also used as diaphoretic, narcotic, hypnotic, anti-arthritis, antirheumatic, and cathartic (Al-Quran, 2008). The leaves of the plant were used as decoction in cold, sore throat, and chest pain (Safa et al., 2013). It has been widely studied to exert antioxidant, antiinflammation, and neuroprotective effects against cerebral ischemia and reperfusion (Al-Snafi, 2014; Torki et al., 2018). *A. italica* possessed many pharmacological effects; these included anticancer, antioxidant, antiviral, central nervous system, endocrine system, and many other effects (Al-Snafi, 2014).
Methods of cultivation: Seeds

DOI: 10.1201/9781003263500-39

Andrographis echioides (L.) Nees

Synonyms: *Erianthera echioides* (L.) M.R. Almeida; *Erianthera lobelioides* Nees.; *Indoneesiella echioides* (L.) Sreemadh.; *Justicia echioides* (L.)

Family: Acanthaceae

Common names: English: Bugloss chiretta, False water willow; **Gujarati:** Kalukariyatun; **Kannada:** Godibarsana; **Malayalam:** Gopuramthang, Pitumba; **Marathi:** Ranchimani; **Tamil:** Kopuramtanki Gopuram tangi

Distribution: The species is native to the Indian subcontinent (including Myanmar, Sri Lanka, and the West Himalaya region)

Description: Erect herbs, 10–50 cm high. Whole plant is villous. Leaves: opposite decussate, about 7.5 cm long and 2.5 cm wide, elliptic oblong, obtuse to round at apex, attenuate at base and hairy on both sides. Inflorescence: spike-like racemes, axillary, short simple or 1–2 branched, the stalk carrying the raceme is densely hairy. Flowers: erect. Calyx: deeply 5-lobed; lobes 0.8–1 cm long, linear. Corolla: tube 0.8–1 cm long, white, 2-lipped; lip with violet blotches; upper lip 2-lobed, 0.6–0.8 cm long, lower about 0.8 cm long, 3-lobed. Stamens: 2; filaments broad, hairy at base; anthers 2-celled. Style slender; stigma bifid. Capsules: 0.8–1.2 cm long, 0.25–0.3 cm wide, compressed, broadened toward tip, attenuate at base, sparsely hairy toward tip. Seeds 0.15–0.2 cm across, black, pitted without retinacula, spiny. https://www.flowersofindia.net/catalog/slides/Bugloss%20Chiretta.html

Flowering and fruiting: April–July

Chemical constituents: From leaves extract of *A. echiodies*, various chemical constituents were isolated: dihydroechioidinin, skullcap avone 1 2'-methyl ether, echioidinin, echioidin, skullcap avone 1 and 2'-O-b-D-glucopyranoside. Some of the other chemical constituents include borneol (2.6%); cyclohexanol 2,4 dimethylphenol (1.89%); 3,4 altroson (3.5%); n-deconoic acid (15.29%); squalene (22.57%); vitamin E (7.40%); methoprene (1.53%); 2-nonenlolOxirane, octyl-, 2, 2-cyclopentene-1-undecanoic acid, ketone, 1,5-methylbicyclo [2.1.0] pent-5-ylmethyl (10.61%); and 2,5-cyclohexadiene-1,4-dione, 2, 5-dihydroxy-3-methyl-6-(1-methylethyl) bicycle heptan-3-one, 2,6,6-trimethyl (1à,2á,5à) (1.60%); and (-)-3-á-Acetoxy-5-etienic acid (3.0%; Nirubama et al., 2014).

Used in: Siddha

Parts used: Whole plant and leaves

Uses: Whole plant: pounded in mustard oil and paste applied on snakebite, also whole plant dried and powdered, mixed with water and drunk; whole plant decoction in fever, dysentery, stomachache, and liver disorders. Leaves: juice used as a febrifuge, cathartic, anthelmintic, laxative, alterative, stomachic, antimalarial; leaf paste applied to cure headache; leaves decoction given in fever and jaundice, against intestinal worms and for constipation. Properties and uses of this plant are considered similar to those of *Andrographis paniculata* (Quattrocchi, 2012).

Methods of propagation: Seeds and stem cuttings

DOI: 10.1201/9781003263500-40

Anemonastrum obtusilobum (D. Don.) Mosyakin

Synonyms: *Anemone pulsatilla* Linn. *Anemone discolor*, Royle;*Anemone micrantha* Klotzsch; *Anemone mollis* Walli. *Anemone obtusiloba* D. Don; *Pulsatilloides obtusiloba* (Don) Starod

Family: Ranunculaceae

Common names: English: Windflower, Wood anemone, Pasque flower, Pulsatilla, Himalayan thimbleweed; **Hindi:** Ratanjot

Distribution: Native to mountainous regions of Pakistan, Myanmar, Nepal, Western China, and Tibet; also cultivated. In India, found in Himalayas from Kashmir to Sikkim at an altitude 9000–15,000 ft

Description: A perennial herb, densely tufted, glabrate, or softly hairy. Rootstock: woody, fibrous, clothed with old root sheaths. Leaves: many stalked, sub-orbicular, deeply cordate 2–5 cm across with rounded outline, deeply 3-lobed; the lobes are further lobed or cut rarely shortly petiolate. Flowers are born on short, spreading, tufted stems about 5–15 cm long, about 2.5 cm, are purplish-white or golden; pedicels long, slender, petals 5–7 elliptic silky-haired beneath. Sepals: silky outside, generally lead-colored near the claw. Achenes: strigose, rarely glabrous, very variable in size.

Flowering and fruiting: April–July

Chemical constituents: Pulsatilla contains ranunculin, which hydrolyzes to a toxic, unstable compound, protoanemonin, which readily dimerizes to nontoxic anemonin. Anemonin and protoanemonin exhibit sedative and antipyretic activity (Khare, 2007).

Used in: Folk and Unani

Parts used: Seeds and root

Uses: Nervine and sedative used for anxiety neurosis, nervous exhaustion, tension, headache, migraine, insomnia, antispasmodic (in catarrh); used for amenorrhea, inflammation of ovaries, painful menstruation, and genito-urinary infections. The rootstock is given with milk for concussions. The seeds cause vomiting and purging. The seed oil is used in rheumatism. Pulsatilla is used for infectious diseases of the skin (Chopra et al., 1986; Khare, 2007).

Methods of cultivation: Seeds

DOI: 10.1201/9781003263500-41

Angelica archangelica Linn.

Synonyms: *Angelica commutata* (C.A. Mey. ex Rupr.) M. Hiroe; *Angelica discocarpa* (Fr.) M. Hiroe; *Angelica intermedia* Schult. ex Steud.; Fr.; *Ligusticum angelica* Stokes; *Selinum archangelica* (L.) Vest

Family: Apiaceae (Umbelliferae)

Common names: English: Garden angelica, Wild celery, Norwegian angelica; **Sanskrit:** Chandaa, Chandaam shuka, Kathachoraa

Distribution: Native to Syria; now grown in Kashmir at 1000–3900 m

Description: Perennial plants that often are monocarpic (meaning that they will die after the flowering and fruiting has ended). Leaves are large, 2–3 pinnate with wide lobes the petioles are enlarged and inflated. Inflorescence is convex with a lot of small flowers gathered with white, greenish, or pink petals and even sometimes yellowish. Fruit is compressed, which gives it an oblong to ovate or oblong shape with 4 wings and visual ridges. During its first year, it grows only leaves, but during its second year, its fluted stem can reach a height of about 8 ft. Its leaves consist of numerous small leaflets divided into three principal groups, each of which is again subdivided into three lesser groups. The edges of the leaflets are finely toothed or serrated. The flowers are small and numerous, yellowish or greenish, and are grouped into large, globular umbels that bear pale yellow oblong fruits. https://en.wikipedia.org/wiki/Angelica_archangelica

Flowering and fruiting: July–August

Chemical constituents: *Angelica* seeds have a similar chemical composition to the roots, including α-pinene, β-pinene, camphene, myrcene, β-phellandrene, limonene, caryophyllene, borneol, carvone, and others, such as vaginate, archangelicin, oxypeucedanin hydrate, bergapten, byakangelicin angelate, imperatorin, isoimperatorin, and isopimpinellin (Burdock, 2016). The roots of *A. archangelica* contain osthol, ostruthol, oxypeucedanin, phellopterin, psoralen, and xanthotoxin, (Härmälä et al., 1992).

Used in: Homeopathy

Parts used: Stem, leaves, seeds, and root

Uses: Expectorant, carminative, digestant, cholagogue, antispasmodic, diaphoretic, diuretic, anti-inflammatory, smooth muscle relaxant, antifungal, and antibacterial. Key application: in flatulence and peptic discomforts. The root is a component in numerous gastrointestinal, cholagogue, and biliary remedies. Dosage: root—3–5 g powder (Khare, 2007). The dry extract of root has been shown to have anti-inflammatory activity. The root is reported to inhibit bacterial and fungal growth. Angelicin, a resin, is stimulating to the lungs and skin, contraindicated in bleeding disorders, peptic ulcers, and pregnancy (Khare, 2007). *A. archangelica* has been used in traditional and modern medicine to cure nervous headache, fever, skin rashes, wounds, rheumatism, and toothaches (Sarker and Nahar, 2004).

Methods of cultivation: Seeds

DOI: 10.1201/9781003263500-42

Angelica glauca Edgew

Synonyms: *Angelica nuristanica* Kitam.
Family: Apiaceae (Umbelliferae)
Common names: English: Angelica; **Hindi:** Chora, Gandrayan; **Sanskrit:** Choraka, Chorakaa, Kopanaa, Chorakaakhya, Nishaachara, Dhanhar, Taskara, Kshemaka
Distribution: It is endemic to India. The plant is endangered in status, which is distributed in Western Himalaya from Kashmir to Uttarakhand, in alpine scrub and forest shades between 2700–3700 m
Description: It is a glabrous aromatic perennial or biennial herb, 1–2 m tall glabrous, aromatic. The stem is hollow and ribbed, and root is thick and rhizomatous. Leaves are long stalked triangular 20–30 cm long 15–20 cm broad, unipinnate, bipinnate, or tripinnate; the large pinna is toothed, ovate, or lanceolate. The inflorescence is a compound umbel with numerous rays. Flowers are white, yellow, or purple in color, bracteate; florets are white or purple. Fruit is oblong-ellipsoidal 0.8–1.2 cm long and 4–6 mm broad lateral ridges expanded into wings and bearing single vittae that is the key to the species.
Flowering: June–August **Fruiting:** August–October
Chemical constituents: The root contains furocoumarins, also dimeric, lingusticum, and lactone (Khare, 2007). Roots contain valeric acid, angelic acid, and angelisine resin (Butola and Vashistha, 2003). Steam distillation of the roots yields a pale to brownish-yellow essential oil (0.4%–1.3% dry basis), which contains various compounds (Agnihotri et al., 2004).
Used in: Ayurveda
Parts used: Roots and rhizome
Uses: Root: cordial and stimulant, carminative (used in constipation), expectorant, diaphoretic. Dosage: root—3–5 g powder (Khare, 2007) *A. glauca* is used in dyspepsia, constipations, ulcer of palate, infantile atrophy, dysentery, menorrhagia, and rinderpest. The roots and rhizomes are used as a drug for wounds and gastric pains. It has stomachic, carminative, stimulant, and sudorific properties. It is useful in anorexia, spasms, flatulent colic, and bronchitis. The powdered root is administered in warm water for children's stomach ailments; it also checks vomiting. It is also used for flavoring confectionery and liquors. The powder of the root is applied over affected area and given internally to counter act the poison of snakes and scorpions. The powder of the root about 3–4 g. It is given to treat psychic and mental disorders. The leaves and stem act as a stimulant, cordial, and is used in stomach troubles
Methods of cultivation: Seeds and rootstock split. Seed viability is very low, and moist seeds have a better germination potential than dry seeds. Germination takes 25–40 days.

DOI: 10.1201/9781003263500-43

Annona muricata L.

Synonyms: *Annona bonplandiana* Kunth.; *Annona cearaensis* Barb. Rodr.; *Guanabanus muricatus* (L.) Gomez.

Family: Annonaceae

Common names: English: Soursop, Graviola, Guanabana, Prickly custard apple; **Hindi:** Laxman phal, Hanuman phal; **Kannada:** Laxman phal, Hanuman phal; **Malayalam:** Mullatha; **Marathi:** Laxman phal, Hanuman phal**;** **Sanskrit:** Laxman phala; **Tamil:** Pulippu palam

Distribution: *A. muricata* is native to the warmest tropical areas in South and North America and is now widely distributed throughout tropical and subtropical parts of the world, including India, Malaysia and Nigeria.

Description: *A. muricata* is a small, upright, evergreen tree that can grow to about 30 ft tall. Its young branches are hairy. Leaves are oblong to oval, 8–16 cm long and 3–7 cm wide. They are glossy dark green with no hairs above and paler and minutely hairy to no hairs below. Flower: stalks (peduncles) are 2–5 mm long and woody. They appear opposite from the leaves or as an extra from near the leaf stalk, each with 1 or 2 flowers. The petals are thick and yellowish. Outer petals meet at the edges without overlapping and are broadly ovate, 2.8–3.3 cm by 2.1–2.5 cm, tapering to a point with a heart-shaped base. They are evenly thick and covered with long, slender, soft hairs externally and matted finely with soft hairs within. Inner petals are oval-shaped and overlap. The sepals are quite thick and do not overlap. The fruits are dark green and prickly. They are ovoid and can be up to 30 cm long, with a moderately firm texture. Their flesh is juicy, acidic, whitish, and aromatic

Flowering and fruiting: April–October

Chemical constituents: Extensive phytochemical evaluations on different parts of the *A. muricata* plant have shown the presence of various phytoconstituents and compounds, including alkaloids, megastigmanes, flavanol, triglycerides, phenolics, cyclopeptides, and essential oils. However, *Annona* species, including *A. muricata*, have been shown to be a generally rich source of annonaceous acetogenin compounds (Leboeuf et al., 1980; Matsushige et al., 2012; Moghadamtousi et al., 2015).

Used in: Traditional

Parts used: Leaves, flowers, fruits, and seeds

Uses: The fruit is used as natural medicine for arthritic pain, neuralgia, arthritis, diarrhea, dysentery, fever, malaria, parasites, rheumatism, skin rushes, and worms, and it is also eaten to elevate a mother's milk after childbirth. The leaves are employed to treat cystitis, diabetes, headaches, and insomnia. Moreover, internal administration of the leaf's decoction is believed to exhibit antirheumatic and neuralgic effects, whereas the cooked leaves are topically used to treat abscesses and rheumatism. The crushed seeds are believed to have anthelmintic activities against external and internal worms and parasites. In tropical Africa, the plant is used as an astringent, insecticide, and piscicide agent and to treat coughs, pain, and skin diseases. In India, the fruit and flower are employed as remedies against catarrh, while the rootbark and leaves are believed to have antiphlogistic and anthelmintic activities. In Malaysia, the crushed leaf mixture of *A. muricata*, together with *A. squamosa* and *Hibiscus rosa-sinensis*, is used as a juice on the head to protect against fainting. In South America and tropical Africa, including Nigeria, leaves of *A. muricata* are deployed as an ethnomedicine against tumors and cancer. In addition, the anti-inflammatory, hypoglycemic, sedative, smooth muscle relaxant, hypotensive, and antispasmodic

DOI: 10.1201/9781003263500-44

effects are also attributed to the leaves, barks, and roots of *A. muricata* The unripe fruit along with olive oil is used in neuralgia, rheumatism, and arthritic pain (Adewole and Caxton-Martins, 2006; Adewole and Adewole, 2008; Mishra et al., 2013; Moghadamtousi et al., 2015).

Methods of propagation: grafting and budding

Annona reticulata L.

Synonyms: *Annona excelsa* Kunt.; *Annona laevis* Kunth.; *Annona longifolia* Sesse & Moc.; *Annona riparia* Kunth.

Family: Annonaceae

Common names: Bengali: Nona; **English:** Bullock's heart, Netted custard apple; **Hindi:** Ramphal; **Kannada:** Ramphala; **Marathi:** Ramphal; **Telugu:** Ramphalam

Distribution: *A. reticulata* is widely distributed in tropical and subtropical regions. The plant is indigenous to the West Indies. In India, it is widely cultivated and naturalized as a fruit tree.

Description: It is a small tree growing to a height of 20–30 ft. It contains numerous lateral branches which are glabrous. The stems are cylindrical having lenticels and very short coffee-colored hairs. The stems are cylindrical having lenticels and very short coffee-colored hairs. Leaves: simple alternate, are oblong, lanceolate, membranous, acute, and rounded or curate at the base. The upper surface of leaves is glabrous and on lower surface it contains few spreading hairs. Flowers: bisexual, terminal or leaf opposed, solitary or in fascicles. Sepals: 3, valvate. Petals: 3 or 6, valvate, much larger than the sepals, outer whorl fleshy and triquetrous above with a thinner concave base, inner whorl if present very small, strap shaped. Stamens: numerous, anthers narrow, ovoid, dorsal at the top of connectives. Carpels: many, subconnate, ovule 1 basal, erect, style oblong, carpels partly free in flower, later confluent into large syncarpous fleshy fruit, Fruits are edible, somewhat heart-shaped, rough, and yellow in color which change to yellowish red on ripening (Jamkhande and Wattamwar, 2015;

Flowering: May–July **Fruiting:** September–January

Chemical constituents: Several phytoconstituents have been identified from different parts of *A. reticulata.* Stem bark contains tannins, alkaloid and phenolic compounds. Leaves contain wide range of chemicals like alkaloids, amino acids, carbohydrates, steroids, flavonoids, proteins, tannins, glycosides and phenolics. Root has been identified for the content of acetogenin, alkaloid, carbohydrates, proteins, flavonoids, and tannins (Jamkhande and Wattamwar, 2015).

Used in: Traditional

Parts used: Leaves, fruits, and bark

Uses: *A. reticulata* plants are known to possess wide-ranging medicinal properties. They have great traditional uses in treating epilepsy, dysentery, bacterial infections, fever, cardiac problems, and the like. The fruits are known to have antioxidants and help boost digestive health and prevent blood pressure. The decoction made from the bark is used as a tonic and, due to its astringent qualities, can be used as a therapy for diarrhea and dysentery. The dried unripe fruit is used to treat gastrointestinal issues like diarrhea and dysentery. In extreme cases, a liter of water is boiled with the leaves, bark, and green fruits for 5 minutes to make a highly potent decoction. *A. reticulata* fruits are rich in antioxidants and vitamin C, which can delay the ageing process of the skin

The plant extract is used for the treatment of diarrhea and pediculosis. Fruits are sweet, astringent, and useful in blood complaints (Jamkhande and Wattamwar, 2015).

Methods of propagation: Seeds

DOI: 10.1201/9781003263500-45

Photo: Dr. Srinath Rao

Anogeissus latifolia (Roxb. ex DC.) Wall. ex Guill. & Perr.

Synonyms: *Conocarpus latifolia* Roxb. ex DC.

Family: Combretaceae

Common names: Bengali: Chakwa; **English:** Axlewood, Buttontree, Dindiga tree, Ghattitree, Gum-ghatti; **Bengali:** Dhaoya; **Gugrati:** Dhav, Dhavado; **Hindi:** Bakli, Baajhi, Dhau, Dhawa, Dhawra; **Kannada:** Bejjalu, Dindala Dindiga; **Malayalam:** Malakanniram, Mazhkanjiram, Vellanava; **Marathi:** Dhaora, Dhamora, Dhawada; **Sanskrit:** Baka, Dhava, Dhavala, Dhurandhara Dridhataru; **Tamil:** Naml, NakamVel-naga-maram, Vakkali; **Telugu:** Shrimanu, Vellama, Yelama

Distribution: It is distributed throughout India and Ceylon. The plant is common in dry deciduous forest, except East Bengal and Assam. It is found in sub-Himalayan tract, from the Ravi to Nepal, Bihar, and Chota Nagpur, and ascends to South India.

Description: *A. latifolia* (DC.) is medium sized deciduous tree. It attains height of about 30–40 ft. The trunk is straight and cylindrical branchless up to 8 m, about 80 cm in diameter; branches are drooping. Bark is smooth pale to gray in color and exfoliating in irregular thin scales. Leaves: simple, entire opposite or sub-opposite with grayish yellow or whitish hairs below. Flowers sessile, small in dense spherical heads on stalks in leaf axil or at the end of branches; sepals, 5, are joined together in a stalk-like tube, expanded at top into a 5-lobed cup. Petals are absent; stamens, 10 in 2 rows. Fruit is a 2-winged pseudo-achene, small, packed in a dense head compressed. Sepal: tube survives until fruiting and forms a beak. Seeds: ovoid (http://www.flowersofindia.net/catalog/slides/Axle%20Wood%20Tree.html).

Flowering: June–September **Fruiting:** December–March

Chemical constituents: Reddy et al. (1965) reported tannin, (+) leucocyanidin, and ellagic acid from the bark, sapwood and heartwood, steroid, β-sitosterol, and a triterpenoid, 3-βhydroxy-28-acetytaraxaren, were isolated from stem bark (Rahman et al., 2007).

Used in: Ayurveda, Folk, Siddha, and Unani

Parts used: Leaves, stem bark, sap, and wood

Uses: Reported to be used in the treatment of various disorders including stomach and skin diseases. The bark is effective in anemic conditions, urinary discharges, and piles (Kirtikar and Basu, 1975). Stem bark is astringent, hemostatic, constipating, depurative, and useful in vitiated conditions of kapha and vata; the stem bark is useful in diarrhea, dysuria, cough, colic, liver complaints, snakebite, and skin diseases. Tribes in Udaipur district of Rajasthan use the bark of this tree in the treatment of fever (Nag et al., 2007). The bark is remedy for chronic cough called "Dangya Khokala" (Patil, 2006). Tribal people residing in the forest of Gundla Branhmeswaram Wildlife Sanctuary apply paste of stem bark on scorpion sting (Venkat Ratna and Venkat Raju, 2008). A decoction of bark, two spoons daily, is useful as remedy against cough and leaf decoction is effective in epileptic fits (Pawar and Patil, 2008). Gum is used as tonic and generally consumed after delivery (Pawar and Patil, 2008). According to Jagtap et al., 2009), Pawara tribes of the Satpura hills use the gum with a cup of water or milk during early morning for lactation. Leaf juice is given in purulent discharges from the ear while the fruit is astringent to bowels and cures kapha and biliousness (Kirtikar and Basu, 1975).

Methods of cultivation: Through root suckers and seeds; seed viability is low but increases after very dry seasons. Germination rates can be increased by pre-soaking the seeds for 3–5 minutes in hot water.

DOI: 10.1201/9781003263500-46

Anthemis nobilis Linn

Synonyms: *Chamaemelum nobile* (L.) Allioni

Family: Asteraceae (Compositae)

Common names: English: English chamomile, German chamomile, Roman chamomile, Chamomile, St Anne's flower, Sweet chamomile; **Hindi:** Babuneka phool; **Marathi:** Babuna **Tamil:** Shimai chamantipu; **Telugu:** Bangaraju poovvu

Distribution: It was widely distributed in Asia, Europe, Africa, and Northern America. In India, found in temperate regions of the Himalayas cultivated as well as in wild at several places

Description: A mat-forming perennial grows to a height of about 25 cm, procumbent stems; the stems, hairy and freely branching, are covered with leaves. The leaves are alternate, bipinnate, and downy to glabrous and aromatic, which are divided into thread-like segments, the fineness of which gives the whole plant a feathery appearance. It has daisy-like yellow-centered disk florets and 16–18 white ray florets, about 1.5 cm across. Flowers are solitary, terminal, rising 20–30 cm above the ground, and consist of prominent yellow disk flowers and silver-white ray flowers. The flowering time in the northern hemisphere is June and July, and its fragrance is sweet, crisp, and fruity. The root is perennial, jointed, and fibrous (Al-Snafi, 2016c).

Flowering: July–September; **Fruiting:** September–October

Chemical constituents: The flower heads contain volatile oil (including azulenes and bisabolol); sesquiterpene lactone (nobilin); flavonoids, cyanogenic glycoside, and bitter glucoside (anthemic acid); acetylenic salicylic derivatives, coumarins (including scopolin), and valerianic acid; and tannins (Khare, 2007), flavanoids, apigenin, luteolin, quercetin, patuletin, and essential oil constituents, such as α-bisabolol and its oxides and azulenes (Sharafzadeh and Alizadeh, 2001).

Used in: Ayurveda, Folk, Homeopathy, Siddha, and Unani

Parts used: Leaves and flowers.

Uses: Mild sedative, anticonvulsant, antispasmodic, anti-inflammatory, mild analgesic; used externally for skin disorders, poultice of flowers in sprains and rheumatism. Used mainly for mild spasmodic gastrointestinal disturbances and sluggishness of bowels, also for nervousness. (Newall et al., 1996; Faucounnier, 1996; Khare, 2007). Used in homeopathy. Helpful for the people with sensitivity to cold air and cold things. Effective in treating coryza. Useful in treating gastric disturbances. Helpful in reducing sneezing and coughing. Helps in relieving abdominal pain. Cures itching of anus. Eases frequent urination. Reduces stress, anxiety, and depression. Effective in treating liver-related disorders. Useful in conjunctivitis. Helps in controlling diabetes (https://www.1mg.com/otc/sbl-anthemis-nobilis-mother-tincture-q-otc338820).

Methods of cultivation: Seeds

DOI: 10.1201/9781003263500-47

Apium leptophyllum (Pers.) F. Muell. ex Benth.

Synonyms: *Cyclospermum leptophyllum* (Pers.) Sprague ex Britton and P. Wilson)
Apium leptophyllum; Apium cyclospermam
Family: Umbellifera
Common Names: Bengali: Randhuni, Banyamani; **English:** Oregano; **Gujarati:** Bodi Ajamo; **Hindi:** Ajmuda, Ajmod, Ajwain; **Kannada:** Oma, Ajavana, Omakki; **Malayalam:** Ayamoda, Omam; **Marati:** Ajmoda, Oova; **Sanskrit:** Dipyaka; **Tamil:** Omam; **Telugu:** Vamu
Distribution: Found in India, Sri Lanka, Pakistan, South America, Queensland, and the tropics. In India, cultivated in Andhra Pradesh, Gujarat, Madhya Pradesh, and Karnataka
Description: Low, weak hairless annual herb, reaching up to 50 cm, sometimes branched from the base. Leaves: 3–4 pinnate finely divided. Inflorescences: a compound umbel that appears as multiple simple umbel present in the axils of the upper leaves. Fruit: ovoid to suborbicular, about 1.5 mm long and 1.2 mm broad, glabrous, with prominent ridges.
Flowering and fruiting: June–January
Chemical constituents: Thymohydroquinone dimethyl ether, isothymol methyl ether, thymol methyl ether, cuminaldehyde, p-cymene and γ-terpinene are present in oil (Helal et al., 2015).
Used in: Ayurveda and Folk
Parts used: Fruit
Uses: Traditionally, the fruit is widely used as an antinephritic, antirheumatic, and carminative and is beneficial for prevention of tumor, anorexia, vomiting, colic pain, and mitch (Sahoo et al., 2015). The fruit is useful in bronchitis, cough, and asthma by various traditional practitioners of India. In addition, the fruits possess thermogenic and antispasmodic in asthmatic patients (Warrier et al., 1996; Hardel et al., 2012).
Methods of propagation: Seeds

DOI: 10.1201/9781003263500-48

Aquilaria allocha Roxb.

Synonyms: *Aquilaria malaccensis* Lam.
Family: Thymelaeaceae
Common names: Bengali: Agaru; **English:** Agarwood, Agilawood, Eaglewood; **Gugrati:** Agar; **Hindi:** Agar; **Kannada:** Krishna Agaru; **Malayalam:** Akil; **Sanskrit:** Agaru, Anaryaka, Dhoopavasam, Krumija, Shreshta Vruksha; **Tamil:** Aggaalichandanum; **Telugu:** Agaru
Distribution: Agaru is a native plant of China, India, Indonesia, Malaysia, Philippines, Laos, Thailand, Singapore, Sri Lanka, Tibet, Europe, and Africa. In India, the plant is found growing naturally in evergreen forests of the Eastern Himalayas. It grows at an altitude of 700–1400 m. It is found in abundance in Assam, Tripura, Meghalaya, Nagaland, Sikkim, Manipur, Bhutan, and the Khasi hills.
Description: A large evergreen tree about 70–80 ft in height and 4–5 ft in girth with a straight and fluted stem. Initially, its wood is soft, light, elastic, whitish yellow in color without any characteristic smell. Later as the fungal infection progresses it becomes heavy and black. Leaves: Simple, alternate, ovate-lanceolate to lanceolate-oblong in shape, about 8 to 10 cm long and 2 to 3 cm broad, short petioled, silky glossy surface, and entire margin. Flowers: inflorescence of many small, greenish-yellow-colored flowers on short peduncle umbels occurring on younger branchlets. Fruits: capsule, obovoid shape, slightly compressed, yellowish, and tomentose. **Seeds:** Brownish-black color, ovoid shape with along the tail.
Flowering and fruiting: April–July
Chemical constituents: Agaru's heartwood is rich in essential oil, resins, alkaloids (liriodenine), saponins, steroids, terpenoids, tannins, flavonoids (aquisiflavoside, aquilarisinin, aquilarisin, and aquilarixanthone), and phenolic compounds (Satapathy et al., 2009). Agaru wood predominantly contains 2–(2-phenylethyl)-4H-chromen-4-one derivatives and sesquiterpenes. Sesquiterpenes include agarol, aquilochin, α and β agarofuran, norketoagarofuran, agarspirol, 10-epi-geudesmol, jinkoeremol, jinkohol, kusunol, dihydrokaranone, oxoagarospirol, qinanol A-F, aquilarabietic acid A-K, aquilarin B, aquilacallane A-B, aquimavitalin, abietane ester, gmelofuran apigenin, and 4′,7-dimethyl ether Srivastava et al., 2016). Agaru oil contains selinene, dihydroselinene, agarol, b-agarofuran, vetispira-2(11), valerianol, dihydrokaranone, and tetradecanoic acid (Naf et al., 1992, 1995).
Used in: Ayurveda and Folk
Parts used: heartwood and oil
Uses: It is reported to be used as a folklore medicine in inflammatory diseases, skin diseases, headache, arthritis, vomiting, gout, and the like. In traditional Chinese medicine, Agaru is used as a chi-regulating drug and carminative medicine.to relieve gastric problems, coughs, rheumatism, and high fever. It can promote chi-circulation to relieve pain, warm the middle energizer to arrest vomiting, and regulate respiration to relieve asthma. Agaru paste is applied externally to decrease edema. External application of oil prepared by Agarusaar and Sinshipa are used in chronic nonhealing ulcers and skin diseases like dadru, kitibh, and kustha. Agaru powder mixed with honey twice daily is used as leha to get relief in shwasa and hikka (asthma and hiccough). Incense of Agaru wood can also be used to get relief in respiratory congestion and asthma. Taking 1–3 g of Agaru's heartwood powder with honey gives relief in cough. A decoction prepared with Patha and Agaru is used to cure lavanmeha (a type of diabetes). Daily intake of Agaru powder with milk acts as a rejuvenator.
Methods of cultivation: Seeds

DOI: 10.1201/9781003263500-49

Arctium lappa Linn.

Synonyms: *Arctium majus* (Gaertn.) Bernh.; *Arctium ruderale* Salish; *Arctium vulgare* (Hill) Evans; *Bardana arctium* Hill; Bardana lappa Hil.l; *Lappa glabra* Lam.; *Lappa major* Gaertn.; *Lappa nemorosa* (Lej.) Korn. Ex Griewank.

Family: Asteraceae (Compositae)

Common names: English: Great burdock

Distribution: This species is native to the temperate regions of the Old World, from Scandinavia to the Mediterranean, from the British Isles through Russia, and from the Middle East to India, China, Taiwan, and Japan. Now found in the western Himalayas, Kashmir, and Simla

Description: Greater burdock is a biennial plant, rather tall, reaching as much as 10 ft. It has large, alternating, wavy-edged cordiform leaves that have a long petiole and are pubescent on the underside. The flowers are purple and grouped in globular capitula, united in clusters. The capitula are surrounded by an involucre made out of many bracts, each curving to form a hook, allowing the mature fruits to be carried long distances on the fur of animals. The fruits are achenes; they are long, compressed, with short pappus hairs (http://www.oardc.ohio-state.edu/weedguide/singlerecord).

Flowering and fruiting: July–September

Chemical constituents: Important constituents of burdock roots are fatty acids, organic acids, phenolic acids, lignans, sesquiterpenes, tannin, inulin, and mucilage (Khare, 2007). Burdock roots contain mucilage, sulfurous acetylene compounds, polyacetylenes and bitter guaianolide-type constituents. Seeds contain arctigenin, arctiin, and butyrolactone lignans (Hayashi et al., 2010).

Used in: Folk, Homeopathy, and Traditional Chinese Medicine

Parts used: Leaves, seeds, and roots

Uses: Hypoglycemic. Roots: inhibitory of tumor growth, cardiac stimulant, diuretic, spasmolytic. Leaves and seeds: anti-cutaneous, used in psoriasis and seborrheic eczema (Matsumoto et al., 2006; Khare, 2007).

Methods of cultivation: Seeds

DOI: 10.1201/9781003263500-50

Armoracia lapathifolia Gilib

Synonyms: *Armoracia rusticana* P. Gaertn., B. Mey. & Scherb.; *Armoracia rustica* Schur; *Armoracia sativa* Bernh; *Cardamine armoracia* (L.) Kuntze; *Cochlearia armoracia* L.; *Cochlearia lancifolia* Stokes' *Cochlearia lapathifolia* Gilib.; *Cochlearia rusticana* Lam.; *Cochlearia variifolia* Salisb.; *Crucifera armoracia* E. H. L. Krause; *Nasturtium armoracia* (L.) Fr.; *Raphanis magna* Moench; *Raphanus rusticanus* Garsault; *Rorippa armoracia* (L.) Hitchc.; *Rorippa rusticana* (G. Gaertn., B. Mey. & Scherb.) Godr.

Family: Cruciferae and Brassicaceae

Common names: English: Horseradish

Distribution: Native to Europe; grown in gardens in northern India and in hill stations of southern India

Description: Roots: fusiform or cylindrical, fleshy or woody. Stems: 5–12(–20) dm. Basal leaves: petiole to 60 cm (broadly expanded basally); blade broadly oblong, oblong-lanceolate, or ovate, (10–)20–45(–60) cm × (30–)50–120(–170) mm, margins usually coarsely crenate, rarely pinnatifid. Cauline leaves: proximal shortly petiolate, blade oblong to linear-oblong (lobed), smaller than basal, margins pinnatifid or pinnatisect; distal sessile or shortly petiolate, blade linear to linear-lanceolate, base cuneate or attenuate, margins usually serrate or crenate, rarely entire. Racemes: to 40 cm. Fruiting pedicels: ascending, 8–20 mm. Flowers: sepals 2–4 mm; petals obovate or oblanceolate, 5–7(-8) mm, claw to 1.5 mm; filaments 1–2.5 mm; anthers 0.5–0.8 mm. Fruits (rarely produced), 4–6 mm; style obsolete or to 0.5 mm; stigma well developed. Seeds: compressed (often not produced, rarely to 4 per locule).

Chemical constituents: The root contains glucosinolates, mainly sinigrin, which releases allyl isothiocyanate on contact with the enzyme myrosin during crushing. has high content of vitamin C.

Used in: Ayurveda

Parts used: Stem, leaves, and root

Uses: Circulatory stimulant, diaphoretic, diuretic, antibiotic. Used for both urinary and respiratory tract infections. The roots are antiseptic, aperient, digestive, diuretic, expectorant, rubefacient and liver, spleen, and pancreatic stimulant; urinary antiseptic; diuretic; used in the treatment of gout, increases excretion of uric acid; also for proteinuria and vaginal discharges. An infusion is used for hepatitis. In catarrhs of the respiratory tract and aperient as a supportive therapy for infections of the urinary tract; externally for catarrhs of the respiratory tract and hyperemic treatment of minor muscle aches. The root produces an inhibitory effect on the growth of Gram-negative bacteria of the typhoid-paratyphoid enteritis group Externally a poultice made from the root is used to treat pleurisy, arthritis, and infected roots. It will also relieve the pain of chilblains (https://link.springer.com/referenceworkentry/10.1007/978-0-387-70638-2_150; Khare, 2007).

Methods of cultivation: Root segments

DOI: 10.1201/9781003263500-51

Artocarpus heterophyllus Lam.

Synonyms: *Artocarpus integrifolia* var. heterophylla (Lam.), Pers., nsis *Artocarpus brasilie* Gomez; *Artocarpus integrifolia*, auct.; *Artocarpus jaca* Lam.; *Artocarpus maxima* Blanco; *Artocarpus philippinensis*, Lam.; *Polyphema jaca* Lour.; *Saccus arboreus*, major Rumph; *Sitodium cauliflorum* Gaertn.; Tsjaka-maram Rheede

Family: Moraceae

Common names: Bengali: Kathal; **English:** Jackfruit; **Gujarati:** Phannasa; **Hindi:** Kathal; **Kannada:** Halasina hannu; **Malayalam:** Chakka; **Marathi:** Phanas; **Sanskrit:** Panasah; **Tamil:** Palaa; **Telugu:** Panasa

Distribution: Native to tropical Asia and widely grown throughout the wetland tropics for its large fruits and durable wood.

Description: Evergreen trees to 25 m high, bark 10–12 mm thick, blackish gray, mottled with green and black, exfoliating in large thick flakes, exfoliated surface orange red; blaze pinkish yellow; exudation milky white latex; trunk with warty tubercles; branchlets glabrous. Leaves: simple, alternate; stipules 3–5 cm long, lateral, ovate-lanceolate, sheathing, glabrous, caducous; petiole 20–40 mm long, stout, grooved above, glabrous; lamina 8–23 × 3–13 cm, obovate, obovate-oblong, or elliptic-ovate, base acute, round or cuneate, apex acute or obtuse, margin entire, glabrous and shining above and scabrous beneath; lateral nerves 6–8 pairs, pinnate, prominent, arched, intercostal scalariform, prominent. Flowers: unisexual, minute, yellowish green, in spikes enclosed by spathe-like bracts, male from young branches, catkin narrow-cylindric; perianth 2-lobed, puberulous; stamen 1; filament somewhat flattened, stout; anthers ovate-oblong; female catkins from the trunk and mature branches, more massive, perianth with strongly projecting conical apex; ovary 0.3 mm, superior, globose-obovoid; style exserted; stigma spathulate. Fruit a sorosis 30–45 × 20–25 cm, oblong, tuberculate, tubercles conical yellowish-green, fruiting perianth yellow to light orange, fleshy; seeds 10–12 × 8–10 mm, elliptic-oblong, smooth, and glossy (https://indiabiodiversity.org/species/show/8042).

Flowering: March–April **Fruiting:** June–August

Chemical constituents: Isopentyl isovalerate (28.4%) and butyl isovalerate. The aroma concentrate of "soft jackfruit" was dominated by isopentyl isovalerate, butyl acetate, ethyl isovalerate, butyl isovalerate and 2-methylbutyl acetate. Jackfruit contained 45 volatile components, of which 32 have not been reported previously. Esters represented a high proportion of the jackfruit volatiles and are important contributors to the flavor of the fruit.

Used in: Ayurveda, Folk, and Traditional

Parts used: Leaves, seeds, root, and bark

Uses: Ayurvedic uses: antidiarrheal, astringent, carminative, and tonic. Leaf extract of *A. heterophyllus* was found to cause a hypoglycemic effect. Traditional medicinal uses: Many parts of the plant including the bark, roots, leaves, fruit, and latex are endowed with medicinal properties and used in traditional medicine. In the Philippines, the ash of the leaves, after burning with corn and coconut shells, is applied on wounds and ulcers as cicatrizant. In India, the leaves are used in treating skin disease. The ash of jackfruit leaves, burned, is used alone or mixed with coconut oil to heal ulcers. In Sarawak, ash from the leaves mixed in a little coconut oil and prescribed for scabies or kuris by the Iban; the Melanau used the same for applying on cuts and wounds. In Malaysia, ashes of leaves are used to heal wounds. The root is anti-asthmatic. The decoction of the root is used in diarrhea and for fever. The root is also useful in skin diseases. The latex of the tree is used to promote

 DOI: 10.1201/9781003263500-52

healing in glandular swellings and in snakebites. Mixed with vinegar and applied to these swellings and abscesses, it promotes absorption or suppuration. The wood has a sedative property; its pith is said to produce abortion. The ripe fruit is demulcent, nutritive, and laxative. The pulp envelopes or arils of the seeds are considered by the Chinese to be cooling, tonic, and nutritious. The starch of the seeds is given in bilious colic. The roasted seeds are believed to have aphrodisiac properties. Hot water extract of mature jak leaves (*Artocarpus heterophyllus*) is recommended by Ayurvedic and traditional medical practitioners as a treatment for diabetes mellitus (Lim, 2012). Snakebite: part used—pedunele. Feed 1 cup of juice thrice daily immediately after bite. Bark and root juice is fed to the human and cows, 1/2 a cup juice twice daily, until cured (Patil, 2008). In Burma and China, the latex is applied to heal ulcers and abscesses. The roots are used to stop diarrhea. In the Philippines, the latex is used as it is in Burma, and the leaves as it is in Malaysia. In Vietnam, the wood is used to stop convulsion. A decoction of the leaves is drunk to stimulate the production of milk. The sap is used to treat syphilis and to expel intestinal worms (Wiart, 2006). Used as a tonic to treat discomfort from alcohol influences (Thomas, 2006).

Methods of propagation: Seeds

B

Baccaurea ramiflora Lour.

Synonyms: *Baccaurea cauliflora* Lour.; *Baccaurea flaccida* Mull. Arg.; *Baccaurea oxycarpa* Gagnep.; *Baccaurea sapida* (Roxb.)

Family: Euphorbiaceae

Common names: Bengali: Bhubi, Latok-tenga, Latka; **English:** Burmese grapes; **Hindi:** Leteku, Khattaphal

Distribution: Native of Southeast Asia. It is found growing in sub-Himalayan tract, in the hills from Nepal, Sikkim, and Darjeeling. It is also found in Assam, Arunachal Pradesh, Bhutan, Burma, Malaysia Tripura, and Tibet (Gogoi, 2017).

Description: The generic name is derived from Latin *baccaurea*, referring to the golden-yellow color of the fruits (Raghavan and Ramjan, 2018). This is a semi-evergreen tree about 5–10 m in height. Stem gray brown in color, branches hispid when young. Leaves: simple, alternate, 9–15 cm long and 3–8 cm broad, obovate-oblong, in shape, papery, glabrous, cuneate, margin entire or shallowly repand, apex acute, stipules caducous. Flowers: small, dioecious, apetalous, many-flowered, in raceme panicles. Male and female on different trees. Fruit: is a pericarp, about 2–3 cm in diameter, yellowish and velvety, contains pinkish pulp. The pericarp is leathery. Seeds: 3 in number (Gogoi, 2017).

Flowering: April–July; **Fruiting:** September–October

Chemical constituents: The fruit is abundant of magnesium (504 mg), potassium (730 mg), phosphorous (132 mg), and iron (100 mg) per 100 gm of fruit pulp. It has a significant amount of ascorbic acid (Ragavan and Ramjan, 2018).

Used in: Traditional medicine

Parts used: Leaves, fruits, seeds, and root

Uses: It is used as anti-inflammatory, in treatment of arthritis, abscesses, injuries, and cellulitis. The juice of the fruit is used for the treatment of constipation. The plant is useful to treat abscesses and arthritis (Gogoi, 2017). Wine known as latkan wine is prepared from fruit juice. The fresh bark of the plant is chewed or the juice is taken orally for complaints of constipation in India (Khan, 2008).

Methods of propagation: Seeds

DOI: 10.1201/9781003263500-54

Baliospermum montanum (Willd.) Muell. Arg

Synonyms: *Baliospermum axillare* Bl.; *Baliospermum polyandrum* Wt.; *Baliospermum montana* Muell.; *Croton polyandrus* Roxb; *Jatropa montana* Willd.

Family: Euphorbiaceae

Common names: Bengali: Harun, Haphun; **English:** Red physic nut, Wild castor, Wild croton; **Hindi:** Jangli Jamaalgotaa; **Kannada:** Naga danti; **Marathi:** Danti; **Sanskrit:** Hastidantidanti, Nikumbha, Udumbarparni, Erandphalaa; **Tamil:** Neeradimuthu, Danti; **Telugu:** Cinna nepalum

Distribution: The Himalayas, Assam, Khasi hills to Kashmir, Bengal, Madhya Pradesh, Bihar, and Peninsular and Central India, ascending to 1800 m

Description: It is a stout undershrub 3–4 ft in height with herbaceous branches from the roots. Leaves are simple, sinuate-toothed, upper ones small, lower ones large and sometimes palmately 3–5-lobed. Flowers are numerous, arranged in axillary racemes unisexual with male flowers above and a few females below. In male flowers, the calyx is globose, 2.5 mm long, 4–5 partite, glabrous or slightly pubescent, membranous, finely mottled with a disc of 6 glands. Stamens are about 20 in number. Female flowers have ovate-lanceolate and pubescent sepals, and a disc about 2.5 mm in diameter. Fruits are capsules, 8–13 mm long, and obovoid. Seeds are ellipsoid smooth and mottled.

Flowering: January–February **Fruiting:** March–April

Chemical constituents: Several compounds have been isolated from this plant such as steroids, triterpenoids, diterpenes (baliospermin, montanin, phorbol-12-deoxy-13-O-palmitate, phorbol-12-deoxy-16-hydroxy-13-O-palmitate and phorbol-12-deoxy-5b-hydroxy-13-myristate) glycosides, saponins, alkaloids, flavonoids, and phenolic compounds (Johnson et al., 2010; Ogura et al., 1978).

Used in: Ayurveda, Folk, and Siddha

Parts used: Leaves, seeds, root, and latex

Uses: Root in the medicinal part, which is used in constipation, abdominal pain, general anasarca, piles, calculus, helminthic infections, scabies, dermic affections, suppurative ulcers, and diseases caused by the morbidity of kapha and pitta (Kurup et al., 1979). It is purgative (also used in dropsy), anti-asthmatic (decoction is given in asthma). Latex is used for body ache and joint pain. Root and seed oil is cathartic and anti-dropsical. Along with other therapeutic applications, the *Ayurvedic Pharmacopoeia of India* indicated the use of dried root in jaundice, abdominal lump, and splenomegaly. EtOH extract of roots showed in vivo activity in P-388 lymphocytic leukemia. Dosage: Root—103 g powder (Khare, 2007) Antileukemic and cytotoxic activities have been demonstrated in the esters of both 12-deoxyphorbol and 12-deoxy-16-hydroxyphorbol, isolated from *B. montanum* (Kinghorn, 1979).

The roots have the following properties: purgative, tonic, anodyne, digestive, acrid, thermogenic, anthelmintic, diuretic, febrifuge, diaphoretic, rubefacient and anti-inflammatory. The leaf is anti-asthmatic and wound healing. The root and seed oil are cathartic and anti-dropsical (Bhakuni et al., 1971).

Methods of cultivation: Seeds and terminal cuttings

DOI: 10.1201/9781003263500-55

Balsamodendron myrrha Nees.

Synonyms: *Commiphora molmol* Engl.; *Commiphora abyssinica* (Berg.) Engl

Family: Burseraceae

Common names: English: Myrrh; **Gugrati:** Hirabol; **Hindi:** Bola, hirabol; **Malayalam:** Merra; **Marathi:** Hirabol; **Sanskrit:** Pranpinda, Bola, Hiraabol, Nirloha; **Tamil:** Vellaibolam; **Telugu:** Balintabolu, Valentrapolam.

Distribution: The tree is native to the Arabian Peninsula (Oman, Yemen) and Africa. Its resin is sold in Indian market.

Description: Shrub or small tree, up to 5 m tall, spiny, glabrous; bark whitish or grayish, peeling in papery flakes; all branchlets spine-tipped. Leaves 1(–3)-foliolate, stiff, grayish green, or glaucous; petiole 1–10 mm long; lateral leaflets, if present, usually much smaller than the middle one; middle leaflet elliptic, lanceolate, or obovate, 0.6–4.4 long and 0.3–2 cm, broad, attenuate, cuneate, rounded, or truncate at the base, rounded to acute at the apex, with entire or dentate margins. Male flowers: 2–4 together in up to 4 mm long often glandular cymes, female flowers 1 or 2 together. Calyx: 3–4 mm long. Petals about 4.5 mm long. Fruits: ovoid to ellipsoid with a distinct beak, 10–16 long and 6–8 mm wide, fruit stalk about 10 mm long; pericarp 2-valved; pseudaril a basal cup with 2 broad lobes over the cells of the stone and 2 reduced sutural lobes; stone flattened-ovoid, 6–9 × 4–6 mm.

Chemical constituents: Contain cadinene, commiferin a and b, amyrones, myrrhin, limonene, The gum (30%–60%) contains acidic polysaccharides, volatile oil (2%–10%) including other constituents, heerabolene, eugenol, furanosequiterpenes and monoterpenes (Khare, 2007). It also contain burseracin; commiferin; α-, β-, and γ-commiphoric acids; α- and β-heerabomyrrhols; and β-sitisterol (https://www.easyayurveda.com/2015/07/07/bola-myrrh-commiphora-myrrha).

Used in: Ayurveda, Siddha, and Unani

Parts used: Gum resin

Uses: Oleo-gum resin is emmenagogue (used for irregular menstruation and painful periods), anti-inflammatory (on pharyngitis and gingivitis), antiseptic, bacteriostatic, antiviral, astringent, stimulant, expectorant, stomachic, carminative (in dyspepsia), a leuko-cytogenetic agent (increases number of white cells in the blood). Used externally for treating acne, boils and pressure sores, internally as a blood purifier. Key application: in topical treatment of mild inflammations of the oral and pharyngeal mucosa. Used as a gargle or mouth rinse for the treatment of aphthous ulcers, tonsillitis, common cold, and gingivitis. Dosage: gum resin 3–5 g (Khare, 2007).

Methods of cultivation: Seeds and hardwood cuttings

DOI: 10.1201/9781003263500-56

Bambusa bambos (L.) Voss.

Synonyms: *Bambusa arundinaceae* (Retz.) Roxb.; *Arundo bambos*

Family: Poaceae (Gramineae)

Common names: Bengali: Banas, behar; **English:** Spiny or thorny bamboo, Banas; **Gugrati:** Wans **Hindi:** Bamboo; **Kannada:** Bidaru; **Malayalam:** Illi, Mungil, Mulla; **Marathi:** Bamboo; **Sanskrit:** Vansha, Venu, Kichaka, Trinadhwaj, Shatparvaa Vanshalochana; **Tamil:** Moongil, Mukku Moongil; **Telugu:** Mullabongu, Mullavendru

Distribution: Wild throughout India, especially in the hill forests of western and southern India.

Description: Its tall, 30–40 m and woody with numerous thorny stems curving at the top, grows in clusters. Culms are 15–18 cm across, nodes prominent armed, with 2–3 stout recurved spines. Internodes are 30–45 cm long. Leaves: 17–20 cm long and 2–3 cm broad, linear or linear lanceolate tip stiff, glabrous, margins scabrous, base ciliate, midrib narrow, leaf sheath ending on a thick callus and shortly bristly auricle. Inflorescence with many panicles often occupying the whole stem. Fruit a caryopsis, oblong, 5–8 mm long, grooved on one side.

Flowering and fruiting: Occurs once in its lifetime, often during September–May (Kaikini Aakruti et al., 2013)

Chemical constituents: The leaves of *B. bambos* contain digestible crude proteins, mainly lysine and methionine. Six phenolic acids, namely, chlorogenic acid, ferulic acid, coumeric acid, protocatechuic acid, vanillic acid, and caffeic acid, were identified in the aqueous extract of matured *B. bambos* leaves. The root of *B. bambos* showed the presence of cyanogenic glycosides and this is identical with taxiphyllin (Sanjita et al., 2012).

Used in: Ayurveda and Folk

Parts used: Leaf bud, young shoots, leaves, roots, and resin

Uses: Leaf bud and young shoots: used in dysmenorrhea; externally in ulcerations. Leaf: emmenagogue, antileprotic, febrifuge, bechic; used in hemoptysis. Stem and leaf: blood purifier (used in leukoderma and inflammatory conditions). Root: poisonous. Burnt root is applied to ringworm, bleeding gums, and painful joints. Bark: used for eruptions. Leaf and bamboo manna: emmenagogue, pectoral, expectorant, carminative, cooling, aphrodisiac, and tonic (used in debilitating diseases, urinary infections, chest diseases, cough, asthma). *B. bambos* has been proven to have great pharmacological potential, with wide utility and usage as folklore medicine. It is widely used in folk medicine for its anti-inflammatory, astringent, laxative, diuretic, anti-ulcer, anti-arthritic, anti-obesity, and abortifacient activities (Kaikini Aakruti et al., 2013). Root paste and resin powder are used. The sprouts are acrid, bitter, laxative, and useful in inflammation ulcer and wounds (Sanjita et al., 2012).

Methods of cultivation: Vegetatively. Bamboo is propagated through clump divisions, rhizomes, offsets, layering, marcotting, culm cutting, branch cutting and macro-proliferation.

DOI: 10.1201/9781003263500-57

Barleria prionitis L.

Family: Acanthaceae

Common names: Bengali: Kantajinti; **English:** Procupine flower, Yellow hedge, Barleria; **Gujarati:** Kantashila, Pilikantashelio; **Hindi:** Jinti, Vajradanti; **Kannada:** Mullu Goranti, Muljajee; **Malayalam:** Chemmulli, Kuttivetila, Manjakanakambaram; **Marathi:** Kate koranti, Pivala koranta; **Sanskrit:** Kurantaka; **Tamil:** Kattu kanagaambram, Semulli; **Telugu:** Mullugorinta

Distribution: It is commonly found in tropical Asia include India, Malayasia, Pakistan, the Philippines, Sri Lanka, and tropical Africa. In India, it is found throughout the hotter parts, often cultivated.

Description: Erect armed shrubs, spines about 1.2 cm long. Leaves: simple 5–9 × 2.5–4 cm, elliptic, acute, mucronate, base cuneate, sparsely puberulus, ciliate on the margins, gland dotted beneath; petiole to 2 cm. Cymes: axillary; bracts 2, 1.5 cm, oblong mucronate; outer calyx lobes 1.3 × 0.4 cm, inner, 1.1 × 0.2 cm, mucronate, hirsute; corolla tube 2.5 cm, lobes 2 cm obovate; filaments 1.3 cm, staminodes 2, included at the base of the corolla tube; ovary 2.5 mm, style 2.5 cm (https://indiabiodiversity.org/species/show/32443).

Flowering and fruiting: September–February

Chemical constituents: Preliminary phytochemical analysis of hydro-methanolic extract of *B. prionitis* whole plant indicated the presence of glycosides, saponins, flavonoids, steroids, and tannins. The leaves and flowering tops were reported to rich in potassium salts (Khare, 2007). Several phytochemicals, namely, balarenone, pipataline, lupeol, prioniside A, prioniside B, and prioniside C, have been isolated from the ethanolic extract of *B. prionitis* (Ata et al., 2007). Numbers of glycosides include barlerinoside, verbascoside, shanzhiside methyl ester, 6-O-trans-p-coumaroyl-8-O-acetylshanzhiside methyl ester, barlerin, acetylbarlerin, 7-methoxydiderroside, lupulinoside has been also isolated from the aerial parts (Chen et al., 1998; Ata et al., 2009; Banerjee et al., 2012).

Used in: Traditional

Parts used: Leaves

Uses: In indigenous system of medicine in India, the juice of *B. prionitis* leaves is used in stomach disorders, urinary affections, ulcer, and fever (Khare, 2007). The leaf juice mixed with honey given to children in catarrhal affections and fever (Khare, 2007). Leaves are chewed to relieve from toothache. Some tribal communities are used leaves for the treatment of piles and reduce irritation (Aneja et al., 2010; Shukla et al., 2011). The leaf juice is applied externally in lacerated soles of feet and pimples. The dried stem bark is used as an expectorant in whooping cough and diaphoretic (Khare, 2007; Aneja et al., 2010). The aerial parts of this plant are also used in inflammation and gastrointestinal disorders. The root paste is externally applied to disperse boils and glandular swellings (Aneja et al., 2010). The flowers are used internally for the treatment of migraine, internal abscesses, edema, hemoptysis, urethral discharges, seminal disorders, and reduce obesity (Khare, 2004). The whole plant is also used in stiffness of limbs, enlargement of scrotum, and sciatica. The whole plant, specifically the roots, are used as diuretic and tonic. It is also used in urinary infection, jaundice, hepatic obstruction, and dropsy (Shukla et al., 2011). Ash of the whole plant with honey is given in bronchial asthma. The crude extract of this plant in oil is recommended in arresting the graying of hair, arthritis, and gout (Khare, 2004, 2007). In South India, this plant is widely used in neurological disorders like paraplegia and sciatica, as well as in leprosy and other skin diseases (Khare, 2004). The plant

DOI: 10.1201/9781003263500-58

formulation, available in over the counter is prescribed in dysuria, rheumatic affections, internal abscesses, nervine disorders, and in chronic sinusitis (Khare, 2004; Banerjee et al., 2000).

Methods of propagation: Seeds

Photo: Dr. Srinath Rao

Barringtonia acutangula (L.)

Synonyms: *Barringtonia spicata; Butonica acutangula* Lam.; *Euginia acutangula*; *Huttum acutangulum* Britten.; *Michelia acutangula* Kuntze.; *Stravadium acutangulum* Sweet

Family: Lecythidaceae

Common names: Bengali: Higla; **English:** Indian oak tree, Mango-pine; **Gugrati:** Samudraaphala; **Hindi:** Samudraaphal; **Kannada:** Dhatripala, Mavinkubia, Niruganigily, Samudraapha; **Marathi:** Newar, Samudraaphala, Sathaphala, Tiwar; **Sanskrit:** Nichula, Samudraaphala, Vidula; **Tamil:** Kadambu, Kadappai, Samutrapallam; **Telugu:** Kanagi

Distribution: In Asia, Afghanistan, Indian subcontinent, Myanmar, Thailand, Cambodia, Laos, Vietnam, Malaysia, Indonesia, Philippines, New Guinea, and Australia. Grows near Riverbanks, swampy sites or rocky or sandy shores at elevations from sea level to 400 m

Description: Evergreen trees, about 15 m in height, bark 10–12 mm thick, surface dark brown, rough, rich in tannin; branchlets densely tomentose. Leaves: simple, alternate, clustered toward the tip of branchlets; stipules lateral, caducous; petiole pink about 6–12 mm long, slender, glabrous; lamina 7–18 × 2.5–7 cm, oblanceolate, obovate or elliptic-obovate, base cuneate, or attenuate, apex obtuse, round or subacute, margin finely serrate, glabrous, subcoriaceous; lateral nerves 8–12 pairs, pinnate, slender, prominent, intercostal reticulate, slender, prominent. The plant has a drooping raceme of up to 50 cm long, with numerous large, white flowers. Flowers: bisexual, 6–8 mm across, red, in terminal pendulous racemes; bracts small, deciduous. Calyx: tube campanulate, adnate to the ovary; lobes 4, ovate, acute, imbricate. Petals: 4, ovate, obtuse, imbricate, basally connate, adnate to the staminal tube; stamens many, in several rows; filaments exserted, filiform, 1.5 cm long, pink, connate below; ovary inferior, 2 mm, 2-celled, ovules 6–8 in each cell; style 2 cm, filiform; stigma small. Fruit: a fibrous berry, 2.5–3.8 cm long, oblong-ovoid, bluntly quadrangular. Seed: 1, ovoid in shape.

Flowering and fruiting: Almost throughout the year

Chemical constituents: Its bark contains potent painkillers, structurally unrelated to known opioid painkillers. Also contain 3,3′-di-O-methoxyellagi acid, dihydromyricetin, gallic acid, bartogenic acid, and stigmasterol (Sun et al., 2006), triterpenoids, olean-18-en-3beta-O-E-coumaroyl ester, and olean-18-en-3beta-O-Z-coumaroyl ester (Rahman et al., 2005).

Used in: Ayurveda and Folk

Parts used: Leaves, fruit, seed bark, and root

Uses: *B. acutangula* is commonly used in India by tribal people for the treatment of liver disorders, diarrheal diseases, eye diseases, spleen disorders, and worm infestation. It is one of the useful traditional medicinal plants in India and used in various Ayurvedic formulations for the treatment of various diseases. It is a plant traditionally used for the cure and treatment of many ailments, such as hemolytic disease (various diseases of blood), abdominal colic, lumbar pain, syphilis, blennorrhea, febrifuge, malaria, and diabetes. All parts of the plant—leaves, fruit, seed, bark, and root—are used for their potential medicinal properties (Kaur et al., 2013). Research on this plant has reported a number of medicinal uses, including antitumor (seed extract; Samanta et al., 2010), antibiotic (Rahman et al., 2005), inhibition of growth of *Helicobacter pylori* (Bhamarapravati et al., 2003), antinociceptive activity (Deraniyagala et al., 2003), and antifungal activity (Vijaya Bharathi et al., 2010; Sahoo et al., 2008).

The 1889 book *The Useful Native Plants of Australia* records that "[i] India an extract or juice is obtained from the leaves of this tree which, when mixed with oil, is used in

DOI: 10.1201/9781003263500-59

native practice for eruptions of the skin. The kernels powdered and prepared with sago and butter, are used in diarrhea; mixed with milk they produce vomiting. The root is bitter and is said to be similar to *Cinchona* but also cooling and aperient. The bark in decoction is given as a stomachic. The scraped bark is squeezed with coconut meat, and the juice is drunk daily for treating pneumonia, diarrhea, and asthma. Externally, it is used for poulticing wounds, ulcers, sores, and itches. The root is considered aperient". Powdered seeds, in small doses, are given to children as an expectorant and emetic. It is further used as an anthelmintic. The seed is used to treat ophthalmia (http://proseanet.org/).

The liquid obtained by pounding the wood in water is considered hemostatic and given in menorrhagia. The leaves are used to treat diarrhea.

Methods of cultivation: Seeds and cuttings of half-ripe wood

Bauhinia purpurea Linn.

Synonyms: *Bauhinia castrata* Blanco.; *Bauhinia coromandeliana* DC; *Bauhinia platyphylla* Zipp. ex Spanoghe.; *Bauhinia rosea* Kurz.; *Bauhinia triandra* Roxb.; *Caspareopsis purpurea* (L.) Pittier.; *Casparia castrata* (Blanco) Hassk; *Perlebia purpurea* (L.) A. Schmitz.; *Phanera rosea* Rich.; *Phanera kurzii* (Prain) Thoth.; *Telestria purpurea* (L.) Raf.

Family: Caesalpiniaceae

Common names: Bengali: Kanchan, Koiral; **English:** Purple bauhinia, Camel's foot, Butterfly tree; **Hindi:** Kaniar; **Kannada**: Kanchuvaala, Devakaanchana, Basavana pada; **Malayalam:** Chuvanna Mandaram; **Marathi:** Rakta kanchan; **Sanskrit:** Kovidaara, Rakta Kaanchanaara Deva kanchan; **Tamil:** Mandarai Nilattruvatti, Khairwaal; **Telugu:** Bodanta

Distribution: Native to the Indian subcontinent and Myanmar and widely introduced elsewhere in tropical and subtropical areas of the world

Description: *B. purpurea* is a small to medium-size deciduous tree growing to 17 ft (5.2 m) tall. Leaves are alternate, 7.5–15 cm long and broad, rounded, and bilobed at the base and apex. The flowers are conspicuous, pink, and fragrant, with 5 petals. Pedicels: 5–13 mm long, stout, tomentose, bract and bracteoles small tomentose, deltoid. Calyx: tomentose, tube 7.5–10 mm long. Petals: 3.8–5 cm long, oblanceolate, long-clawed. Stamens: usually 3 fertile, others reduced to antherless filaments. Ovary: downy, long-stalked; style long; stigma large, oblique. Seeds: 12–15. The fruit is a pod 15–25 cm long and 1.5–2 cm broad containing 12–16 seeds suborbicular, flattened, 1.3 cm. wide and dark brown and smooth (Kirtikar and Basu, 1991).

Flowering: September–February; **Fruiting:** January–April

Chemical constituents: The flowers contain astragalin, isoquercitrin, and quercetin, also anthocyanins. Seeds contain chalcone glycoside (Khare, 2007). A wide range of chemical bioflavonoids, bibenzyls, dibenzoxazepine, mixture of phytol fatty esters, lutein, β-sitosterol, isoquercitrin, and astragalin (Yadav and Bhadoria, 2005; https://en.wikipedia.org/wiki/Bauhinia_purpurea).

Used in: Ayurveda, Folk, Siddha, and Unani

Parts used: Flowers and bark

Uses: Bark: used as astringent, antidiarrheal. Flower buds and flowers: fried in purified butter and given to patients suffering from dysentery. Extract of stems are used internally and externally for fractured bones. Plant is used for goiter (Khare, 2007) and used to make poultices for treating swelling, bruises, boils, and ulcers. Various parts of the plant are also used in decoctions to treat fever and stomach ailments, as well as being used as an astringent. In Indian traditional medicine, the leaves are used to treat coughs while the bark is used as an antidote for poisons and glandular diseases. The flowers are also used in pickles and curries and is regarded as a laxative (Kurian, 2004).

Methods of cultivation: Seeds

DOI: 10.1201/9781003263500-60

Photo: Dr. Srinath Rao

Bauhinia racemosa Lamk

Synonyms: *Bauhinia parviflora* Vahl.; *Piliostigma racemosa* (Lam.) Benth.; *Piliostigma racemosum* (Lam.) Benth.

Family: Fabaceae (Caesalpiniaceae)

Common names: Bengali: Banraj; **English:** Sona tree, Beedi leaf tree; **Gugrati:** Apto; **Hindi:** Katmaul, Jhinjheri; **Kannada:** Aralukadumandara; **Malayalam:** Arampaali, Kutabuli, Malayaththi; **Marathi:** Kachnaala, Apta, Sona; **Sanskrit:** Ashmantaka, Kanchini, Yamalapatrakah, Yugmapatra; **Tamil:** Kokku mandarai, Tataki; **Telugu:** Tellarecettu

Distribution: Found in Ceylon and China, to an altitude of 1650 m from sea level. This species is commonly found across India in sub-Himalayan tracts from Ravi eastward, ascending to 1000 m in the Uttar Pradesh, West Bengal, and Central and South India.

Description: This is a deciduous tree, 3–5 m tall. The bark is rough dark brown and longitudinally fissured. The leaves are simple petiolate, sub-orbicular, cordate at the base, bilobed, glabrous above and pubescent beneath, shaped like cow's hooves, broader than long about 3–6 cm broad and 2–4 cm long. The flowers are creamy white 2–4 cm. In long terminals and racemes about 5–10 cm long. Petals are 5, narrow lance-like stamens, 10. The pods are linear oblong, 13–25 cm by 1.8–2.5, woody slightly curved, dark brown or brownish black, constriction between all or few seeds. Seeds: 12–20, glabrous, dark reddish brown or black, compressed, 8 mm long (Anonymous, 1985).

Flowering: January–June **Fruiting:** June–August

Chemical constituents: The different parts of this plant contain β-sitosterol and β-amyrin, flavanols (kaempferol and quercetin) and two coumarins (scopoletin and scopolin), tannins (Soni et al., 2015). Also contain octacosane, beta-amyrin, and beta-sitosterol have been isolated from the bark (Khare, 2007). The phytochemical studies have identified more than 37 chemicals from *B. racemosa*, including phenolics, flavonoids, saponins, glycosides, terpenoids, steroids, propanoids, coumarin, lipids, alkaloids, and tannins (Prakash and Khosa, 1976; Sharanabasappa et al., 2007). β-sitosterol and β-amyrin were isolated from the stem bark of this plant (El-Hossary et al., 2000).

Used in: Ayurveda, Folk, Siddha, and Unani

Parts used: Leaves, flowers, seeds, and bark

Uses: Bark is highly astringent, anti-inflammatory; used in glandular inflammations, ulcers, in the treatment of headache, fever, skin diseases, blood diseases, dysentery, and diarrhea. A decoction of the bark is recommended as a useful wash for ulcers (Panda et al., 2015). Leaves are anthelmintic used with onion for diarrhea. Flowers are used in hemorrhages, piles; also in cough. Seed is antibacterial. Ethanol (50%) extract of seeds exhibited anticancer activity (Khare, 2007). In Ayurveda, the plant is commonly used for the treatment of initial stage of cancer (Kirtikar and Basu, 1975). Methanol 80% extract of *B. racemosa* was the most active as an anti-HIV-1 agent. methanol extract (50, 100, and 200 mg/kg) of *B. racemosa* stem bark showed antitumor activity against Ehrlich ascites carcinoma tumor in mice (Soni et al., 2015).

Methods of cultivation: Seeds

DOI: 10.1201/9781003263500-61

Bauhinia tomentosa Linn.

Synonyms: *Alvesia bauhinioides* Welw.; *Alvesia tomentosa* (L.) Britton & Rose.; *Bauhinia pubiscens* DC.; *Bauhinia Volkensii* Taub.; *Bauhinia wituensis* Harms.; *Pauletia tomentosa* (L.) A. Schmitz

Family: Fabaceae (Caesalpiniaceae)

Common names: English: Yellow bauhinia, St. Thomas tree, Bell bauhinia; **Gugrati:** Pido Kachnar, Pido Abondaro; **Hindi:** Kural, Gurial, Padrian, Gwiar, Kachnar, Koliar; **Kannada:** Mandaara, Vanasampige, Aachaly; **Sanskrit:** Ashvamantaka, Pita kovidaara, Pita Kanchana phalgu; **Tamil:** Kokkumandarai, Tiruvaatti, Kanjani

Distribution: Found in South Africa, Mozambique, Zimbabwe, tropical Africa, India, Sri Lanka, southern India, Assam, and Bihar

Description: Yellow bauhinia is a small tree with a maximum height of 13–15 ft. It has drooping slender branches with multiple scrambling stems. The bark is grayish, smooth, and sometimes hairy, which gives its specific name *tomentosa*. Greenish leaves are deeply divided and elliptic in nature, margin entire. Flowers are bell-shaped with large, yellow petals with a dark maroon patch at the base. Usually in pairs, rarely 1–3, on short axillary or leaf—opposed peduncles, bracts, and bracteoles present, calyx hairy, broadly ovate, spathaceous, petals up to 5 cm long, yellow or creamy yellow, obovate—spatulated, imbricate, the upper petals with a purple blot on a face, stamens 10 fertile, the filaments long, unequal, ovary stalked, hairy, stigma peltate. The fruit is a pod, black, veined, flattened, 10–12 cm long and 1.2–1.8 cm broad, slightly puberulous when fully ripe, pointed at the apex. Seeds: 8–12, oblong, smooth, glabrous.

Flowering: November–January **Fruiting:** January–June

Chemical constituents: Seeds yield fatty oil called ebony oil, a water-soluble mucilage and saponins. Flowers gave isoquercitrin (6%), rutin (4.6%) and quercetin (small amounts) (Khare, 2007).

Used in: Ayurveda, Folk, Siddha, and Unani

Parts used: Root bark, fruits, and seeds

Uses: Antidysenteric. Fruit is diuretic. Bark: astringent. Root bark: vermifuge. A decoction of the root bark is prescribed for liver diseases. It is used internally to treat conditions of the large intestine. An infusion of the root bark is used as an external application to treat inflamed glands, abscesses, and skin conditions. Root bark is used in inflammation of liver. Seed used for wound healing (Khare, 2007). Flowers, buds, and dried leaves are used in dysentery. Seeds are tonic and aphrodisiac. Leaves have antidiabetic action. Plant is used in snakebite and scorpion string. Infusion of stem bark is useful as an astringent gargle (Kaur et al., 2009; https://tropical.theferns.info/viewtropical.php?id=Bauhinia+tomentosa).

Methods of cultivation: Seeds

DOI: 10.1201/9781003263500-62

Bergenia ciliata (How) Stemb.

Synonyms: *Bergenia ligulata* var. *ciliata* (Royle) Engl.; *Bergenia milesii* Stain.;*Bergenia thysanodes* (Lindl.) C.K. Schneid.; *Saxifraga ciliata* Lindl.; *Saxifraga schmidtii* Regel.; *Saxifraga thyrsoidea* Tausch.; *Saxifraga thysanodes* Lindl.

Family: Saxifragaceae

Common names: Bengali: Patharkuchi, Himasagara; **English:** Fringed elephant's ears, Winterbegonia, Hairy bergenia; **Gugrati:** Pashanbheda; **Hindi:** Pashanbheda; **Kannada:** Alepgaya, Hittaga, Pashanbheda, Silpbheda; **Malayalam:** Kalluvanchi, Kalluvanni; **Marathi:** Pashanbheda; **Sanskrit:** Pashanbheda, Shailagarbhaja; **Tamil:** Sirupilai; **Telugu:** Kondapindi

Distribution: The plant is endemic to Northern and Eastern temperate Himalayan region in Himachal Pradesh, Jammu and Kashmir, Uttarakhand, and North-Eastern hilly states between altitudes of 1200–3000 m in the cold or glacial mountain rocky slopes in stone crevices. It is also found in adjoining countries like Nepal, Pakistan, and Afghanistan up to Tibet and China at higher altitudes. It is mainly found on rocks, and it grows in the months of February and March.

Description: This is a rhizomatic herb with fleshy leaves, growing up to 30 cm tall, having a stout creeping rhizomatous rootstock with scars and intermittent axillary buds. Stem is short, and the rhizome comes out from the cervices of rocks and hangs in the air in sloppy areas. Leaves are 5–30 cm long, glabrous, sparsely hairy in margins, broadly obovate or elliptic, finely or sparsely denticulate or shallowly sinuate-denate. The flowers are bisexual, white, pink, or purple, with long cymose panicles 4–10 cm long. Petals: 5, sepal cup hairless, styles long. The fruit is a capsule and rounded in shape. Seeds are grayish in color, minute, and numerous in capsule (https://vikaspedia.in/agriculture/crop-production/package-of-practices/medicinal-and-aromatic-plants/bergenia-ciliata).

Flowering and fruiting: April–June

Chemical constituents: Bergenin, catechin, gallic acid, gallicin, catechin-7-O-glucoside and β-sitosterol (Dhalwal et al., 2008). Other chemicals reported are paashanolactone, sitoindoside, quercitin, (+) afzelechin, tannic acid, albumen, mucilage, glucose, wax, metarbin, and mineral salts (Ahmad et al., 2018).

Used in: Ayurveda, Unani, and Folk

Parts used: Rhizome

Uses: Used as litholytic agent for urinary calculi. It is widely used in the treatment of dysuria, cystitis, crystalluria and renal failure, vertigo, and headache. The rhizomes and roots of the plant act as an astringent and tonic, have anti-inflammatory effect, and are applied as poultice for stiff joints, boils, abscesses, and skin infections. The root powder is considered to be a mild diuretic, but in higher doses, it exhibited antidiuretic action. Various Ayurvedic classical drugs such as Pashanabhedadi kwath, Pashanabhedadi ghrit, Pashanabhedadi Churan, and others are prepared from Pashanbhed rhizome (https://vikaspedia.in/agriculture/crop-production/package-of-practices/medicinal-and-aromatic-plants/bergenia-ciliata). In the Unani and Ayurvedic systems of medicine, *Bergenia* spp. rhizomes and roots have been used for curing kidney and bladder diseases, dysuria, heart diseases, lung and liver diseases, spleen enlargement, tumors, ulcers, piles, dysentery, menorrhagia, hydrophobia, biliousness, eyesores, cough, and fever. The burns or wounds may be treated with rhizome paste for 3–4 days. The paste can be applied on dislocated bones after setting; it is consumed to treat diarrhea along with honey in fevers (Raina et al., 2008; Chowdhary

DOI: 10.1201/9781003263500-63

et al., 2009; Shakya, 2016; Koul et al., 2020). For century's rhizome of *B. ciliata* has been used for curing pulmonary infections, leucorrhea, and piles and for dissolving bladder and kidney stones (Ahmad et al., 2018).

Methods of cultivation: Seeds and rhizome. Rhizome: Segments about 8–14 cm long are used for direct planting; annular segments of 2-cm thickness are preferred for nursery raising. Seed germination is low, and seed viability is very poor.

Blepharis maderaspatensis (L.) B. Heyne ex Roth.

Synonyms: *Acanthus ciliaris* Burm.f.; *Acanthus maderaspatensis* L.; *Blepharis abyssinica* Hochst. ex A. Rich.; *Blepharis boerhaviifolia* Pers.; Blepharis rubifolia Schum

Family: Acanthaceae

Common names: English: Creeping blepharis, Madars blepharis; **Gujarati:** Utangan; **Hindi:** Dudhiya choti, Utangan; **Kannada:** Kodali soppu; **Malayalam:** Murikootipacha; **Marathi:** Madarasi katemark; **Sanskrit:** Chatushpatri; **Tamil:** Kooravaalchedi, Naalilai naagam; **Telugu:** Antrinta pulu.

Distribution: The species is native to continental Africa, Arabia and tropical parts of Asia: the Indian subcontinent, Southeast Asia, and Hainan in China.

Description: Prostrate herbs. Leaves simple, ternate, unequal, 3–5 × 2 cm, obovate, apex apiculate, puberulous, margins distantly serrate; petiole to 3 cm. Flowers: axillary; bracts 4 pairs, obovate, margins dentate with stiff elongated hairs, bracteoles 1.5 cm, spathulate; outer calyx lobes 1.3 cm and 1.8 cm, oblong, 2 and 3 veined, inner 8 mm; corolla tube urceolate, 5 mm, upper lip reduced, lower lip 3 lobed, 18 × 12 mm, white with pink nerves; filaments 4 and 5 mm, anthers 1-celled, apiculate, bearded; ovary 2.5 mm, oblong, style 1 cm, with a basal ring (http://medplants.blogspot.com/search/label/Blepharis%20maderaspatensis).

Flowering: November–February **Fruiting:** February onward

Chemical constituents: The plant contain steroids, cardiac glycosides, flavonoids, saponins, and phenolic compounds (Neelambika and Leelavathi, 2014).

Used in: Traditional and Folk

Parts used: Leaves and seeds

Uses: Plant ash is used for dropsy, swellings, edema, and gout; dry alcoholic extract of the plant is a potent diuretic; plant parts are crushed, mixed with sheep milk, and given for venereal diseases. Leaf juice is used for throat troubles and asthma; leaves are ground with egg and onion and applied for bone fracture; leaf paste is applied on wounds. Seeds are used as dysuria, are aphrodisiacs, and are diuretics used in diseases of nervous system. Roots are used as an antidote to snakebites. The flat branches of the plant are heated and tied in the case of joint pains (Neelambika and Leelavathi, 2014). The plant is mainly helping to prevent and/or cure inflammation, wound, diabetic wound, ulcer, as an antioxidant, and antihyperlipidemic types of the diseases (Joshi et al., 2021).

Methods of propagation: Seeds

DOI: 10.1201/9781003263500-64

Blepharispermum subsessile DC.

Synonyms: *Leucoblepharis subsessilis* (DC.) Arn.

Family: Asteraceae (Compositae)

Common names: English: Stalkless eyelashpod; **Hindi:** Rasanajadi, Raasnaa; **Kannada:** Naama banta; **Marathi:** Raasnaa; **Sanskrit:** Rasanajadi, Raasnaa; **Telugu:** Adavi banti

Distribution: Distributed in Africa, Madagascar, the Arabian Peninsula, India, and Sri Lanka. In India, it is distributed in Odisha, Karnataka, Madhya Pradesh, and Maharashtra.

Description: It is a shrub and small tree. Leaves: alternate, entire or toothed with short petiole. The leaf blades are sometimes borne on petioles, which may have spines. Inflorescence is a head. The flower head is white about 10 mm, contains 2 female florets, tubular about 3 mm, glandular 3–4 lobed lobes, ovate obtuse style, 3–4 mm, exerted bisexual florets 4, about 4–5 mm. Five-lobed, lobes lanceolate. Stamens: 5 anthers about 3 mm linear sagittate, style about 5 mm recurved. The fruit is a rough-edged cypsela 3–5 mm, black with a pappus.

Flowering and fruiting: August–January

Chemical constituents: The main phytoconstituents isolated from ethyl acetate root extracts were betulin; dronabinol; benzoic acid, 4-heptyl-, 4-cyanophenyl ester; benzamide; 4-chloro, o-(2,6-dimethoxybenzoyl) oxime; 9,12-Octadecadienoic acid (Z,Z)-; n-Propyl 9, 12-octadecadienoate; n-Propyl 9, 12-octadecadienoate; 2-amino-4–(2-cyclohexyl-ethyl)-7-methyl-5-oxo-4H,5H-Pyranol(4,3-b)pyran-3-carbonitrile; carbamult; and carvacrol (Das and Suresh Kumar, 2014).

Parts used: Root, stem, and leaves

Used in: Ayurveda and Folk

Uses: In common cold and rhinitis, as a wormicide, as a tonic for diarrhea, eye troubles, backache, rheumatism, and irregular menstruation. Decoction, as well as powder of root, is used in nervous disorders (Jain, 1965). The whole plant is used in treating worms, stomachache, rheumatic affections, diarrhea, skin diseases, eye troubles, backache due to rheumatism, and irregular menstruation. It is also used for the treatment of diarrhea (Hemadri et al., 1996; Dash and Padhy, 2006), skin diseases, irregular menstruation (Prusti, 2007), and anti-inflammatory diseases. Its roots are being used as *Rasna* in treating rheumatic, gynecological, and nervous disorders (Jadhav et al., 2015).

Methods of cultivation: Seeds

DOI: 10.1201/9781003263500-65

Borassus flabellifer L.

Synonyms: *Borassus flabelliformis, Borassus Sundaicus*
Family: Asteraceae
Common names: Bengali: Taala; **English:** Doub palm, Great fan palm, Toddy palm, Ice apple; **Gujarati:** Taad; **Hindi:** Taad, Tala Tadgola, Trinaraja Munjal; **Knnada:** Olegari, Taalegari, Taati nungu; **Malayalam:** Karimpana, Nonku; **Sanskrit:** Mahatah, Talah, Tantuniyosah, Tranam-ketuh; **Tamil:** Panai, Nungu; **Telugu:** Tatichettu, Munjal
Distribution: The native range of this species is India to Indochina and Jawa to Lesser Sunda Islands. It is a tree and grows primarily in the wet tropical biome.
Description: *B. flabellifer* is a robust tree and can reach a height of 30 m (98 ft). The trunk is gray, robust, and ringed with leaf scars; old leaves remain attached to the trunk for several years before falling cleanly. The leaves are fan-shaped and 3 m (9.8 ft) long, with robust black teeth on the petiole margins. *B. flabellifer* is dioecious with male and female flowers on separate plants. The male flowers are less than 1 cm (0.39 in.) long and form semicircular clusters, which are hidden beneath scale-like bracts within the catkin-like inflorescences. In contrast, the female flowers are golf-ball-sized and solitary, sitting on the surface of the inflorescence axis. Fruits: fleshy, 15–25 cm (5.9–9.8 in) wide, each containing 1–3 seeds. The fruits are black to brown with sweet, fibrous pulp, and each seed is enclosed within a woody endocarp (https://en.wikipedia.org/wiki/Borassus_flabellifer).
Flowering and fruiting: March–April, and again in September–October
Chemical constituents: Contains albuminoids, fats and the fresh pulp is reportedly rich in vitamins A and C. The fresh sap is reportedly a good source of vitamin B-complex. Male inflorescence constitutes spirostane-type steroid saponins like borassosides and dioscin. It also contains 20 known steroidal glycosides and carbohydrates like sucrose. It also contains a bitter compound called flabelliferrins; these are steroidal saponins (Gummadi et al., 2016).
Used in: Fruit, bark, and root
Parts used: Folk and Traditional
Uses: The different parts of the plant are being used for medicinal properties like anthelminthic and diuretic. Young roots are diuretic, and a decoction is given in certain respiratory diseases. The bark decoction, with salt, is used as a mouthwash, and charcoal made of the bark serves as a dentifrice. The fruit pulp helps to cure skin inflammations. It is used to treat nausea and vomiting as well as worm infestation. It is used as an expectorant and a liver tonic. A thin layer of palm fruit jelly applied to the affected area has a soothing effect and immediately alleviates the itchiness associated with prickly heat. The fruit pulp of *B. flabellifer* has been used in traditional dishes and the sap and as a sweetener for diabetic patients (Gummadi et al., 2016).
Methods of cultivation: Seeds

DOI: 10.1201/9781003263500-66

Breynia androgyna (L.) Chakrab. & N.P. Balakr.

Synonyms: *Aalius androgyna* (L.); *Clutia androgyna* L.; *Phyllanthus androgynus* (L.) Chakrab. & N. P. Balakr; *Sauropus androgynus* (L.) Merr.; *Sauropus albicans*

Family: Phyllanthaceae

Common names: English: Sweet leaf, Star gooseberry

Distribution: A plant of the tropics, where it is well adapted to lowland conditions but can also be cultivated at elevations up to 4000 m. Found in India also

Description: A shrub, perennial about 3–6 ft in height. Stem: long and upright. Leaves: biseriate, alternate, oblong to orbicular, 2–7.5 cm long and 1–3 cm broad, entire, glaucous, above dark green, often with vague grayish spots, beneath light green. Flowers: apetalous, red, small but prominently produced in leaf axils. Female flowers have 6 red sepals surrounding a single creamy-colored pistil; male flowers have 6 creamy sepals enclosing 3 reddish stamens. Fruits: a globose to the faintly 6-angular capsule, about 1.5 cm in diameter, white or purplish, dehiscing with 3 valves. Seeds: black, when fully mature, they split open to reveal up to 6 black angular seeds.

Flowering and fruiting: Summer and autumn

Chemical constituents: Contains alkaloids, steroids/triterpenoids, saponins, tannins, polyphenol, glycoside, flavonoids, papaverine, sterols, catechol, cardiac glycosides, and acidic compounds.

Used in: Traditional

Parts used: Leaves and roots

Uses: The leaves are used as a medicine for coughs and to soothe the lungs, as a tonic, and as a febrifugal to relieve internal fever. Plant acts as an antioxidant, antimicrobial, wound healing, anti-inflammatory, antidiabetic, and anti-obesity agent; it also has the potential to increase breast milk production. A decoction of the roots is used against fever and urinary problems and to relieve congestion. The leaves of this plant have been used to treat certain diseases, for weight loss, and in vegetable dishes. The leaf juice is used to treat cholecystitis, diarrhea, and other forms of fever and rhinitis. Decoction of the leaves and roots is a remedy for epistaxis and oriental sores. Another application of the leaves is for oral thrush in infants. A paste of the leaves is applied over nasal ulcers and yaws, erythema, and measles (https://www.socfindoconservation.co.id/plant/341?lang=en#:~:text=Traditional).

Methods of propagation: Can be propagated through leaf cuttings, stem cuttings, and seeds

DOI: 10.1201/9781003263500-67

Photo: C.T. Chandralekha

Breynia retusa (Dennst.) Alston

Synonyms: *Breynia angustifolia* Hook. f.; *Breynia patens* (Roxb.) Rolfe.; *Breynia turbinata* (Oken) Cordem.; *Melanthesa retusa* (Dennst.) Kostel.; *Phyllanthus retusa*

Family: Euphorbiaceae

Common names: Bengali: Chitki; **English:** Cup saucer plant; **Hindi:** Kangli, Retund; **Kannada:** Bilisull; **Malayalam:** Aattacherukola, Ekdania, Perin niruri; **Marathi:** Dalfodi

Distribution: Sri Lanka to Indochina; in India, throughout semi-evergreen and deciduous forests, also in the plains

Description: Shrubs, 1–1.5 m high. Leaves: 1.3–2.5 × 0.7–1.3 cm, broadly elliptic or ovate, base rounded or acute, apex obtuse, glaucous below, black on drying; petiole to 2 mm long. Flowers: unisexual, solitary, axillary. Male flowers: about 3 mm across; pedicels about 1 mm long; perianth turbinate, 3–4 mm long, truncate; lobes 5–6; stamens 3. Female flowers: 5–6 mm across; pedicels 3–4 mm; perianth campanulate, 3–4 mm long, 5–6-lobed. Ovary: globose; stigma bifid. Capsule: 1.2–1.5 cm across, depressed globose, orange red on ripening. Fruiting: perianth enlarged (https://indiabiodiversity.org/species/show/228949).

Flowering and fruiting: February–September

Chemical constituents: Various phytochemical reported from this species, such as reducing sugars, phenolics (most abundant phenols are gallic acid, ellagic acid, coumaric acid, ferulic acid, and vanillic acid), alkaloids, tannins, glycosides, flavones, saponins, and proteins (Kumar and Balakrishnan, 1996; Bhagyasri et al., 2017).

Used in: Traditional

Parts used: Leaves, stem, and fruits

Uses: *B. retusa* leaf juice is used to cure body pain, skin inflammation, hyperglycemia, diarrhea, and diuretic. The fruits have been used for dysentery and twigs used for toothache. The juice of the stem is used in conjunctivitis dysentery, toothache, skin inflammation, hyperglycemia, and diarrhea and for its astringent, diuretic, and anti-arthritic effect (Bhagyasri et al., 2017).

Methods of propagation: Seeds

DOI: 10.1201/9781003263500-68

Brugmansia suaveolens (Humb. & Bonpl. ex Willd.) Bercht. & Presl

Synonyms: *Datura aborea* Mart.; *Datura gardneri* Hook.; *Datura suaveolens* Humb. & Bonpl. ex Willd.

Family: Solanaceae

Common names: English: Angel's trumpet, White angel's trumpet; **Hindi:** Dhatura; **Malayalam:** Ummata; **Tamil:** Madulam

Distribution: The species was originally a native to the coastal rainforests of Southeast Brazil, growing up to an altitude of about 1000 m, along riverbanks and forest edges with warm temperatures, high humidity, and heavy rainfall. Now it can now be found growing in residential areas throughout the world. It is widely cultivated and naturalized in India, and can be seen growing up to altitudes of 1300–1700 m.

Description: *B. suaveolens* is a semi-woody shrub or small tree, growing up to 10–15 ft tall, with a many-branched trunk. The leaves are large, oval, 25 cm long and 15 cm wide. The flowers are very beautiful and sweetly fragrant, slightly pink, about 20–30 cm long, and shaped like trumpets. Calyx: about 10 cm long, 5 teethed, corolla body is slightly recurved to 5 main points about 1–2.5 cm long. The flowers are usually creamy white to pinkish shade but may be yellow or pink and hang downward, lower half of corolla pipe-like and upper half funnel-shaped. Fruits are rarely formed and are short spindle-shaped about 6–8 cm long and 3 cm broad, with large brown seeds.

Flowering: January–November **Fruiting:** May–June and October

Chemical constituents: The plant is rich in atropine, scopolamine, hyoscyamine, and several other tropane alkaloids. It also contains cuscohygrine, an alkaloid unique to this species.

Used in: Homoeopathy

Parts used: Leaves, stem, flowers, and seeds

Uses: Fresh flowers are soaked in milk or hot water to prepare a tea, which is aphrodisiac in nature. Leaf paste is applied to treat wounds, rashes, ulcers and snakebite. Tea prepared from the plant is taken to cure menstrual pain, infections, and mental weakness. The plant is highly effective in curing asthma, bronchitis, inflammation Parkinson's disease, piles, paralysis, epilepsy, convulsions, spasm, and rheumatism; it is analgesic, anesthetic, and antispasmodic. It is used to calm toothache and treat inflammation from trauma (De Feo, 2004) and for treating pain in general (Parker, 2007), especially chest pain, and treating abscesses, dermatitis, and fungal infections of the skin (Reis et al., 2019). Flower scent induces sound sleep. However, overdose of any preparation from this plant is serious and may lead to delirium, physical discomfort, and even death.

Methods of propagation: Stem cuttings

DOI: 10.1201/9781003263500-69

Photo: Dr. Srinath Rao

C

Callicapra pedunculata R.Br.

Synonyms: *Callicarpa formosana*; *Callicarpa dielsii* (H. Lev.) C. Pei; *Callicarpa lasiantha* Lem.; *Callicarpa panduriformis* H. Lev.; *Callicarpa purpurea* Lem.

Family: Lamiaceae (Verbenaceae)

Common names: English: Velvet leaf

Distribution: Bangladesh; Cambodia; China South-Central; China Southeast; Hainan; India, Jawa; Laos; Malaya; Myanmar; Philippines; Sumatera; Thailand; Vietnam. In India found in Assam and Eastern Himalayas

Description: An evergreen shrub or small tree growing occasionally to 4–5 m tall. The twigs, the petioles, and the underside of the leaf blade have a covering of stalked stellate hairs, while the upper surface of the leaf has a covering of stellate and simple hairs, which become sparse when older. The leaf blades are about 6–18 cm long and 3–6 cm broad, and there are small, pale yellow glands on the underside of the leaf. The bottom part of the leaf has smooth margins, but the remainder is toothed. The inflorescences are 2–3 cm long, inserted a little above leaf axil. The flowers have stalks, which are 0.5–1 mm long, while the calyx is 1–1.5 mm long, and the purple or mauve corolla is 2–3 mm long. It produces whitish to purple berries that are drupes (Conn, 1992; Zich et al., 2020).

Flowering and fruiting: June–October

Chemical constituents: Numerous phytochemicals have been isolated from species in *Callicarpa*, including representatives from the following structural classes: clerodane and phyllocladane diterpenes, fatty acids, flavones, lignans, monoterpenes, phenylpropanoids, phytosterols, sesquiterpenes, and triterpenes (Kapoor et al., 1975; Sinha and Malik, 2015).

Used in: Folk

Parts used: Whole plant, leaves, and roots

Uses: The whole plant is analgesic, antiphlogistic, and hemostatic a decoction is used in the treatment of hemoptysis, hematemesis, nosebleed, hematuria, traumatic bleeding, traumatic injuries, and rheumatoid arthritis. The whole plant can be crushed for external applications to wounds (Duke and Ayensu, 1985). An infusion of the leaves, together with leaves of basil (*Ocimum basilicum*), is drunk after parturition and as an emmenagogue. The leaves are used to relieve difficulty in breathing. They are rolled into cigarettes, which are considered to have a sedative effect on asthma. The crushed flower and leaf buds are applied as a styptic on wounds. The roots are used as an antidote for poisonous fish, crabs and toadstools, as a remedy for gonorrhea, and as an emmenagogue. A decoction of the roots is an ingredient of a poultice used to mature boils and ulcers. The whole plant can be crushed for external applications to wounds (http://proseanet.org/).

Methods of cultivation: Seeds

DOI: 10.1201/9781003263500-71

Callicarpa macrophylla Vahl

Synonyms: *Callicarpa dunniana* H. Lev.; *Callicarpa incana* Roxb.; *Callicarpa roxburghii* Wall.; *Callicarpa salviifolia* Griff.; *Callicarpa tomentosa* K.D. Koenig ex Vahl

Family: Lamiaceae (Verbenaceae)

Common names: Bengali: Massandari, Matara; **English:** Beauty berry, French mulberry of Western Ghats, Perfumed cheery; **Gugrati:** Latapriyangu Priyangu; **Hindi:** Bastara, Dahiya, Phalini; **Kannada:** Ibbani, Rushipatri; **Malayalam:** Cerutheekk, Naaykkubis, Naayththeekk; **Marathi:** Aesar, Kanphuli; **Sanskrit:** Gandhaphali, Gochandana, Karambha, Priyangu; **Tamil:** Ittauduga, Seembakulthu, Vettilai pattai; **Telugu:** Kodauduga

Distribution: Bhutan, China, India, Indonesia, Myanmar, Nepal, Pakistan, Papua New Guinea, Sri Lanka, Thailand, Vietnam. In India. Andhra Pradesh, Arunachal Pradesh, Assam, Bihar, Himachal Pradesh, Jammu and Kashmir, Maharashtra, Manipur, Meghalaya, Mizoram, Nagaland, Odisha, Punjab, Sikkim, Tripura, Uttar Pradesh, and West Bengal

Description: *Callicarpa* species are under-shrubs, shrubs, or trees, about 5–15 ft tall. Stem: acutely or obtusely 4 angular. Leaves: opposite-decussate, simple, ternate, exstipulate, subsessile, or petiolate. Inflorescence: axillary, cymose, solitary, pedunculate or sessile. Flowers: bisexual, actinomorphic, fragrant. Calyx: 4 or rarely 5, toothed, pubescent, corolla infundibular or hypo-crateriform, 4 or rarely 5, lobed. Stamens: 4, exserted, anthers oblong, 2 loculed, dorsifixed, pistil 2-carpellary, ovary superior, 4 lobed, 4 loculed, 1 ovule in each locule, style short, stigma 2 lobed. Fruit: a drupe, globose or subglobose, about 2–2.5 mm in diameter, succulent, green and white or purple when ripe.

Flowering and fruiting: July–November

Chemical constituents: Calliterpenone and its acetate are present in seeds, leaves, and aerial parts (https://echarak.in/echarak/templates/Callicarpa%20macrophylla%20Vahl.pdf.A). Crategolic acid, 2alpha-hydroxyurosolic acid, urosolic acid, beta-sitosterol, beta-D glucoside, apigenin, and luteolin are obtained from the leaves. A new diterpenoid-Isopropyl idenocalliterpenone is isolated from essential oil of leaves, along with calliterpenone (https://www.easyayurveda.com/2017/03/14/priyangu-callicarpa-macrophylla/).

Used in: Ayurveda

Parts used: Leaves, root, and oil

Uses: In Ayurvedic literature, Priyangu is described as cooling, refrigerant, a deodorant and antipyretic, tones up the digestive system, checks excessive perspiration, disinfects intestines, controls diarrhea and allays burning sensation during fevers. Leaves are applied in rheumatic joints. Oil from the root is aromatic (https://echarak.in/echarak/templates/Callicarpa%20macrophylla%20Vahl.pdf.A). A paste of the bark is applied over forehead to treat headaches. The powder of the bark is used for rubbing the gums to treat gingivitis and used in face packs to improve the complexion of face. The powder of the bark is used to control the bleeding from wounds. The decoction of the bark is given (30–40 ml) to stop internal bleeding in peptic ulcer and internal piles. The dried powder is given with milk to improve the strength of the body. The paste of the bark is applied externally to treat skin disease. The cold infusion of the bark or root is given (40 ml) for blood purification. Oil of the leaf or bark is applied over joints to treat swelling and pain (https://www.easyayurveda.com/2017/03/14/priyangu-callicarpa-macrophylla/).

Methods of cultivation: Seed

DOI: 10.1201/9781003263500-72

Capparis zeylanica L.

Synonyms: *Capparis acuminate* Roxb.; *Capparis formosa* Wall.; *Capparis latifolia* Craib.; *Capparis ovalifolia* Zipp. Ex Miq.; *Capparis rufescens* Turcz.

Family: Capparidaceae

Common names: Bengali: Kalokera; **English:** Ceylon caper, Indian caper; **Gujarti:** Govindakal, Karrallura; **Hindi:** Ardanda; **Kannada:** Mullukattari; **Malayalam:** Elippayar, Gitoran, Karthotti; **Marathi:** Govindi, Vaghanti; **Sanskrit:** Karambha, Vyaghranakhi; **Tamil:** Adondai; **Telugu:** Arudonda.

Distribution: Very common in the deciduous forests from foothills to 400 m. India, Sri Lanka, to Myanmar, Thailand, Indochina, and Malesia. In India, very common in the deciduous forests from foothills to 400 m

Description: It is a large shrub, 10–15 ft in height; stems have climbing and drooping branches. Leaves: alternate, elliptic, ovate, 6–9 cm long, 5–6 cm broad, apex mucronate, base cuneate-obtuse, petiole about 1 cm long; stipular spines present; flowers: solitary in leaf axis, 3–4 cm across, white, buds densely pubescent, pedicels 2–4 cm long; stamens: numerous long, exerted, white, gynandrophore as long or longer than filaments; ovary: 2–3 mm ellipsoid. Fruit: an ovoid berry, pendulous, smooth, bloodred when ripe. Seeds: many.

Flowering: February–April **Fruiting:** April–June

Chemical constituents: Preliminary phytochemical screening of the extracts showed the presence of alkaloids, flavonoids, saponins glycosides, terpenoids, tannins, proteins, and carbohydrates (Ghulea et al., 2007). Phytochemical analysis of the whole plant has shown the presence of alkaloids, fatty acids, flavonoids, tannins, E-octadec-7-en-5-ynoic acid, saponins glycosides, terpenoids, saponin, p-hydroxybenzoic, syringic, vanillic, ferulic, and p-Coumanic acid. Leaves and seeds showed presence of β-carotene, thioglycoside, glycocapparin, n-triacontane, α-amyrin, and fixed oil, whereas root bark showed presence of an alkaloids (Raghunathan and Mitra, 2003; Padhan et al., 2010).

Used in: Folk

Parts used: Leaves and bark

Uses: Leaves used as cataplasm for boils, swellings, and hemorrhoids and to increase appetite. Root bark is ground with water, boiled, and taken orally to treat indigestion (Muthu et al., 2006). Traditionally, it is used as an antidote for snakebite and to cure swelling of testicle, smallpox, boils, cholera, colic, hemiplegia, neuralgia, sores, pneumonic, and pleurisy (Kirtikar and Basu, 1993; Chopra et al., 1992).

Methods of propagation: Seeds

DOI: 10.1201/9781003263500-73

Carapichea ipecacuanha (Brot.) L. Andersson.

Synonyms: *Callicocca ipecacuanha* Brot.; *Cephaelis acuminate* H. Karst.; *Cephaelis ipecacuanha* (Brot.) A. Rich.; *Psychotria ipecacuanha* (Brot.) Stokes; *Uragoga ipecacuanha* (Brot.) Baill.

Family: Rubiaceae

Common names: English: Ipecac; **Hindi:** Gurmarbuti; **Kannada:** Ipikaakyuna oushadhi **Tamil:** Ipika, ipikakku

Distribution: It is native to Brazil; it has been cultivated in Darjeeling hills of West Bengal, India, and elsewhere in Malaysia and Burma.

Description: It is a small shrub, growing about 50 cm tall. The plant bears opposite leaves above. Leaves are petiolate, entire, oblong, dark green with rough texture. Roots are branched and covered with a thick, transverse ringed bark, that becomes a diagnostic character for the drug. Color is reddish brown, with characteristic round ridges that are linked to a subterranean stem by a ground-distinct filament. At the base of leaf stalks, a pair of whitish stipules are seen. The inflorescence is capitate, enclosed by a large one-leafed involucres. Flowers are small, white, funnel-shaped, enclosed within four large ovate bracts, corolla white with reflexed limb segment, stamens 5, slightly exserted. The stamens and pistils are dimorphic with some flowers bearing long stamens and short pistils and in contrast, other flowers short stamens and long pistils. Fruit is a berry, dark violet, crowned by the limb of the calyx, 2-celled, 2-seeded (Mondal and Moktan, 2020).

Flowering: June–July and October–January **Fruiting:** February–April and July–August

Part used: Root

Uses: The roots are used internally in the treatment of coughs, bronchitis, whooping cough, and amoebic dysentery. The plant is used in homeopathy in the treatment of nausea (Brown, 1996). It is especially useful in the treatment of drug overdoses by inducing vomiting. Smaller doses are strongly expectorant, and it is a common ingredient cough medicines (Chaevallier, 1996). The plant needs to be used with caution because an excess causes severe vomiting and diarrhea (Brown, 1996). Also used in treatment of blood disorders, leukemia, teething children, and cancer (DeAlbuquerque et al., 2007).

Chemical constitutions: Roots contain the alkaloids emetine (methylcephaeline) and cephaeline. It also contains the pseudo-tannin ipecacuanha acid or cephaëlic acid (Roberts et al., 1998).

Methods of propagation: Propagated through stem or root cuttings; growth is slow from seeds.

DOI: 10.1201/9781003263500-74

Cassia tora (L.)

Synonyms: *Senna tora* (L.) Roxb

Family: Fabaceae (Cesalpiniaceae)

Common names: Bengali: Panevar, Chakunda; **English:** Stinking cassia, Chinese senna, Java bean; **Hindi:** Chakund, Chakramard; **Kannada:** Chagate Chakramard; **Malayalam:** Sakramardakam; **Marathi:** Takala, Tankala, Godhadi; **Sanskrit:** Chakramardak, Dadrughna, Uranaakhya; **Tamil:** Senavu

Distribution: Distributed in India and other tropical countries

Description: An erect, usually annual often profusely branched herb or under-shrub, 30–120 cm tall. Leaf: approximately 5–10 cm long, leaflets 3 pairs, membranous, mostly obovate, nearly equal sided, about 0.5–5 cm long and 0.5–2.5 cm wide, both surfaces hairy, or glabrous, base oblique, cuneate to rounded, apex delicately acute, petiolules 1.0–2 mm long, glands between 2 lower pairs of petiolules. Racemes: short, peduncle hairy, about 3–4 mm long, with a single pair of flowers, or axillary solitary. Bracts: linear-acute, 2–4 mm long, pedicel 4–10 mm long, in fruit up to 1.5 cm, hairy. Sepals: ovate, 5–6 mm long. Petals: unequal, ovate, top rounded, 8–10 mm long. Stamens: 7, developed, 3 largest, 4 lateral, 3 upper absent or staminodial. Ovary: appressed, densely hairy, ribbed, style glabrous, stigma truncate. Pod: glabrescent, often curved, indehiscent, 10–20 cm long, 4–6 mm broad. Seeds: 20–30, rhomboidal, glossy, dark, 5 mm across.

Flowering and fruiting: July–December

Chemical constituents: Several anthraquinones have been isolated from the seeds. The leaves showed mainly the presence of anthraquinone glycosides and flavonoids. The anthraquinone glycoside includes rhein, emodine, physcion, chrysophanol (marker), obtusin, chryso-obtusin, chryso-obtusin-2-O-β-D-glucoside, obtusifolin, and chryso-obtusifolin-2-O-β-D-glucoside. Sennosides have also been detected in the leaves of the plants (Shibata et al., 1969; Raghunathan et al., 1974; Lohar et al., 1975; Pawar and D'mello, 2011).

Used in: Folk

Parts used: Leaves, seed, and root

Uses: Different parts of the plant (leaves, seed, and root) are reputed for their medicinal value. The leaves of *C. tora* are reported to have antirheumatic activity in folklore practice. Decoction of the leaves is used as laxative. The seeds have been used in Chinese medicine as aperients, anti-asthmatic, and diuretic agent. It is well recognized in traditional medicine as laxative and is useful for treatment of leprosy, ringworm infection, ophthalmic, skin diseases, and liver disorders. The leaves have been found to possess significant hepatoprotective activity and anti-inflammatory activity (Pawar and D'mello, 2011). Emodine, an anthraquinone, present in root and bark of *C. tora* possess antitumor activity (Choi et al., 1998).

Methods of propagation: Seed

DOI: 10.1201/9781003263500-75

Cerbera manghas L.

Synonyms: *Cerebera linnaei* Montr.; *Cerbera venenifera* (Poir.) Steud; *Odollamiamanghas* (L.) Raf.; *Tanghinia manghas* (L.) G. Don.; T*abernaemontena obtusifolia* Poir.

Family: Apocynaceae

Common names: English: Sea mango, Madagascar ordeal bean, Odollam tree, Pink-eyed cerbera, Dogbane; **Kannada:** Chande, monde; **Malayalam:** Chattankaya, Utalam; **Marathi:** Sukanu; **Tamil:** Kodalma, Kattarali, Kottuma, Caat aralie

Distribution: Sea mango is native to Madagascar, Southeast Asia, and many Pacific islands. In India distributed in Maharashtra, Ratnagiri, Karnataka, Mysore, North and South Kanara, and Odisha

Description: Sea mango is a small evergreen coastal tree growing up 12 m tall. Its bark is gray and fissured. The branches are thick and succulent. The stem has large lenticels and soft wood. The leaves alternate, shiny, dark green, and ovoid in shape and arranged spirally. The flowers are fragrant, possessing a white tubular 5-lobed corolla about 3–5 cm in diameter, with a pink to red throat. They have 5 stamens, and the ovary is positioned above the other flower parts. The fruits are egg-shaped, are 5–10 cm long, and turn bright red at maturity; they have single seed.

Flowering and fruiting: April–August

Chemical constituents: The leaves and the fruits contain the potent cardiac glycoside cerberin. *C. manghas* contains reducing sugars, tannins, steroids, alkaloids, glycosides, polyphenols, terpenoids, flavonoids, and saponins (Hossain et al., 2013; Anwar et al., 2013). p-Hydroxybenzaldehyde, benzamide, n-hexadecane acid monoglyceride, loliolide, b-sitosterol, cerberin, cerleaside A, daucosterol and neriifolin, are the compounds present in the leaves of *C. manghas* (Chan et al., 2016; Zhang et al., 2010). Fresh leaves contain digitoxigenin, oleagenin glucosyl-thevetoside, gentiotriosyl-thevetoside, glycosides, and glucosyl-thevetosides of digitoxigenin and tanghinigenin (Yamauchi et al., 1987).

Used in: Folk

Parts used: Leaves, seeds, stem, and bark

Uses: The seed oil in plasters applied to the skin is effective for scabies and prurigo and applied to the hair kills head lice. The glycosides extracted from the seeds are active on heart failure. A decoction of the inner bark is drunk with cold water as an abortifacient. The trunk bark or the leaves are occasionally used as a purgative and antipyretic and in the treatment of dysuria and ringworm, while the flowers are used to treat hemorrhoids (https://tropical.theferns.info/viewtropical.php?id=Cerbera+manghas). This plant has applications in traditional folk medicine as analgesic, anticonvulsant, a cardiotonic, and for hypotensive activity (Maharana, 2021). It is also used for the treatment of fever, malaria, pain, diabetes, gastrointestinal ailments, skin diseases, and ectoparasitic diseases (Wiart, 2006).

Methods of cultivation: Vegetative propagation of this species takes place through stem cuttings and air layering.

DOI: 10.1201/9781003263500-76

Chenopodium ambrosioides L.

Synonyms: *Dysphania ambrosioides* (L.) Mosyakin & Clemants; *Teloxys ambrosioides* (L.) WA Weber,

Family: Chenopodiaceae

Common names: Bengali: Bathu sag; **English:** Mexican tea, Spanish tea, American wormseed; **Hindi:** Khatua, Sughandha vastuka; **Kannada:** Kaadu oma, Guddeoma; **Malayalam:** Eechchekkolli, Kattayamodakam Naatachcheti; **Marathi:** Chandanbatva; **Tamil:** Pulukkolliched; **Telugu:** Kaadu Omu.

Distribution: An herb native to Central America, South America, and southern Mexico. Also in India (Sathyavathi and Janardhanan, 2007)

Description: It is an annual or short-lived perennial plant, growing to 1.2 m tall, irregularly branched, with oblong-lanceolate leaves up to 12 cm long. The flowers are small and green, produced in a branched panicle at the apex of the stem.

Flowering and fruiting: Throughout the year

Chemical constituents: The essential oil contains ascaridole (up to 70%), limonene, p-cymene, and smaller amounts of numerous other monoterpenes and monoterpene derivatives (α-pinene, myrcene, terpinene, thymol, camphor and trans-isocarveol). Ascaridole (1,4-peroxido-p-menth-2-ene) is rather an uncommon constituent of spices; another plant owing much of its character to this monoterpene peroxide with a characteristic aromatic odor. Ascaridole is toxic and has a pungent, not very pleasant flavor; in pure form, it is an explosive sensitive to shock. Allegedly, the ascaridole content is lower in epazote from Mexico than in epazote grown in Europe or Asia.

Used in: Traditional

Parts used: Leaves and oil

Uses: It is commonly believed to prevent flatulence. It has also been used in the treatment of amenorrhea dysmenorrhea, malaria, chorea, the now-discredited diagnosis of hysteria, catarrh, and asthma. Some of its chemical constituents have been shown in the laboratory to affect certain cancer cell lines. The oil of *Chenopodium* is derived from this plant. It is a colorless or pale yellow toxic essential oil of unpleasant odor and taste, formerly used as an anthelmintic. In the early 1900s, it was one of the major anthelmintics used to treat ascarids and hookworms in humans, cats, dogs, horses, and pigs. Usually, the oil of *Chenopodium* was used. *Chenopodium* is still used to treat worm infections in humans in many countries. In Honduras, as well as other Latin American countries, the whole plant or the leaves are ground and added to water; this mixture is then consumed.

Methods of cultivation: Seeds. Sown indoors in April/May and transplant or sown directly outdoors once soil warms up in early June. Optimal temperature for germination: 21°C Seeds will sprout in 7–14 days.

DOI: 10.1201/9781003263500-77

Chromolaena odorata (L.) R. M. King and H. Robinson

Synonyms: *Eupatorium odoratum L.; Eupatorium conyzoides* Vahl; *Eupatorium brachiatum* Siam weed. ex Wiestr; *Eupatorium atriplicifolium* Vahl and *Osmia odorata* (L.) Schultz-Bip.

Family: Asteraceae (Compositae)

Common names: English: Christmas bush, Communist weed, Siam weed, Devil weed; **Hindi:** Bagh dhoka, Tivra gandha; **Sanskrit:** Ropani, Seekhrasarpi

Distribution: India, China, and Thailand

Description: *C. odorata* is a scrambling perennial shrub that grows 2–3 m in height with straight, pithy, brittle stems that branch readily. The arrowhead-shaped leaves are 6–12 cm in length and 3–7 cm in width, with three veins in a pitchfork appearance. The leaves grow in opposite pairs along the stems and branches. There are 15–25 tubular florets per head, each 10 mm long and in several colors, such as white, purple, pink, or blue. The color of seeds is brown gray to black and is 4–5 mm long with a pale brown pappus that is 5 or 6 mm long.

Flowering and fruiting: December–March

Chemical constituents: Its phytochemical components are alkaloids, flavonoids, flavanone, essential oils, phenolics, saponins, tannins, and terpenoids. The other important constituents of this plant are eupolin, chromomoric acid, quercetagetin, and quercetin, acacetin, chalcones, eupatilin, luteolin, naringenin, kaempferol, quercetin, quercetagetin, and sinensetin, all of which contribute to its remedial properties (Sirinathipapora and Jiraungkoorskul, 2017). It also contain terpenes and terpenoids alkaloids, including pyrrolizidine, saponins, and tannins; phenolic acids, including ferulic acid and protocatechuic acid; and phytoprostane compounds, including chromomoric acid (Wafo et al., 2011; Biller et al., 1994; Phan et al., 2001; Heiss et al., 2014).

Used in: Traditional

Parts used: All parts

Uses: It is a traditional medicinal plant that is widely used for its wound-healing property. In particular, the several parts of this herb have been used to treat wounds, burns, and skin infections. Furthermore, it has also been shown to possess anticancer, antidiabetic, antihepatotoxic, anti-inflammatory, antimicrobial, and antioxidant properties. Traditionally, fresh leaves or a decoction of *C. odorata* have been used throughout Vietnam and other tropical countries for the treatment of leech bites, soft-tissue wounds, burn wounds, skin infections, rashes, diabetes, and periodontitis and as an insect repellent. A poultice of the leaves is traditionally applied to cuts or wounds to stop bleeding and promote healing. Eupolin, a product made from *Chromolaena* spp., has already been licensed for use in Vietnam for the treatment of soft tissue burns and wounds. A decoction of the roots is used as an antipyretic and analgesic remedy, and a leaf extract with salt is used as a gargle for sore throats and colds. The fresh leaves and extracts of *C. odorata* are traditional herbal treatments in certain developing countries, including Thailand, India, and Vietnam, for burns, soft tissue wounds, and skin infections. A decoction is used in traditional medicine. The leaf is used as a cough treatment and as a component with guava leaves and lemongrass

DOI: 10.1201/9781003263500-78

therapy for malaria. In certain underdeveloped nations, it is used to treat soft-tissue burns, infections of the skin, and wounds (Suksamrarn et al., 2004; Vital and Windell, 2009; Vaisakh and Pandey, 2012; Sirinathipapora and Jiraungkoorskul, 2017).

Methods of propagation: Seeds

Photo: Dr. Srinath Rao

Cissampelos pareira L.

Synonyms: *Cebatha oriculata* (Thunb.) Kuntze; *Cebatha villosa* Christensen; *Cissampelos accuminata* Benth.; *Cissampelos auriculata* Miers.

Family: Menispermaceae

Common names: Bengali: Akanadi; **English:** Abuta, Velvet leaf, Ice vine; **Gujarati:** Venivel; **Hindi:** Padhi, Padha; **Kannada:** Padavali; **Malayalam:** Kattuvalli, Patathali; **Marathi:** Padaveta; **Sanskrit:** Ambasthtaki, Laghu patha; **Tamil:** Vatta tiruppi; **Telugu:** Chiruboddi

Distribution: Mainly native of India found in tropical and subtropical regions; also distributed in Asia, East Afrca, and the Americas

Description: It is a slender tomentose climber. The leaves are peltate, 2.5–12 cm long, 2.5–11.5 cm broad, triangularly broad-ovate, or orbicular, obtuse, mucronate, base cordate or truncate, ± tomentose on both sides; petiole pubescent. Flowers are small in size, pedicels filiform. Male flowers are clustered in the axil of a small leaf; sepals are 4 in number, obovate-oblong, hairy outside; petals 4 in number, united to form a 4-toothed cup, hairy outside; stamens 4, column short, anthers connate, encircling the top of the column. Female flowers are clustered in the axils of orbicular, hoary imbricate bracts, on 5–10 cm long racemes; sepal 1, petal 1; carpel 1, densely hairy; style shortly 3-fid. Drupe are 4–6 mm long, 3–4 mm broad, subglobose, compressed, hairy-pubescent, red when fresh, black when dry, endocarp transversely ribbed, tuberculate. Seeds are horseshoe-shaped.

Flowering and fruiting: July–September

Chemical constituents: Chemical compounds present in this plant are berberine, hayatin, hayatinin, menismine, methanolic acid, cissamine, cycleanine, hayatidin, and quercitol.

Used in: Ayurveda

Parts used: Leaves and roots

Uses: It has been extensively used in the traditional medicinal system since ancient times for the treatment of numerous diseases such as ulcer, wound, rheumatism, fever, asthma, cholera, diarrhea, inflammation, snakebite, malaria, rabies, and is recommended for blood purification (Kumari et al., 2021). Patha root powder, 1–2 gm, is taken twice daily mixed with rice washed water to treat nonspecific leucorrhea and backache. Leaf juice is used for washing eyes in case of conjunctivitis and burning eyes. Leaf paste is placed in the middle of the scalp or over closed eyes to reduce burning of the scalp and induce sleep. Five grams of patha and fenugreek powder is mixed in butter milk and administered for breast milk cleansing. The herb is also used in fracture healing, wounds, ulcer, and female reproductive ailments. This herb is also used to treat edema, cystitis, fever, and kidney ailments.

Methods of propagation: Seeds

DOI: 10.1201/9781003263500-79

Clerodendrum paniculatum L.

Synonyms: *Clerodendrum diversifolium* Vahl.; *Clerodendrum pyramidale* Andrewa.; *Volkameria angulata* Lour.; *Volkameria diversifolia* Vahl.

Family: Lamiaceae (Verbenaceae)

Common names: English: Pagoda flower; **Hindi:** Hanuman kireet; **Kannada:** Rathapushpa, Krishnana kirita; **Malayalam:** Krishnakireetam, Hanuman kireetam; **Marathi:** Deepmal; **Sanskrit:** Krishnakireeta; **Tamil:** Krishna kireedam

Distribution: Distributed in countries such as India, Sri Lanka, Malaysia, Australia, China, Taiwan, Laos, Cambodia, Vietnam, Indonesia, Bangladesh, and many other Asian countries

Description: It is an erect, semi-woody shrub, reaching a height of 1–2 m. Leaves are simple, large, opposite, up to 30 cm in diameter. Leaves are membranous, cordate-ovate, 3–7 lobed (lobes shallow), margins minutely denticulate, glandular beneath and petiole is up to 30 cm long with acuminate apex. Flowers are in large terminal panicles. Calyx is red to orange-red in color and divided nearly to the base. Corolla are orange-red to scarlet with slender tube, which is up to 2 cm long, and the lobes are approximately 7 mm long. Fruit is a drupe, globose, bluish black in color, and approximately 1 cm across.

Flowering and fruiting: More or less throughout year

Chemical constituents: Various phytochemicals such as rutin, quercetin, β-sitosterol, β-amyrin, lupeol, oleanolic aldehyde acetate, stigmasta-4, 25-dien-3-one, and (3β)-stigmasta-4,22,25-trien-3-ol have been identified in this plant (Prashith Kekuda and Sudharshan, 2018).

Used in: Traditional and Folk

Parts used: Leaves, flowers, and roots

Uses: The plant is reported to have ethnomedicinal importance as the plant is used as a remedy for ailments and disorders such as wounds, typhoid, snakebite, jaundice, giddiness, malaria, anemia, and hemorrhoids. The plant is shown to exhibit biological activities such as antimicrobial, antioxidant, anthelmintic, anti-inflammatory, antimutagenic, cytotoxic, hypolipidemic, insecticidal, and anti-aging activity (Prashith Kekuda and Sudharshan, 2018). Traditionally, it has been used in the treatment of inflammation, rheumatism, arthritis, diabetes, dropsy, swellings, edema, and gout.

Methods of propagation: Seeds

DOI: 10.1201/9781003263500-80

Photo: Dr. Srinath Rao

Clerodendrum phlomidis Linn.f.

Synonyms: *Clerodendrum multiflorum* (Burm. f.) O. Kuntze. Volkameriamultiflora Burm.f.
Family: Lamiaceae (Verbenaceae)
Common names: English: Indian headache tree, Sage glory bower; **Gugrati:** Arani, Aranel, Airanamul; **Hindi:** Arni, Agathu, Ganlyarl, Tekar; **Kannada:** Arani, Taggiberu; **Malayalam:** Munja; **Marathi:** Eirani, Eran, Chamari Takali; **Sanskrit:** Agnimantha, Arani, Tarkaari, Vikraantaa, Jayanti, Jai, Jayaa, Ganikaarikaa, Vaijayanti; **Tamil:** Tazhuthaazhai, Taludalal, Munnay; **Telugu:** Gabbunulli, Nelli
Distribution: Throughout India, in the drier parts
Description: A large shrub/small tree, 1.5–3 m tall. Stem: ash gray, branches velvet. Leaves: opposite, deltoid ovate, obtuse or acute, crenate, dentate, hairy below, wavy. 1.5–5cm long, 1–3 cm broad. Inflorescence: axillary cyme. Flowers: creamy white or pale yellowish, about 1.5 cm across in dichotomous axillary cymes forming a rounded terminal panicle, fragrant, bracts small, leafy lanceolate. Calyx: 5-lobed lobes ovate veined. Corolla: white or pinkish, tube long, lobes 5, unequal, elliptic, obtuse, veined. Stamens: 4, didynamous, much exserted. Ovary: glabrous, style slender, stigma bilobed. Fruit: a drupe obovoid, 4-lobed, with one pyrene in each lobe.
Flowering and fruiting: August–February
Chemical constituents: Flavonoids, scutellarein, and pectolinarin have been isolated from the leaves. Stems gave d-mannitol, beta sitosterol, its glucosides, and ceryl alcohol. The roots contain ceryl alcohol, clerodin, clerosterol, and clerodendrin (Khare, 2007).
Used in: Ayurveda, Folk, and Siddha
Parts used: Leaves, roots, and bark
Uses: Plant parts used in dyspepsia, stomachache, colic, cholera, dysentery, postnatal fever, during convalescence from measles. Root and bark: used to create a bitter tonic, used in debility and nervous disorders. Use of the root in dysuria leaves exhibited hepatoprotective activity. The aqueous extract of leaves exhibited anthelmintic activity. The plant also exhibited antidiabetic activity and retention of urine. Dosage: root—12–24 g for decoction (Khare, 2007). Root is used to create a bitter tonic and given in convalescence of measles. Juice of leaves is alterative and given in syphilitic complaints. The root is given as a demulcent in gonorrhea.
Methods of cultivation: Seeds

DOI: 10.1201/9781003263500-81

Clerodendrum serratum (Linn.) Moon

Synonyms: *Rotheca serrata* (Linn.) Steane & Mabb. *Volkameria serrata*; *Cyclonema serratum*

Family: Lamiaceae (Verbenaceae)

Common names: English: Blue fountain bush, Blue-flowered glory tree, Beetle killer; **Hindi:** Bharangi; **Kannada:** Gantubharangi, Kirutaggi; **Malayalam:** Cherutekku; **Marathi:** Bharang; **Sanskrit:** Bhaargi, Bhaarangi, Angaarvalli, Phanji, Braahmanyashtikaa, Kharshaak, Padma, Bhragubhavaa, Brahmayashtikaa; **Tamil:** Cheruthekku, Kandoorbarang; **Telugu:** Gantubharangi

Distribution: A shrub distributed throughout the country, especially common in Assam and Bengal

Description: It is a small shrub, 2–4 m tall, bearing opposite leaves and having woody rootstock. The plant has quadrangular, glabrous branches. Leaves are large and ovate or oblong, usually ternate whorled, coarsely and sharply serrate, glabrous, and pale beneath with six pairs of lateral nerves. Flowers are large pinkish white in color and numerous, appearing from May to August. It has a stout deflexed compressed pedicel in lax, dichotomous, long terminal panicles. It has leafy bracts. Calyx is cup-shaped, 5 mm long. Corolla is pale to pinkish blue with tube about 6–7 mm long; the lower larger lip like lobe is sky blue in color. Stamens are long, exerted, curved, and bluish. The drupes are 1–4 lobed, bluish black, and glossy.

Flowering and fruiting: May–August

Chemical constituents: The bark contains triterpenoids and serratagenic, oleanolic, and queretaric acids; leaves contain alpha-spinasterol and flavonoids, including luteolin, apigenin, baicalein, scutellarein, phenolic acids, caffeic, and ferulic acids (Khare, 2007). Isolated bioactives from roots like icosahydropicenic acid and ursolic acid (Patel et al., 2014).

Used in: Ayurveda, Siddha, Unani, and Traditional

Parts used: Leaves, bark, and roots

Uses: Anti-asthmatic, antihistaminic, antispasmodic, antitussive carminative, febrifuge. Leaf: febrifuge. In Ayurveda, the use of the dried roots in cough, bronchitis, dyspnea, chest diseases, and sinusitis is indicated (Khare, 2007). Uses include treatment of pain, inflammation, rheumatism, and fever, especially malarial fever. Scientific studies on root extracts and formulations revealed anti-asthmatic, mast cell stabilization and anti-allergic effects. Also includes hepatoprotective, antioxidant, anti-inflammatory, and anticancer potential of the drug (Patel et al., 2014).

Methods of cultivation: Seeds

DOI: 10.1201/9781003263500-82

Cocculus hirsutus (L.) Diels

Synonyms: *Cocculus aristolochiae* DC.; *Cocculus hastatus* DC.; *Cocculus villosus* Lam.; *Holopeira laeviuscula* Miers; *Holopeira torrida* Miers; *Holopeira villosa* (Lam.) Miers.; *Menispermum hirsutum* L.

Family: Menispermaceae

Common names: English: Broom creeper, Inkberry; **Hindi:** Patalgarudi, Jaljamani; **Kannada:** Sogadiballi, Dusariballi; **Malayalam:** Patalagarudakkoli; **Sanskrit:** Patalagarudah, Chilihindah; **Tamil:** Kottukotti; **Telugu:** Dusaratiga

Distribution: Throughout India in dry localities

Description: A perennial climber shrub with softly villous young parts; leaves simple, alternate, ovate-oblong, obtuse, apiculate, subcordate, or truncate at the base, softly villous on both surfaces, petioles densely villous: male flowers small, in axillary cymose panicles, female flowers 2–3 together in axillary clusters; fruits: drupes, green when raw purplish black. The leaves are simple, alternate, truncate at the base, and ovate-oblong in upper parts. The flowers are very small, unisexual, and green. Fruits: a drupe, green when raw and purplish black on ripening.

Flowering: November–April; **Fruiting:** April–May

Chemical constituents: Several isoprunaline alkaloids are present in roots and leaves. The important ones are cohirsile, cohirsinine, cohirsitinine, hirsutin and jamtine, and shaheenine. It contains essential oil, β-sitosterol, ginnol, glycosides, sterols, and alkaloids (Marya and Bothara, 2011).

Used in: Ayurveda, Siddha, Tradtional, and modern medicine

Parts used: Roots and leaves

Uses: The roots are bitter, acrid, thermogenic, laxative, alternate, emollient, alexeteric, depurative, demulscent, digestive, carminative, diuretic, aphrodisiac, expectorant, anodyne, antipyretic, and tonic and are useful in vitiated conditions of Kapha and Vata, poisonous bites, leprocy, skin diseases, prurites, dyspepsia, flapulence, colic, stranguary, cough, bronchitis, gout, and cephalalgia. Crushed leaves for the bleeding of the lower digestive tract, dysentery, gonorrhea, inflammation, and rheumatism; the leaves are crushed in castor oil, and the paste applied on sprain; the leaf paste applied for skin burns and skin diseases; the leaf extract is mixed with curd and given to overcome spermatorrhea and bleeding; it is also drunk for scorpion sting. The aqueous extract of stem and root is a sedative, anticonvulsant and as a hypotensive, bradycardiac, cardiotonic, and sapasmolytic.

Methods of propagation: Seeds and vegetative method

DOI: 10.1201/9781003263500-83

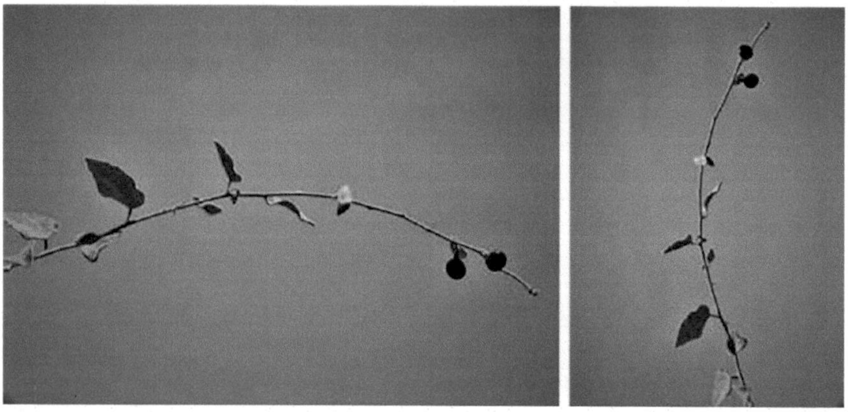

Photo: Dr. Kotresh

Cocculus pendulus (J. R. Forst. & G. Forst) Diels.

Synonyms: *Cocculus leaeba*; *Cabetha esculenta* Forsk. ex Chiov.; *Cabetha pendula* (J. R. & C Forst) Kuntze.; *Cocculs cebatha* DC.; *Cocculs glabra* Roxb.

Family: Menispermaceae

Common names: Gujrati: Parwatti; **Hindi:** Parwatti; **Malayalam:** Villumbi; **Tamil:** Villumbi; **Telugu:** Dusaratiga

Distribution: Distributed throughout the tropical and subtropical countries of the world (India, Pakistan, and Africa)

Description: A dioecious, much-branched liana or scandent shrub; stem up to 15 cm in diameter at base, striped, dark gray; branchlets long, slender, terete, hairy. Leaves are simple, arranged spirally, petiole 2–10 mm long; blade is oblong-lanceolate, in lower leaves sometimes ovate, 1.5–5 × 0.5–2 cm, base cuneate, rounded or sometimes spear-shaped, apex obtuse, sometimes notched, leathery, and glabrous. Inflorescence: a small axillary cyme, solitary or clustered; male inflorescence: few to many flowers, up to 2 cm long; female inflorescence: 1–2-flowered, up to 1.5 cm long. Flowers: unisexual, small; sepals 6, ovate-elliptical, fleshy to membranous, 3 outer ones 1–1.5 mm long, slightly hairy, 3 inner ones larger, finely hairy to glabrous; petals 6, ovate-obovate, 0.5–2 × 0.5–1 mm, apex notched; male flowers: sessile or with short pedicel, stamens 6–9, up to 1.5 mm long, free; female flowers: with pedicel up to 1 cm long, staminodes 6, about 1 mm long, ovary superior, consisting of 3(–6) free, ovoid, laterally compressed carpels about 1 mm long, stigma about 0.5 mm long. Fruit: composed of 1–3 obovoid, flattened drupes, each drupe 4–7 × 4–5 mm, dark red, stone ribbed on lateral faces, 1-seeded. Seed: horseshoe-shaped, laterally flattened.

Flowering and fruiting: October–January

Chemical constituents: Cocsulinin is the main alkaloid other alkaloids present are cocsoline, penduline, tetradine, and isotrilobine. The stems and leaves contain a great variety of bisbenzylisoquinoline dioxine alkaloids, including cocsuline, cocsoline, cocsulinin, siddiquine, penduline, tetradine, isotrilobine, siddiquamine, kohatine, telobine, pateline, kurramine, isotrilobine, and tricordatine, and many derivatives of these. Little investigation has been made from the chemical components of the root (https://uses.plantnet project.org/en/*Cocculus pendulus_*).

Used in: Folk and Traditional

Parts used: Leaves, stem, and roots

Uses: The roots of the plant are used in the indigenous system of medicine for the treatment of intermittent fevers and as a tonic. Hypotensive and anticancer activity has been attributed to the alkaloid fractions from the leaves and stem, 4. The juice of the leaves contains mucilage, which when mixed with water forms a jelly. This is taken as a cooling medicine for gonorrhea. The roots and leaves are used in rheumatic pain. The leaves and stems of *C. pendulus* are used in the indigenous system of medicine for the treatment of intermittent fever and biliousness, febrifuge, vermifuge, antipyretic, diuretic, cholagogue, menstrual cycle troubles, internal parasites, antiblennorrhagic, and rheumatic pain (Oliver-Bever, 1986).

Methods of propagation: Seeds

DOI: 10.1201/9781003263500-84

Cochlospermum gossypium DC.

Synonym: *Cochlospermum religiosum* (L.) Alston; *Bombax gossypium*; *Maximilianea gossypium.*

Family: Bixaceae (Cochlospermaceae)

Common names: Bengali: Sonali simul; **English:** Golden silk cotton; **Hindi:** Galgal; **Kannada:** Arasinaburuga; **Tamil:** Kattupparutti; **Telugu:** Kondagogu; **Marathi:** Ganeri

Distribution: Native of India, Burma and Thiland. In India found in Andhra Pradesh, Karnataka, Tamil Nadu, Madhya Pradesh, and Bihar; cultivated at Agartala in Tripura

Description: Deciduous small tree grows to a height of 7–8 m, bark is smooth and pale gray, sparsely clothed with leaves, and sheds them at the height of the flowering season. Leaves alternate, appear at the tips of the branches and are palmately lobed, lobes 5, lanceolate, acute smooth, green above and tomentose and white beneath, veins prominent. Petiole cylindrical 5–6 in. long. Inflorescence: a panicle, terminal. Flowers are large bright yellow, growing up to about 10 cm, butter cup–shaped stamens and orange. Fruits are brown and oval-shaped. They come in the form of a capsule made up of 5 segments. Capsule splits open to release the seeds which are embedded in the silky cotton contained within

Flowering: January–March **Fruiting:** April

Chemical composition: Uronic acids to be the major component of the polymer (approximately 26 mol%). Furthermore, analysis of the gum by gas chromatography-mass spectrometry (GC-MS) indicated the presence of sugars such as arabinose (2.52 mol%), mannose (8.30 mol%), alpha-D-glucose (2.48 mol%), beta-D-glucose (2.52 mol%), rhamnose (12.85 mol%), galactose (18.95 mol%), d-glucuronic acid (19.26 mol%), beta-d-galactouronic acid (13.22 mol %), and alpha-d-galacturonic acid (11.22 mol%; Vinod et al., 2008).

Parts used: Bark

Uses: A gum extracted from the bark is used in curing dysentery, cough, asthma, diarrhea, gonorrhea, syphilis, hoarse throat, hemoptysis, eye troubles, and scalding urine. It also has sedative and cooling effect.

Methods of propagation: Seeds

DOI: 10.1201/9781003263500-85

Coleus zeylanicus (Benth). L. H. Cramer

Synonyms: *Plectranthus zeylanicus* Benth.; *Plectranthus hadiensis* (Forssk), Schweinf. Ex Sprenger.

Family: Lamiaceae (Labiatae)

Common names: English: Hribera; **Hindi:** Hariveera, Valaka; **Malayalam:** Iruvely; **Sanskrit:** Hruberam, Ambu, Ambas

Description: It is an annual herb growing up to 1 m tall, commonly found in fertile soil. The leaves are opposite, ovate to cordate retuse or emarginate at apex, rounded or truncate at base, margin crenate—serrate, surface bullate, about 2–8cm long and 4–6 cm wide, bearing well defined veins, slightly fleshy. They are grayish green stained with red purple above and darker purple beneath. Stems are violet in color; it is aromatic. Flowers are small, 1–1.5 cm diameter, 5-lobed, bell-shaped with imbricate bracts. These are white in color with faint purple marks and appear within the terminal 2–10 cm long spikes. Calyx: bi-lipped, glandular; corolla bi-lipped, upper lip 4-lobed, lower entire, boat-shaped; stamens: 4, didynamous, filaments connate below; ovary: 4-lobed; style slender, stigma bifid. Capsules are small, slender, oval, linear, and light green in color. Seeds are small, flat, and white in color (Lakshmi and Nandagopal, 2017). Closely similar to *C. ambonicus* Lour. but can readily be recognized by its characteristically different odor and less succulent leaves.

Flowering and fruiting: August–October

Chemical constituents: This plant contains number of medicinally important compounds such as alkaloids, tannins, flavonoids, and phenol sesquiterpene volatile compounds. The main compounds from dried leaves identified include geroniol/lemanol, geraniol, eremophylene/valencene), m-cymene, ledol α-bergamotene, fenchyl acetate, and α-Pinene (Lakshmi and Nandagopal, 2007).

Used in: Traditional

Parts used: Whole plant

Uses: To treat indigestion, diarrhea, fever, rheumatic arthritis, cough, and cold. Its extract oil is used to repel mosquitos. Leaves boil in water and inhale the vapor is better to reduce the cough and cold. Leaves are crushed and the paste is applied on head to get relief in severe headache. Its utilized for treating of inflammatory conditions, *Psilopogon zeylanicus* is also known as a medication for microbial infections, such as dysentery and diarrhea, and against smallpox, the common cold, and eye diseases (Jayaweera, 1982; Lakshmi and Nandagopal, 2007). It also used as anti-hepatotoxic, antiulcerogenic, anti-inflammatory, digestive, stomachic, liver tonic, and cardiotonic (Lakshmi and Nandagopal, 2007; Rebecca et al., 2015).

Methods of propagation: Stem cuttings

Further reading: https://www.google.co.in/search?q=coleus+zeylanicus

DOI: 10.1201/9781003263500-86

Convolvulus arvensis L.

Synonyms: *Convolvulus ambigens* House.; *Convolvulus incanus* Vahl.; *Strophocaulos arvensis* (L.) Small

Family: Convolvulaceae

Common names: English: Field bindweed, Creeping jenny, European bindweed; **Hindi:** Hiran khur

Distribution: It is distributed throughout the world.

Description: A perennial vine (0.4–2 in. in height) arising from deep, persistent, spreading roots. Leaves: dull green with readily visible veins. The petiole is flattened and grooved on the upper side. The leaves of this species are extremely variable. The most common leaf type is hastate or sagittate; they have distinctive arrowhead shapes with pointed lobes at the bases. The leaf margin is entire and divided into three lobes. Stem length ranges from 1 to 6 ft longer and are trailing, twinning, herbaceous, and glabrous to pubescent. Due to twisting, they appear to be 4–5-angled. Inflorescence is axillary cyme composed of 1–3 flowers. The pedicels are significantly longer than calyx, with hairy sepals being 2.5–5mm long. Flowers: have five fused petals forming a 2–2.5-cm-long funnel-like corolla. Petals are generally white to very pale pink. and occasionally has a pinkish and whitish mid-petaline bands. Five stamens of unequal lengths are attached to the base of the corolla. The pistil is compound with two thread-like stigmas. Fruits: about 8 mm wide, rounded, light brown. Seeds are dark brown to black and have rough surfaces. They are 0.5–1.2 cm long, and their shapes vary, depending on the number produced in the fruit; they are rounder when only one is produced and progressively thinner as more are produced (Al-Snafi, 2016c).

Flowering and fruiting: June–September

Chemical constituents: *C. arvensis* was found to contain the tropane alkaloids, tropine, pseudotropine, and tropinone, as well as cuscohygrine, meso-cuscohygrine, and calystegines. Aerial parts of the plant contained 8 alkaloids, including plantyneciene 211.07, fagomine 111.82, swainsonine 438.79, 1-deoxynojirimycin 50.61, austaline 1-epiaustaline, 6-epicastanospermine, and castanospermine-N-oxide. Polyphenolic compounds in the plant, such as coumarins and phenolic acids, showed that it contained umbelliferone and scopoletin in coumarin fraction. Protocatechuic, caffeic, chlorogenic, gentisic, p-Coumaric, p-hydroxybenzoic, p-hydroxyphenylacetic, ferulic, vanillic, syringic, benzoic, and salicylic acids were detected in the phenolic acids fraction. The plant contained four coumarins, esculetin, scopoletin, scopoletin7-O-glucoside umbelliferone. There are 10 phenolic compounds, their total concentration is 1479.12 (μg/ml), including caffeic acid 53.53, thujone 197.55, cymene, ferulic acid, isoferulic acid, cimiracemoside, β-phellandrene, gentisic acid, leutoline, and coumaric stearic, oleic linoleic, linolenic, arachidic, and behenic acid. It also contain steroids, including campesterol, stigmasterol, and β-sitosterol (Al-Snafi, 2016c).

Used in: Leaves, stem, roots, flowers, and fruits

Parts used: Leaves, roots, flowers

Uses: Aerial parts of convolvulus arvensis are used as laxative, wound healing, antispasmodic antihemorrhagic, and antiangiogenetic and for the treatment of parasites and jaundice. In addition, it was used as diuretic and in skin disorders such as anti-furunculosis, antidandruff, and in spider bites. *C. arvensis* is also used traditionally as decoction in cough and flu, to treat the painful joints, inflammation and swelling. All five parts (root, stem, leaf, flower, and fruit) used in preparations to support urinary function, increase libido, alleviate

DOI: 10.1201/9781003263500-87

chronic anemia and coughs, and treat a swollen penis. To relieve bone and joint aches, all five parts are mashed, wrapped in cloth, and placed on the painful area. For mouth sores, liquid from boiling the five parts is held in the mouth; the liquid is also used as a wash for old sores. Leaf: mashed and applied with a bandage to bumps, cysts, and other skin sores. The juice is used for rashes and itching. Root: used in laxative medicines.

Methods of propagation: Rhizome and seed

Couroupita guianensis Aubl.

Synonyms: *Couroupita membranacea* Miers.; *Couroupita peruviana* O. Berg.; *Couroupita saintcroixiana* R. Knuth.; *Couroupita surinamensis* Mart. ex Berg

Family: Lecythidaceae

Common names: Bengali: Kaman gola; **English:** Cannonball tree; **Hindi:** Nagalinga, Top gola; **Kannada:** Nagalinga; **Marathi:** Shivalingam; **Tamil:** Nagalingam

Distribution: It is native to the tropical forests of Central and South America, and it is cultivated in many other tropical areas throughout the world because of its fragrant flowers and large fruits.

Description: A deciduous tree reaching about 15–20 m height. Stem: soft-wooded, straight, stout trunk and rough brownish-gray bark. Leaves: alternate, oblong, oblanceolate, entire, glabrous, nerves prominent, up to 30 cm long. Inflorescence: racemes produced from the upper part of the trunk or from the main branches. Flowers: large, pleasing combination of rose pink, white, and yellow colors, arranged in long racemes. Calyx: turbinate. Petals: 6, unequal, concave, pink inside but with a mixture of white, yellow, and pink outside. Fertile stamens are numerous, monadelphous, arranged on a band that arise from the base of the ovary and bends over it like a hood so that the anthers press down on the ovary. Some abortive stamens (staminodes) also present at the base of the band, surrounds the ovary. This structure looks like a hooded snake sheltering the Shivalinga. Hence, the tree is also called snake-flower tree. Fruit: remarkably large berry and round, with a brown hard shell, containing pulp, about the size of a man's head, up to 25 cm in diameter.

Flowering and fruiting: January onward

Chemical constituents: The studies of chemical constituents show the presence of α-amirin, β-amirin, β-sitosterol, nerol, tryptanthrine, indigo, indirubin, isatin, linoleic acid, carotenoids, and sterols [3–8]. Singh et al. [9] reported that aqueous extract of *C. guianensis* is endowed with flavonoids, alkaloids, phlorotannins, glycosides, tannins, steroids, and terpenoids. Isatin is also a chemical constituent of *C. guianensis* (Wong and Tie, 1995; Anjaneyulu and Rao, 1998; Rane et al., 2001; Famina Jasmine and Vinayaga Moorthi, 2017).

Used in: Ayurveda, Folk, and Traditional

Parts used: Leaves, flowers, and bark

Uses: Medicinal uses constitute antipyretic, antidepressant, analgesic, antiseptic, anti-inflammatory, antiprotozoal, anticancer, and antiulcer activities. In Ayurveda, they are used extensively for its anti-inflammatory property. The flower shows antibacterial activity and antifungal property. The tree contains antibiotic, antifungal, antiseptic, and analgesic qualities. It is used as an ingredient in many ways for curing gastritis, scabies, bleeding piles, dysentery, and scorpion poison. The various parts such as leaves, flowers, and bark are used to treat hypertension, tumor, pain, and inflammatory reactions. Leaves are used to make juice which is used to treat skin diseases and Malaria. It has a chemical constituent of *C. guianensis* that has antioxidant activity, it is cytotoxic to human promylocytic leukemia (HL60) cells, and it is used as a prophylactic agent to prevent free radial-induced cancer and as a chemotherapeutic agent to kill cancer cells (Famina Jasmine and Vinayaga Moorthi, 2017).

Methods of propagation: Seeds and nodal segments

DOI: 10.1201/9781003263500-88

Photo: Dr. Srinath Rao.

Crataeva nurvala Buch.-Ham

Synonyms: *Crataeva lophosperma* Kurz.; *Crataeva magna* Lour.; *Capparis magna* Lour.; *Tricianthera corymbosa* Raf.

Family: Capparidaceae

Common names: Bengali: Trikosha, Varun; **English:** Three-leaved creeper, Sacred lingam tree; **Gujarati:** Barno, Kagdakeri; **Hindi:** Varun, Barun; **Kannada:** Adhiraaja, Bipatri, Mattamavu, Neervala mara; **Malayalam:** Kili, Niravila; **Marathi:** Bhatavarna, Hadavarna; **Sanskrit:** Setuvruksha, Kumarka, Tamalaka Varuna; **Tamil:** Adhicharam, Marlingam; **Telugu:** Bilvaram, Ulimiri chettu

Distribution: Distributed widely in India and other tropical and subtropical parts of the world.

Description: This plant is a medium-sized, deciduous tree about 30 m height, much branched. Leaves: three-foliated with petioles 3.8–7.6 cm, bear leaflets that are ovate, entire, lanceolate/obovate, acute, or acuminate and reticulately veined, the lateral leaflets with petiole 69 mm long. The bark is generally 6–15 cm in length and 3–10 cm in width, with a thickness of 5–15 mm. The bark on the outer side is rough and appears gray to grayish brown, whereas the inner side is suave and whitish brown to pale in appearance. Flowers: whitish cream, pale yellow, or reddish yellow in color, appearing either just before or with the leaves. Fruits are spherical or oval, hard, slim, or scaly berry. Fruits have multiple reniform seeds, which are implanted in the yellow and fleshy pulp of the fruit (Kumar et al., 2020).

Flowering and fruiting: March–June

Chemical constituents: The plant is rich in alkaloids, saponins, triterpenes, tannins, flavonoid glycosides, glucosinolates, and phytosterols. The bark and leaves contain alcohol, lupeol, friedin, diosgenin, and betulinic acid (https://www.planetayurveda.com/library/varuna-crataeva-nurvala/).

Used in: Ayurveda, Folk, and Traditional

Parts used: Leaves, bark, and root

Uses: This herb is used since the time of the Vedas. The bark of this herb is used to treat kidney stones, urinary tract infection, and prostate related disorders, and bengib prostatic hypertrophy. The bark is anti-inflammatory, antiseptic, laxative, and carminative. The herb is used to clear blockage of arteries (https://www.planetayurveda.com/library/varuna-crataeva-nurvala/). The plant is also beneficial as an antipyretic, anthelminthic, demulcent, diuretic, stomachic, and as a contraceptive for females. Root and bark of this herb are laxative and lithotriptic, promote appetite, and increase biliary secretion. It is also used extensively in migraine, dysuria, renal calculi, and intestinal worms (https://www.easyayurveda.com/2012/08/30/varuna-benefits-usage-dose-ayurvedic-details/).

Methods of propagation: Seeds

DOI: 10.1201/9781003263500-89

Crescentia cujeta L.

Synonyms: *Crescentia accuminata* Kunth.; *Crescentia angustifolia* Willd. ex Seem.; *Crescentia arborea* Raf.; *Crescentia fasciculata* Miers.; *Crescentia latifolia* Raf.

Family: Bignoniaceae

Common names: English: Begger's bowl tree, Calabash tree; **Kannada**: Sokeburde; **Marathi:** Fakirancha wadga; **Tamil:** Tiruvottukkay

Distribution: Native to the Americas, that is grown in Africa, Central America, South America, the West Indies. It is naturalized in India and found in almost all the states.

Description: Small tree with a crooked growth form, low and much-branched, able to grow up to able 10 m tall. Trunk: fissured and gray bark. Green oblanceolate leaves, alternate arrangement, measuring about 4–26 cm long and 1–7.5 cm wide, subsessile; yellow tubular flowers with purple veins, solitary or sometimes paired and about 5 cm long, musty fragrance. Fruit is a large, green globular berry about 13–20 cm long and up to 30 cm wide, attached to the trunk and branches, contains flat seeds that are embedded in the pulp (https://indiabiodiversity.org/species/show/229336).

Flowering and fruiting: June–October.

Chemical constituents: The ripe fruit pulp contains crescentic, tartaric, citric, tannic, chlorogenic, and hydrocyanic acids. The seed oil contains oleic acid.

Used in: Traditional

Parts used: Leaves, fruit, and bark

Uses: Calabash tree has a wide range of medicinal uses. The fruit is used in the treatment of colds, diarrhea, pneumonia, and intestinal irregularities. It is also used for relief from menstrual pains and to ease childbirth and procure an abortion. The leaves, by comparison, can be used in the treatment of dysentery, colds, lung diseases, toothache, wounds, and headache. The bark is used to clean wounds. The pulp of the fruit is astringent, emollient, expectorant, and laxative. It is used in domestic medicines. The fruit is abortifacient, emetic, emmenagogue, purgative and vermifuge. A syrup made from the pulp of the fruit is a popular remedy for colds. The juice of the fruit is used to treat diarrhea, pneumonia, and intestinal irregularity. It is made into a strong tea and drunk to procure an abortion, to ease childbirth, and is used in a mix to relieve severe menstrual pains by eliminating blood clots. A syrup made from the fruit is used to treat consumption. The leaves are cholagogue, emetic (in larger doses), and purgative. An infusion is sometimes administered for treating dysentery. It is boiled with sugar, soft grease, or Buckley's white rub to make a syrup that is used to treat colds. The juice of young leaves is drunk to remedy colds and lung diseases. The leaf is chewed to treat toothache. The leaves are used as a wash to cleanse dirty wounds. The whole plant is used as a diuretic against hydropsy and diarrhea (https://pfaf.org/User/Plant.aspx?LatinName=Crescentia+cujete#:~:text).

Methods of propagation: Seeds

DOI: 10.1201/9781003263500-90

Photo: Dr. Srinath Rao

Crinum defixum Ker. Gawl.

Synonyms: *Crinum viviparam* (Lam.) R. Ansari & V. J. Nair.; *Crinum ensifolium* Roxb. ex Gawl.; *Crinum roxburghi* Dalzell & Gibson.; *Amaryllis coenosa* Hook. f.; *Amaryllis vivipara* Lam.

Family: Amaryllidaceae

Common names: Bengali: Sukdarshan; **English:** River crinum lily, Spider lily; **Gujarati:** Nagrikand; **Hindi:** Jangli lahsun; **Kannada:** Kaadu irulli; **Manjina:** naaru; **Sanskrit:** Kandali; **Tamil:** Visha moongil; **Telugu:** Kaesarachaettu, Kaja gadda

Distribution: Generally found in wet habitats along with the riverbanks throughout India up to 900-m elevation

Description: It is an erect perennial and bulbous herb. Stem: herbaceous, underground, bulbos, long, and columnar. Leaves: simple, alternate, ex-stipulate sessile, few to many, 60–90 cm long and 2–4 cm wide, linear, concave or channeled, thick; the margins are slightly rough, apex acute or obtuse, with parallel venation. Inflorescence: racemose umbel of 2–20 or more flowers produced above the foliage over a well-developed and stout scape and subtended by two large spathaceous bracts. Flowers, hermaphrodite, actinomorphic, white to reddish color, bracteates, bract filiform, sessile or shortly stalked, large. The perianth is made up of 6 tepals arranged in 2 whorls of 3 each, salver-shaped cylindrical tube, 6–13 cm long, narrow, linear-lanceolate, reflexed or drooping. Stamens: 6, epiphyllous, inserted at the throat of perianth tube, filaments long purplish red with white base. Ovary: inferior, carpels 3 syncarpous, trilocular, stigma capitate or 3 lobed. Fruits: a capsule, subglobose, 2–3 cm in diameter, short-stalked, beaked by the perianth tube. Seeds are rugose.

Flowering and fruiting: August–November

Chemical constituents: It is reported to contain the active constituents such as caranine, crinamine, crinine, galanthamine, galanthine, haemanthamine, and hippestrine (Kirtikar and Basu, 1975). A new alkaloid 5α-hydroxyhomolycorine has also been reported (Boit et al., 1957).

Used in: Ayurveda and Siddha

Parts used: Leaves, rhizome, and corm

Uses: The bulb is considered emollient, emetic, laxative, nauseant, and diaphoretic. The juice of the leaves is used as an eardrop to treat the disease of otitis. The leaf, rhizome, and corm are used to treat gout, edema, leprosy, and fever. The leaves are used to treat dropsy, earache, carbuncles and paronychia. Seeds are used as purgative and diuretic (Shilpa et al., 2012). Leaf extract is used for treatment for vomiting and for earaches. The bulbs are crushed and applied onto piles and abscesses to cause suppuration. In addition, the roasted bulbs are used as a rubefacient in rheumatism (Jayaweera, 1981; Sumona Afroz et al., 2018).

Methods of propagation: Bulbs

DOI: 10.1201/9781003263500-91

Photo: Dr. Srinath Rao

Crinum latifolium L.

Synonyms: *Crinum asiaticum*; *Crinum bracteatum*; *Crinum toxicarium*

Family: Amaryllidaceae

Common names: Bengali: Bada kanod, Sukh darshan, Gaerhonar patta; **English:** Poison bulb, Wide-leaved crinum; **Hindi:** Chinder, Kanwar, Pindar; **Kannada:** Visha mungali; **Malayalam:** Karimulli, Vishamulla; **Marathi:** Gadambikanda; **Sanskrit:** Sudarshana, Dudarshana, Madhuparnika, Nagdaman, Vrishakarni; **Tamil:** Vishamungali; **Telugu:** Kesarachettu

Distribution: Throughout the tropics, subtropics, and warm temperate regions of the world

Description: A stout herb of 40–60 cm in height, with bulb-form trunk (like big onion) of diameter 10 cm or more. Leaves: around root and have linear form of 60–90 cm in length, 5–11 cm in width, with parallel veins and slightly waved margins. Leaf: blade wide. Bundle of flowers rise from one main pedicle and form a halo of 30–60cm in length; white fragrant 10–20-flowered umbels; flower has form of funnel, with 6 stamen and 1 stigma.

Flowering and fruiting: June–August.

Chemical constituents: Various parts of plant contain following metabolites. Alkaloids latisolin, latisodin, ambellin, 11–0-acetylambellin, 11–0 acetyl 1, 2-ß epoxyambellin, crinafolin, crinafolidin, lycorin, epilycorin, epipancrassidin, 9–0-demethylhomolycorin, lycorin-1, 0-glucosid, pratorin (hippadin), pratorinin, pratorimin, pratosin, beladin, latindin, and latifin; aminoacids like phenylalanine, L-leucin, DL-valin, L-arginin monohydrochloride; Glucans A and B; organic acids; and saponin.

Used in: Ayurveda and Siddha

Parts used: Leaves, bulbs, and rhizome

Uses: It is a medicinal herb that is used from time immemorial. The leaves of plant are non-toxic and have anti-inflammatory, analgesic, antitumor, immune stimulating, antibacterial, antifungal effects, uterine fibroids, detoxification, and tissue-regeneration activities. The bulb is said to be poisonous. The leaf juice is used in earache. For swelling, pain, sprain, inflammation of joints and inflamed toes, and fingers, the heated leaves are tied at affected body part. The leaves juice is applied topically for skin diseases, anti-lice and anti-scabies treatments, and on piles to reduce pain and swelling; it is also used to induce vomiting to help poison to be thrown out of body in case of poisoning. Roasted bulbs are applied for discharge of pus and to cure rheumatism. Poultice of heated, pounded fresh bulb is used to relieve pain in bones and rheumatism. Rhizome extract is used to treat prostate inflammation or enlargement; the extract is also used to control the hot flushes in symptoms of menopause. Plant extract is used in the treatment of allergic conditions.

Methods of propagation: Bulbs

DOI: 10.1201/9781003263500-92

Crotalaria burhia Buch.-Ham

Synonyms: *Crotalaria burhia* var.tomentosa Boiss.
Family: Fabaceae
Common names: English: Burhia rattlepod; **Hindi:** Saniya
Distribution: Found in Afghanistan, Pakistan, and India
Description: It is a bushy, appressed-pubescent under shrubs, 1–2 ft tall. Leaves: simple alternate, elliptic-oblong, or lanceolate; appressed silky pubescent, sessile, about 0.6–2.5 cm long and 3–10 mm broad and deciduous. Inflorescence: 6–12 flowered terminal racemes. Flower: yellow in terminal racemes with two bracteoles, calyx about 8–9 mm long, velvet-hairy teeth lance-shaped, style bearded at the top. Fruits: pods 8–10 mm long and 4–5 mm broad, appressed pubescent to almost villous 3–4 seeded. Seeds: subreniform, dark brown, compressed, shining, smooth, and glabrous.
Flowering and fruiting: January–March, may be throughout the year also
Chemical constituents: There are several active compounds in this plant, including triterpenoids, flavonoids, anthraquinones, phenols, polyphenols, steroids, alkaloids, and tannins (Kataria et al., 2011; Kumar et al., 2011; Bibi et al., 2015).
Used in: Ayurveda, Folk, and Traditional
Parts used: Leaves, branches, and root
Uses: In the Ayurvedic system, this plant is identified as having great medicinal potential. Anticancer and soothing properties are found in the leaves, roots, and branches of *C. burhia*, while fresh plant juice can be used to treat eczema, gout, hydrophobia, pain, and edema. The root extract with sugar is used to alleviate chronic kidney pain and treat typhoid fever. It has a wide range of medical properties, cooling medication can be made from the plant's leaves, branches, and roots. Gout, eczema, hydrophobia, pain and swelling, wounds and cuts, infection, renal pain, stomach disorders, rheumatism, and joint pain can all be treated using plant juice in traditional medicine (Katewa and Galav, 2006; Sandeep et al., 2010; Bibi et al., 2015).
Methods of propagation: Seeds

DOI: 10.1201/9781003263500-93

Cuminum cyminum L.

Synonyms: *Cuminum aegyptiacum* Merat ex DC; *Cuminum odorum* Salisb.; **Cuminum officinale** Garsault.; *Ligusticum cuminum* (L.) Crantz.

Family: Apiaceae (Umbelliferae)

Common names: English: Cumin; **Gujarati:** Jeeru; **Hindi:** Jeera; **Kannada:** Jeeregae; **Marathi:** Jeera; **Sanskrit:** Jeeraka, Svetajiraka; **Tamil:** Seeragum; **Telugu:** Jilkara

Distribution: Cultivated in the Middle East, India, China, and several Mediterranean countries

Description: It is an annual herbaceous plant, with a slender, glabrous, branched stem that is 20–30 cm (8–12 in) tall and has a diameter of 3–5 cm. Each branch has 2–3 sub-branches. All the branches attain the same height, so the plant has a uniform canopy. The stem is colored gray or dark green. The leaves are 5–10 cm long, pinnate or bipinnate, with thread-like leaflets. The flowers are small, white or pink, and borne in umbels. Each umbel has 5–7 umbellets. The fruit is a lateral fusiform or ovoid achene 4–5 mm long, containing two mericarps with a single seed. Cumin seeds have 8 ridges with oil canals. They resemble caraway seeds, being oblong in shape, longitudinally ridged, and yellow brown in color (https://en.wikipedia.org/wiki/Cumin).

Flowering: June–July; **Fruiting:** August–September

Chemical constituents: *C. cyminum* contains alkaloid, coumarin, anthraquinone, flavonoid, glycoside, protein, resin, saponin, tannin, and steroid (Al-Snafi, 2016c).

Used in: Ayurveda, Folk, and Traditional

Parts used: Seeds

Uses: Cumin is widely used in traditional system of medicine as an antioxidant, digestive, antispasmodic, anti-genotoxic, antihyperglycemic, galactagogue, anticonvulsant, diuretic, and stomachic (Mann et al., 2008). It also fights diabetes, epilepsy, and tumors and increases immunity. Another health benefit is that it promotes digestion. Cumin is one of the most gut-friendly herbs and promotes pancreatic enzymes, assisting in the digestion process; treats hemorrhoids; helps in weight loss; fights cancer; treats insomnia; relieves asthma and cold; betters memory; soothes the skin; and treats acne. In India, the seeds are powdered and used in different forms such as *kashaya* (decoction), *arishta* (fermented decoction), and *vati* (tablet/pills; https://www.google.com/search?q=cuminum+cyminum+benefits&oq=Cuminum+cyminum).

Methods of propagation: Seeds

DOI: 10.1201/9781003263500-94

Photo: Dr. Srinath Rao

Curcuma amada ROXB.

Synonym: *Curcuma mangga* Velton&Van zipp

Family: Zingiberaceae

Common names: Bengali: Amada; **English:** Mango ginger, Wild turmeric; **Gujarati:** Aambaa haldhar; **Hindi:** Aamaa-haldi, Amiyaa haldi; **Kannada:** Ambarasini, Huli Arsin; **Malayalam:** Mangayinji; **Marathi:** Ambehalad; **Sanskrit:** Amragandha, Amragandhiharidra, Karpuraharidra, Padmapatra; **Tamil:** Mankayinci; **Telugu:** Maamidi allam

Distribution: A native of Indonesia and China. Also widely distributed in Asia, Africa, and Indo-Malya regions. Cultivated in Karnataka, Kerala, Maharashtra, Gujarat, Uttar Pradesh, and West Bengal

Description: The herb attains 60–90 cm height; leaves are long, petiolate, oblong-lanceolate, tapering at both ends, glabrous, green on both sides; sheathed, petiolate, and in tufts. Each plant bears 5–6 pairs of leaves. The flowers are white or pale yellow in spikes that occur in the center of the leaves, lip is semi-elliptic, yellow, three-lobed, with the middle lobe emarginated. Mango ginger rhizomes are fleshy, buff-colored, and demarcated into nodes and internodes. The rhizome of the plant about 10–12 in. and has pleasant smell of raw mango. At the rhizome nodes, scaly leaves are arranged circularly giving the appearance of growth rings with scars on the surface. The rhizomes are branched, and the branching is sympodial.

Flowering and fruiting: Springtime

Chemical constituents: The ethanol extract of rhizome showed the presence of hydroxyl, carbonyl, ester, and olefin functional groups in it and methyl, methylene, methionine proteins and olefinic proteins (Rao et al., 1989; Mujumdar et al., 2000). Three bioactive molecules, difurocumenonol, amadannulen and amadaldehyde, were isolated and characterized from mango ginger, which are responsible for the medicinal properties of mango zinger (Policegoudra et al., 2007a, 2007b, 2010).

Used in: Ayurveda and Unani

Parts used: Rhizome

Uses: The rhizomes are very similar to common ginger but lack its pungency and instead have a raw mango flavor. The rhizomes are used as appetizer, carminative, digestive, demulcent. stomachic, expectorant, antipyretic, anti-inflammatory, febrifuge, alexeteric, aphrodisiac, laxative, diuretic, and antipyretic and used in the treatment of anorexia, dyspepsia, flatulence, colic, bruises, wounds, chronic ulcers, skin diseases, pruritus, fever, constipation, hiccough, cough, bronchitis, sprains, gout, halitosis, otalgia, and inflammations (Kirtikar and Basu, 1984; Hussain et al., 1992; Warrier et al., 1994). Specific in rheumatism and inflammation of liver, in contusions, and sprains. It is useful against inflammation of the mouth, ear and gleet, ulcers on the male genitalia, scabies, lumbago, and stomatitis (Kirtikar and Basu, 1984; Hussain et al., 1992; Warrier et al., 1994). Its rhizomes are used as flavoring for pickles and other dishes and valued for their medicinal properties. The fresh as well as dried rhizomes are used for flavoring curries. The fresh cut rhizomes have the flavor and the color of mango, hence the name mango ginger (Policegoudra et al., 2010).

Methods of propagation: Rhizome

DOI: 10.1201/9781003263500-95

Cyperus scariosus R. Br.

Family: Cyperaceae

Common names: English: Umbrella's edge, Nut grass; **Hindi:** Nagaramotha; **Kannada:** Tungegadde; **Malayalam:** Korakizhanna; **Marathi:** Lawala; **Sanskrit:** Nagaramustaka, Charukesara; **Tamil:** Koraikkilangu; **Telugu:** Kolatungamuste

Distribution: Widely distributed in India, especially in South India. It is found in damp areas of West Bengal, Uttar Pradesh, and eastern and southern parts of India.

Description: *C. scariosus* is an erect glabrous herb, up to 60 cm high. Leaves are usually shorter than the stem; linear inflorescence an umbel of more or less condensed spikes; bracts foliar, generally exceeding the umbel. Spikelets are pale yellow, or brown, often with reddish tinge. Glumes decurrent below as hyaline wings. Nuts are obvoid ellipsoid. Flowers are borne in contracted umbels; spikelets linear, pale straw-colored. Fruits are trigonal nuts.

Flowering and fruiting: Rainy season

Chemical constituents: Steroid, terpenoid, phenolics and resin, respectively, main active components appear to be the sesquiterpenes. Main sesquiterpenes identified in Cyperus rhizomes. These are cyperone, selinene, cyperene, cyperotundone, patchulenone, sugeonol, kobusone and isokobusone, pinene (monoterpene) derivatives of sesquiterpenes such as cyperol, isocyperol, and cyperone. These active constituents are found in the volatile oil of *Cyperus* rhizomes, which make up only about 0.5%–1% of the dried rhizome (Srivastava et al., 2016).

Part used: Bulb, Rhizomes and Roots

Uses: Its roots are used to make perfume and in aromatherapy. The herb's tubers are credited with astringent, diaphoretic, diuretic, desiccant, cordial, stomachic diuretic, emmenagogue, tonic, and vermifuge properties. It is prescribed to treat various diseases, including diarrhea, epilepsy, gonorrhea, syphilis, and liver damage. It is an anti-inflammatory, central nervous system stimulant, antimicrobial, and hypotensive.

Methods of propagation: Rhizomes

DOI: 10.1201/9781003263500-96

D

Decalepis hamilteionii Wght & Arn.

Synonym: *Decalepis arayalpathra*

Family: Asclepiadaceae

Common names: English: Swallow root; **Kannada:** Maakali beru; **Malayalam:** Mahannikizhangu; **Sanskrit:** Sariba, Svetasariva; **Tamil:** Peru nannari; **Telugu:** Maredu kommulu, Barresugandhi

Distribution: This plant is endemic to peninsular India. It has been recorded in the dry and moist deciduous forests of Karnataka, Andhra Pradesh, and Tamil Nadu. It grows between rocks and places where there is thick vegetation.

Description: It is a climbing shrub, branchlets jointed, entire plant consists of milky latex. Leaves: petiolate, obovate-elliptical, subcoriaceous, longitudinally folded, tip of the leaves blunt, base wedge shaped, size 6 × 4.5-cm leaf stalk about 1.5 cm. Inflorescence: cyme, axillary peduncled, trichotomously branched. Flower tube is 1 mm. Calyx: deeply 5 lobed, oblong about 2 mm in size. aestivation valvate. Corolla: creamy, campanulate lobes, spreading, aestivation valvate; petals 3 × 2 mm, oblong, recurved, white pubescent inside. Stamens: connivent; pollinia horizontal. Corona: double with 10 scales, outer scales truncate, inner flat, adhering to gynostegium. Glands: 5 at the base of corolla, alternating with stamens, bifid. Anthers: attached to the style apex. Ovaries: subglobose, style present, stigma obtuse, about 1 mm. Fruit: a follicle oblong-lanceolate, cylindric; epicarp thick, crinkled; woody when dry. Seeds are many, egg-shaped with long silky hairs (Reddy et al., 2013).

Flowering: January–March **Fruiting:** May–July

Chemical constituents: The root has shown to contain aldehydes, alcohols, ketones, sterols, and triterpenes such as amyrin and lupeols derivatives and 2-hydroxy 4-methoxy benzaldehyde and vanillin. Analysis of this oil showed that, along with the major component, the presence of benzaldehyde (0.017%), salicylaldehyde (0.018%), methyl salicylate (0.044%), benzyl alcohol (0.016%), 2-phenylethyl alcohol (0.081%), ethyl salicylate (0.038%), p-anisaldehyde (0.01%), and vanillin (0.45%) in minor quantities, which are biologically significant (Nagarajan et al., 2001).

Used in: Ayurveda and Folk

Parts used: Roots

Uses: Its tuberous roots are widely used as a health drink and are well known for its medicinal properties. It is used in curing various diseases like stomach disorders, gastric ulcers and to stimulate appetite. It is also used in treatment of epilepsy and central nervous system disorders (Murti and Seshadri, 1942). The tubers have antimicrobial, antipyretic, antiulcer, antidiabetic, antioxidant, anti-inflammatory, chemoprotective, cytoprotective, insecticidal, neuroprotective, and hepatoprotective activities (Reddy et al., 2013). This plant has been used in the preparation of several herbal drugs like Amrutamataka taila, Drakshadi churna, Shatavari rasayana, and Yeshtimadhu taila (George et al., 1999a). A drink is prepared from the roots of this plant along with roots of *Hemidesmus indicus*, which is called Nannarior sughandapala. The drink has medicinal value which cools the system gives good appetite and purifies blood (Vedavathy, 2004).

DOI: 10.1201/9781003263500-98

Methods of propagation: By seeds or stem cuttings. Loamy soil intermixed with stones encourage the production of long, thick roots.

Photo: Dr Srinath Rao

Desmodium laxiflorum Sensu Miq., p.p.

Synonym: *Desmodium incanum; Desmodium macrophyllum* Desvaux; *D. recurvatum* (Roxburgh) Graham ex Wight; Arnott; *Hedysarum recurvatum* Roxburgh; *Hylodesmum laxum* subsp.*laxum*

Family: Leguminosae (Fabaceae)

Common names: English: Loose-flowered desmodium; **Hindi:** Jangli-ganja; **Kannada:** Otte Gida; **Malayalam:** Unda-orila **Marathi:** Jangli-ganja, Asud; **Sanskrit:** Parsniparni; **Tamil:** Atotti; **Telugu:** Cheppa Thatla

Distribution: India to Taiwan, Malaysia, moist deciduous forests, also in the plains of Maharashtra: Karnataka, Kerala, and Tamil Nadu

Description: It is an undershrub, about 60–100 cm tall; the stem is woody, angled, and adpressed and has minute hooked hairs. Leaves are elliptic in shape and are hairy. Leaves: trifoliate, terminal leaflet 13 × 10 cm, ovate, acute at apex, rounded at base, densely hairy below, glabrate above; petiole 1.5–4 cm long; stipules 8 mm long, lanceolate. Flowers are born in racemes 15–25 cm long, pink; calyx 2 mm long, densely villous. Corolla: white to violet standard, broadly obovate 6 × 3 mm, emarginated, vexillum white wings and keel blue, wings auriculate and clawed; keel clawed. Pods: 6–9 jointed; joints 4 × 1.25 mm, oblong, longitudinally striate, with dense, minute hooked hairs.

Flowering: August–September **Fruiting:** September–January

Chemical constituents: The plant contains apigenin, taxifolin, pentacosanayl-D-glucopyranoside (Sing et al., 2015a).

Used in: Ayurveda and Tribal

Parts used: Leaves and fruits

Uses: Mixture of roots and dried spikes along with roots of *Solanum anguivi* and leaves of *Desmodium triflorum* are given for curing cough, bronchitis, and fever. A decoction of leaves is taken orally along with 100 ml of milk twice a day for 2 days to get relief from dysentery. Leaf antidiarrheal, galactogenic and used in skin disease. Fresh leaf juice is applied to wounds and sores. The leaf juice is used to cure wounds. The fruit is consumed in the liver complaints. A leaf paste is used on wounds and abscesses. The plant is pounded, mixed with water, and given for stomach pain and fever (Varma et al., 1999; Singh et al., 2015a). If the leaves and stem are boiled with water and taken, menstrual cycle irregularity disappears; it helps fight uterine infection (Taid, et al., 2014). The roots, 50–100 g, are crushed and boiled in water; 50 ml extract mixed with powder of three fruits of *Piper longum* is taken daily for 10–15 days to cure jaundice (Maheshwari, 1996).

Methods of propagation: Seeds

DOI: 10.1201/9781003263500-99

Desmostachya bipinnata (L.) Stapf

Synonyms: *Briza bipinnata* L.; *Eragrostis bipinnata* (L.) K. Schum.; *Leptochloa bipinnata* (L.) Hochst.

Family: Poaceae (*Gramineae*)

Common names: Bengali: Derbha; **English:** Halfa grass, Tadch grass, Sacred kusha grass; **Gujarati:** Darbha; **Hindi:** Kushta, Kusa, Durva; **Malayalam:** Darbha, Kusha veru; **Marathi:** Darbha, Godhadi; **Sanskrit:** Barhi. Darbha, Kusha; **Tamil:** Teruppai; **Telugu:** Darbha

Distribution: *D. bipinnata* is native to northeast, west tropical, and northern Africa (in Algeria, Chad, Egypt, Eritrea, Ethiopia, Libya, Mauritania, Somalia, Sudan, Tunisia); countries in the Middle East; and temperate and tropical Asia (in Afghanistan, China, India, Iran, Iraq, Israel, Myanmar, Nepal, Pakistan, Saudi Arabia, and Thailand).

Description: Harsh tussocky rhizomatous perennial; culms stout, up to 1.2 m high. Leaf blades up to 65 cm long, 3.8–10.5 mm wide when unrolled; lower leaf sheaths leathery, often densely flabellate toward the base of the culm. Inflorescence up to 60 cm long; spikes clustered or spaced, 14 cm long. Spikelets: 3–17-flowered, narrowly ovate to linear-oblong, 3–10 mm long; lower glume 0.7–1.5 mm long, upper glume 1.1–2.0 mm long; lemmas straw-colored or suffused with purple, 1.8–2.7 mm long.

Flowering and fruiting: Throughout the year

Chemical constituents: The roots contain cylindrin, arundroine, and fenniole iso-arboricole, β-sitosterol-D-glucopyranoside, quercetin, apigenin, and luteolin have been isolated from this plant species (Sathasivampillai and Pholtan Rajamanoharan, 2021).

Used in: Ayurveda and Folk

Parts used: Leaves

Uses: In folk medicine, it has been used to treat many diseases. This plant has some activities as a diuretic, antidiarrheal, analgesic, and antipyretic and has an anti-inflammatory effect. It was also suggested to have hepatoprotective ability, help against dysentery, menorrhagia, and jaundice, and has antioxidant effect, as well as anti-ulcerogenic properties (Ibrahim et al., 2018). In Ayurveda, it is used for the treatment of diarrhea, skin diseases, renal calculi, dysmenorrhea, and for improving breast milk during lactation. A mat is prepared with this grass and used by sages for sitting and sleeping on it. It acts as a natural coolant, keeps the body cool, and has some protective action against radiation.

Religious significance: *D. bipinnata* has long been used in various traditions (Hindus, Jains, Buddhists) as a very sacred plant. According to early Buddhist accounts, it was the material used by Buddha for his meditation seat when he attained enlightenment. It is mentioned in the *Rig Veda* for use in sacred ceremonies and as a seat for priests and the gods. Kusha grass is specifically recommended by Krishna in the *Bhagavad Gita* as part of the ideal seat for meditation.

Methods of propagation: Rhizome and seeds

DOI: 10.1201/9781003263500-100

Dialium Coromandelica (Houtt.) Merr.

Synonyms: *Lannea coromandelica* (Houtt.) Merr.; *Calesium grande* Kuntze.; *Lannea grandis* Engl.; *Lannea wodier* (Roxb.) Adelb.; *Lannea wodier* (Roxb.) Parker.; *Odina pinnata* Rottler.; *Odina wodier* Roxb.; *Rhus odina* Buch.-Ham. ex Wall.

Family: Anacardiaceae

Common names: Bengali: Jiyol; **English:** Indian ash tree, Moya, Wodier; **Gugrati:** Mavedi; **Hindi:** Mohin; **Kannada:** Godda, Gumpina, Kuratige, Nirmajja, Udimara; **Malayalam:** Aladam kaarilav, Otiyan maram; **Marathi:** *S*hinti, Moi; **Sanskrit:** Ajashringi, Jingini; **Tamil:** Kalasan, Odiyan, uthiyan Oti; **Telugu:** Agasrngi, Appriyada, Oddimanu

Distribution: Throughout the greater part of India to an altitude of 1500 m, in sub-Himalayan "sal"-dominated forests

Description: It is deciduous tree with a spreading crown and stout branches. Growing up to 14 m tall. Branchlets are minutely covered with starry hairs. The color of the bark is gray or whitish, smooth, exfoliates in irregular and has rounded plates. The leaves are alternately arranged pinnate and are 25–45 cm in length, crowded at the ends of branches. There are 7–9 leaflets and are oblong-ovate, 7.5–15 cm long and 2.5–5 cm wide. The flowers are small, yellowish or greenish in color, and unisexual. The male flowers are fascicled in slender compound racemes; female flowers are in simple pubescent racemes. The color of the fruit is red; it is compressed, kidney-shaped, and single-seeded.

Flowering: January–March **Fruiting:** April–June

Chemical constituents: The plant was reported to contain various compounds like carbohydrates, including gums, proteins, glycosides, terpenoids, and polyphenols. Preliminary phytochemical analysis of the bark revealed the presence of terpenoids and flavonoids (Manzur-ul-Kadir Mia et al., 2009). The flowers contain isoquercitin (G). The leaves were proven to contain ß-sitosterol, ellagic acid, quercetin (A), quercetin-3-arabinoside, leucocyanidin (C), and leucodelphinidin (D). The bark contained β-sitosterol and physcion (B), and physcion anthranol-B (Sankara Subramanian et al., 1971).

Used in: Ayurveda, Siddha, and Unani,

Parts used: Leaves, stem bark, and resin

Uses: The bark is used to prepare a lotion for relieving bruises, wounds, sores, ulcers, and sore eyes. The fresh juice of the bark is used externally to relieve sore eyes and promote the healing of obstinate ulcers. The gum derived from the plant is known as "Jingini Gum," which is given in the treatment of asthma. This gum is also considered as a cordial to women during lactation. The bark has stimulant properties and that can be used as the curative product for gout. It is used in the form of a decoction for gargling to cure mouth ulcers and toothache. A dilute decoction of the bark of the Indian medicinal plant Jingini is used as an expectorant among the tribal inhabitants of western Maharashtra. The dried, powdered bark, mixed with neem oil, is regarded as an application for chronic ulcers and skin diseases such as impetigo. The leaves of the plant are boiled in oil and applied to relieve sprains and bruises, local swellings, body pains, and elephantiasis. For the treatment of rheumatism, a paste of the leaves is mixed with black pepper, and the paste is applied externally. In Bengal, the juice of the green branches of Jingini is given as an emetic in

DOI: 10.1201/9781003263500-101

cases of coma or insensibility produced by opium or other narcotics. In Ayurveda, the bark and resin of Jingini are used to treat wounds, cardiac disease, and diarrhea. It is also often used for treating the diseases of the nervous system. The latex from leaf and stem bark is used to treat heel cracks. Juice of leaves is taken orally to prevent white discharge in women (Senthamizhselvi, 2013).

Methods of cultivation: Seeds (the viability of seeds is very short) and stem cuttings

Diospyros melanoxylon Roxb

Synonyms: *Diospyros dubia* Wall.; *Diospyros exsculpta* Bedd; *Diospyros tupru* Buch-Ham. *Diospyros wightiana* Wall.

Family: Ebenaceae

Common names: Bengali: Kendu; **English:** Coromandel ebony; **Hindi:** Tendu; **Kannada:** Abanasi, Beedi elae; **Malayalam:** Kari; **Marathi:** Timburni, Tendu; **Sanskrit:** Dirgha-patraka; **Tamil:** Bedi elai Tumbi; **Telugu:** Beedi Aaku

Distribution: Native to India, Pakistan, and Sri Lanka. It is normally found in dry deciduous forest.

Description: Diospyros melanoxylon is a medium-sized tree or shrub up to 25 m and 1.9 m in girth. A key characteristic is its bark, which is pelican in color and peels off in rectangular scales. The primary root is long, thick and fleshy at first and, afterward, woody, grayish, often swollen in upper part near ground level. The roots form vertical loops in sucker-generated plants. Leaves opposite or alternate and coriaceus, up to 35 cm long, tomentose on both sides when young, becoming glabrous above when fully grown. Male flowers are mauve in color, tetramerous to hexamerous, 1–1.5 cm long, sessile or nearly sessile in short peduncles, mostly 3-flowered. Female flowers mauve, mostly extra-axillary or sometimes solitary, axillary generally 2, opposite each other, larger than the male flowers. Fruits olive green, ovoid or globose 3–4 cm across, 1-, 2-, 3-, 4-, 5-, 6-, or 8-seeded berries. Pulp yellow, soft and sweet. Seeds compressed, oblong, shiny, and often banded. The generic name is derived from the Greek *dios* (divine) and *pyros* (fruit), referring to the excellent fruit of the genus. The specific name is Greek and means "dark wood".

Flowering: April–June

Chemical constituents: Stem bark contains tannins (19%), ceryl alcohol, lupeol, betulin, β-sitosterol, sequoyitol, carboxylic acid, diospyric acid, and naphthoquinones (Rastogi, 1970).

Used in: Folk healing practices, which include therapy for hemorrhage, incontinence, insomnia, hiccough, diarrhea etc.

Parts used: Leaves, stem bark, flowers, and seeds

Uses: The leaf of the tree contains valuable flavones. The pentacyclic triterpenes found in the leaves possess antimicrobial properties. Leaves are used as styptic, in the treatment of scabies and old wounds, and as laxative and carminative medicine. while the bark shows anti-hyperglycemic activity. The bark has been determined to have significant anti-plasmodial effects against *Plasmodium* which causes malaria in humans. Stem bark is used in treatment of diarrhea, dyspepsia, astringent (eye), dysentery, ulceration of cornea, and postnatal pain. In India the leaves are wrapped around tobacco to create *beedi*, a cheap alternative to cigarettes. The seeds can be intoxicating; they have been prescribed in India as a cure for mental disorders, nervous breakdowns, and palpitations of the heart. The fruits have a cooling and an astringent effect. Dried flowers are reportedly useful in urinary, skin and blood diseases, leucorrhea, urinary discharge, and anemia and as a diuretic (Kirtikar and Basu, 1987).

Methods of cultivation: Natural regeneration is through seedlings, coppice, and root suckers. Artificial propagation is through either direct seeding or planting nursery-raised seedlings. Soaking seeds for about 12 hours in cold water improves germination. Stump planting has been found to be as good as direct seeding.

DOI: 10.1201/9781003263500-102

E

Elephantopus scaber L.

Synonyms: *Asterocephalus cochinchinensis* Sprengel.; *Elephantopus carolinensis* G. Mey.; *Elephantopus sordidus* Salisb.; *Scabiosa cochinchinensis* Lour.

Family: Asteraceae (Compositae)

Common names: English: Elephant foot; **Hindi:** Samdudri, Bantambakhu; **Kannada:** Hakkarike, Nela mucchilu, Aane kaalu gida; **Malayalam:** Anayatiyan; **Marathi:** Hastipata; **Sanskrit:** Gojivha; **Tamil:** Anashovadi; **Telugu:** Enugabira

Distribution: Native to tropical Africa, eastern Asia, the Indian subcontinent, Southeast Asia, and northern Australia

Description: Scapigerous herbs; whole plant scabrous. Leaves: radical, obovate to oblong or oblanceolate; cauline leaves small, narrow, sessile, base amplexicaule Heads: homogamous, non-rayed, sessile, aggregated. Flowers: bisexual purple Corolla: purple; tube 5–6 mm long, lobes 5. Stamens: 5, auricled at base. Ovary: 3–4 mm, oblong; style arms subulate. Achenes: 4–5 mm long, 10-ribbed. Pappus of 4–6 spiny bristles, approximately 5 mm long.

Flowering and fruiting: October–January

Chemical constituents: *E. scaber* contains elephantopin, 17, 19-dihydrodeoxyelephantonpin, iso-17, 19-dihydro-deoxy elephantopin and 8-hydroxyl naringenin are the most important bioactive compounds.

Used in: Traditional

Parts used: Young leaves and roots

Uses: Different parts of the plant are used in traditional medicine of India as an astringent agent, cardiac tonic, and diuretic and used in treating asthma, coughs, pulmonary diseases, dyspepsia, diarrhea, dysentery, edema, urethral discharges, venereal diseases, skin diseases, eczema, and rheumatism. Root is prescribed to prevent vomiting; powered with pepper, it is applied for toothache. Leaves are used in applications for eczema.

Methods of propagation: Seeds

Photo: C.T. Chandra Lekha

DOI: 10.1201/9781003263500-104

Elytraria acaulis (L. fil.) Lindau

Synonyms: *Elytraria amara* Blanco.; *Elytraria acrenata* Vahl.; *Elytraria indica* Pers.; *Justicia acaulis* L. fill.

Family: Acanthaceae

Common names: English: Asian scalystem; **Hindi:** Patharchatta; **Tamil:** Pumikatampa; **Telugu:** Yeddadugu

Distribution: Widely distributed in South Africa and India. The plant is mostly found in hills.

Description: Stem less perennial herb with one to several unbranched flowering stems up to 30 cm tall. Leaves in a basal rosette, subsessile, elliptic to obovate, up to 18 cm long, hairy, particularly on the veins. Flowers are arranged in spikes held in tight apiculate, overlapping bracts. Bracts and flowering stem are bluish green. The corolla is white, lower lip and lateral lobes spreading, 2-lobed flowers often not opening, capsule 5.5–6.5 mm long and hairless.

Flowering and fruiting: Throughout the year

Chemical constituents: The phytochemical screening of methanol extract of the leaves of the plant showed the presence of chemical compounds like alkaloids, amino acid, carbohydrates, flavonoids, glycosides, phenol, protein, saponins, steroids, and tannins (Kiruthika et al., 2012).

The compound 4H-1-Benzopyran-4-one, 3-((6-O-(6-deoxy—beta—L—mannofuranosyl)—betaOgalactofuranosyl)oxy)-7—((6-deoxy-beta-Lmannopyranosyl)oxy)—5—hydroxyl—2—(4-hydroxyphenyl) was isolated from the leaves of the plant *E. acaulis* using column chromatography and Nuclear Magnetic Resonance (NMR), Infrared (IR), spectroscopy and mass spectrum (Kumudhavalli et al., 2011). A novel compound (Z)-3–(2-methyl-3-oxoprop-1-en-1-yl) phenyl heptanoate was isolated from the plant (Priya et al., 2022).

Parts used: Root

Uses: *E. acaulis* is being used as a traditional medicine for the treatment of leucorrhea in which the root is made into a paste and given once in a day for 21 days. Root paste with black pepper is applied on snakebite and is also applied on piles in the morning and evening. It is traditionally used in treatment of abscess of mammary glands (Sathish Babu et al., 2015). *E. acaulis* has shown an effective antidiarrheal property (Singh et al., 2013). The plant extract and an isolated compound have shown a variety of pharmacological activities like antidiabetic, antidiarrheal, anti-inflammatory, antiseptic, antioxidant, hypoglycemic, and hepatoprotective activities (Sathish Babu et al., 2015). Priya et al. (2022) reported the anticancer effect of a compound extracted from this plant, against PA 1 cell lines.

Methods of cultivation: Seed

DOI: 10.1201/9781003263500-105

Eranthemum roseum (Vahl) R. Br.

Synonyms: *Amphiscopia rosea* Nees; Daedalacanthus roseus (Vahl) T. Anders; *Ecbolium roseum* (Nees) Kuntze; *Justicia rosea* Vahl

Family: Acanthaceae

Common names: English: Blue eranthemum; **Hindi:** Dasmuli, Gulsham; **Marathi:** Jungli Aboli, Dasmuli; **Tamil:** Nila-mulli; **Telugu:** December poou

Distribution: Throughout India

Description: A shrub up to 1.8 m tall. Leaves opposite, oblong-lanceolate, acuminate, 12–20 cm long and 3.8–7.5 cm wide, lineolate on both sides, margins entire or crenulate, base tapering into petiole. Flowers are blue or rose-colored with a strong, fetid odor, borne in linear, lax, axillary, and terminal spikes up to 15 cm long. Bracts 1.6 cm long, obovate, white with very prominent raised green nerves. Fruit is (capsule) 1.3 cm long, clavate, pointed, glabrous. Seeds are 3–4 mm in diameter, hygroscopically hairy, densely so on the margins.

Flowering and fruiting: November–May

Chemical constituents: It mainly consists of steroid, saponin, amino acid, protein, and carbohydrates.

Used in: Folk

Parts used: Root

Uses: Tuberous root boiled in milk is popular remedy for leucorrhea. Few Ayurvedic manufacturing units have formulated the products by using this plant for body massage oil and as mother tonic. Tanumardan massage oil and Mothertone tonic are typical commercial products available in Indian market (Surana et al., 2008). The plant is reported as gastroprotective, anti-inflammatory, and antioxidant (Patil and Surana, 2010).

Methods of cultivation: Seeds

DOI: 10.1201/9781003263500-106

Eryngium foetidum L.

Synonyms: *Eryngium antihystercum* Roxb.; *Eryngium foetidum* f. comosum Urb.; *Eryngium foetidum* f. nudum H. Wolff

Family: Apiaceae

Common names: English: Culantro, Mexican coriander; **Hindi:** Jangali dhania, Bilati dhania; **Malayalam:** Mallichera

Distribution: India, indigenous to America, introduced in Africa and Asia. In India, found in Assam, Karnataka, Kerala, Odisha, and Meghalaya, mostly cultivated

Description: Perennial herbs, 25–40 cm high, aromatic; roots tuberous; stems dichotomously branched, deeply striate. Leaves simple; lamina about 5–10 × 0.6–2.5 cm, oblanceolate-spathulate, cuneate at base, obtuse at apex, margin spinous-toothed; cauline sessile, palmately 3–5-partite; lobes oblong, ca. 0.1–2.5 × 0.2–0.5 cm, spinous toothed. Flowers: white, in oblong-cylindrical umbels; bracts spinulose, stellate-pubescent; calyx teeth rigid, acute; petals white. Fruit ellipsoid.

Flowering and fruiting: August–December

Chemical constituents: Contains a range of metabolites, such as aldehydes, carotenoids, phenolic compounds, and anthraquinones (Bhavana et al., 2013)

Used in: Traditional and Folk

Parts used: Leaves

Uses: This species is also used in ethnomedicine because it has shown potential application in diseases related to the gastrointestinal tract and acts as an antibacterial, analgesic, anti-inflammatory, anthelmintic, anticonvulsant, and anticancer agent. The tea of the plant is recommended to treat female reproductive problems, promote menstruation, relieve cramps, treat infertility, and facilitate labor, and it is considered to have aphrodisiac action. In South American countries such as Peru, Colombia, and Ecuador, the plant is used to treat diseases and ailments related to the digestive and gynecological tract, such as flatulence, diarrhea, indigestion, and stomach pain (Bussmann et al., 2018; Rodrigues et al., 2022).

Methods of propagation: Seeds

DOI: 10.1201/9781003263500-107

F

Ferula foetida L.

Synonyms: *Ferula scorodosma* Bentley & Trimen; *Scorodosma foetidum* Bunge

Family: Umbiliferae (Apiaceae)

Common names: English: Asafoetida, Devil's dung; **Hindi:** Hinga; **Kannada:** Hingu; **Malayalam:** Kayam; **Marathi:** Hing; **Sanskrit:**Hingu; **Tamil:** Perunkayanm; **Telugu:** Inguva

Distribution: Kashmir region of India, Afghanistan, and Iran

Description: *F. foetida* is a woody perennial growing to 2 m (6 ft) by 1.5 m (5 ft). The species is hermaphrodite (has both male and female organs). Roots: huge, fleshy and tapering. Leaves: large, bipinnate, pubescent; segments oblong, entire, obtuse. Inflorescence: densely pubescent. Umbels: compact. Petals: whitish yellow, persistent. Fruit: broadly oblong to suborbicular, 1 cm long, 8 mm broad; vittae inconspicuous, numerous, minute; fruit wing nearly as broad as the seed.

Flowering: May–July

Chemical constituents: The volatile oil is the main active constituent of the Hing, which is responsible for its therapeutic significance. Its main phytochemical is feluric acid. It varies from 40% to 60%. The components of essential oil were separated into four classes, which were monoterpene hydrocarbons, oxygenated monoterpenes, sesquiterpene hydrocarbons, and oxygenated sesquiterpenes (Estekhdami and Nasiri Dehsorkhi, 2019).

Used in: In Ayurveda, it is used in nervous and neurotic diseases.

Parts used: Resinous exudates of the root

Uses: The pungently flavored gum resin that is obtained from the root is an alterative, anthelmintic, antiperiodic, antispasmodic, carminative, deobstruent, deodorant, expectorant, laxative, sedative, and stomachic. It is used in the treatment of simple digestive problems, such as wind, bloating, indigestion, and constipation, and for respiratory problems, such as bronchitis, bronchial asthma, and whooping cough. It is also used as a circulatory stimulant, lowering blood pressure, and thinning the blood. It is used in tincture as a mild cardio-tonic. It is also used in treating convulsion, depression, epilepsy, facial paralysis, hemiplegia, hysteria, and schizophrenia. It also shows anticancer properties.

Methods of propagation: Seeds

DOI: 10.1201/9781003263500-109

Flacourtia jangomas (Lour.) Raeusch.

Synonyms: *Flacourtia cataphracta* Roxb. ex Willd., *Stigmarota jangomas* Lour.

Family: Flacourtiaceae (Salicaceae)

Common names: English: Puneala plum, Indian plum, Indian coffee plum; **Gugrati:** Taalispatra **Hindi:** Paniyaalaa Taalispatri; **Kannada:** Chankali, Goraji; **Malayalam:** Taalisam Lubica, Lovlolika, Vayyamkaitha; **Marathi:** Champeran; **Sanskrit:** Praachinaamalaka, Sruvavrksha, Vikankatah, Paaniyaamalaka; **Tamil:** Vayangarai; **Telugu:** Kuragayi Taalispatramu

Distribution: It is said to have originated from India and distributed throughout tropical regions of East Africa and tropical Asia. It is indigenous to the northeastern Terai region of Utter Pradesh, Bihar Maharashtra, Bengal, Assam, and Orissa states and some parts of South India. It is often found in the Brahmaputra Valley of Assam and adjoining areas of the Northeast region of India.

Description: It is a small deciduous tree growing up to 6–10 m. Trunk and branches of old trees are thornless, whereas woody thorns are present when young. The bark is light brown to copper red. Leaves are alternate, deciduous, pale pink when young, spirally arranged, rarely ovate-lanceolate, long point toothed, very thin, both surfaces glossy, leaf blade elliptic, serrate, 7.0 cm–11.0 cm × 3.5 cm–4.0 cm, papery. Inflorescence is axillary racemes 1.0–2.0 cm long, sub-corymbose, glabrous, few flowered; the male flowers are 1.5–3.0 cm, and the female 1.0–1.5 cm long. Flowers are dioecious, white to greenish in color, comprising 4 or 5 ovate triangular petals bearing fragrance of honey. Pedicels are very slender, 0.5–1.5 cm. Sepals are 2 mm, 4–5 in number, ovate, obtuse, greenish, and pubescent on both sides. Disk is fleshly entire or slightly lobed and is white, yellow or orange in color. Male flowers are filaments, glabrous and solitary or in clusters while female flowers are solitary. Male and female flowers are on separate trees. The androecium consists of many anthers, which are ovate to suborbicular. Ovary is 4–6-celled, with 2 ovules per locule with 4–6 styles connate into a distinct, 1 mm-high column, each bearing a dilate, bilobed, recurved stigma. Fruits are ellipsoid berries, subglobose, 1.5–2.5 cm in diameter, dull brownish red or purple and then blackish, with greenish-yellow pulp, enclosing 4–5 flat seeds. Seeds are tipped with the single, short-style column with 4–6 minute stigma points (Lim, 2013; Baruh, 2016).

Flowering and fruiting: December–April

Chemical constituents: The bioactive compounds including corymbulosine, tremulacin, hydnocarpic acid, and chaulmoogric acid have been reported. The fruit stem bark and bark yielded a coumarin, ostruthin, limonoids, jangomolide, and limonin (Khare, 2007).

Used in: Ayurveda, Folk, Siddha, and Unani

Parts used: Leaves, fruit, and bark

Uses: Leaf decoction is used as an astringent, antidiarrheal, and stomachic; it is used in chronic bronchitis and in treating piles. The roots are said to suppress toothache and are sweetly refrigerant, depurative, alexipharmic, and diuretic. They are useful for asthma and anemia (Sasi et al., 2018). The fruit is used in affections of the liver. The bark and the fruits are used to overcome digestive disorders, allay thirst, biliousness, fevers, nausea, and diarrhea. They are also given for jaundice and enlarged spleen. An infusion of bark is used as

DOI: 10.1201/9781003263500-110

a gargle and treatment of intermittent fever biliousness, bleeding gums, toothache, piles, and weakness of limbs (Rai and Mishra, 2020). Dried leaves are reportedly effective for bronchitis and roots are said to suppress toothache.

Methods of cultivation: The tree is propagated through seeds. However, seeds are slow to germinate.

Flacourtia montana Graham.

Synonyms: *Flacourtia inermis* Miq. ex Hook. Fil. & Thomas

Family: Salicaceae (Flacourtiaceae)

Common Names: English: Mountain sweet thorn; **Kannada:** Hennu sampige, Attak; **Malayalam:** Chaliparzham, Vayyamkaitha, Muripacha, Painellicka; **Marathi:** Raan-tambut

Distribution: It is a tree mainly seen in evergreen forest to semi-evergreen forest, slopes, and wet deciduous forest. Common in Western Ghats and Western subalpine Himalaya

Description: The tree attaining height of 24–30 feet, trunk branched with thorns Bark is brownish smooth. Leaves: simple, alternate in arrangement, distichous, 7–8 cm long, 4–8 cm in width, leaves are narrow elliptic-oblong, toothed margin, tapering tip, coriaceous. Flowers: Borne in cymes in leaf axils, unisexual, small, yellowish, male and female flowers on separate trees, sepals 4 minute ovate, petals 0. Male flowers: stamens numerous. Female flowers: ovary urceolate, sometimes surrounded by a few staminodes at base; styles 5(–6), bifid at apex. Fruit: a berry, round, red when ripe, globose, and fleshy 1-seeded.

Flowering and fruiting: November–April

Parts used: Leaves, bark, and root

Used in: Ayurveda

Chemical constituents: Ripe fruits are delicious to taste and include the following nutritional value: moisture: 77.10%, total sugars: 64%, reducing sugars: 9.88%, nonreducing sugars: 54.11%, calcium: 0.30%, magnesium: 0.60%, potassium: 0.89%, sodium: 57.10 mg/100 g, phenolics: 1.63%, flavonoids: 0.66%, and ascorbic acid: 23.30 mg/100 g (Mundaragi and Thangadurai, 2015). *F. montana* also contains the following nutritional components: vitamins A and C, riboflavin, thiamine, iron, phosphorus, potassium, and sodium.

Uses: Its antibacterial, antidiabetic, anti-inflammatory, and hepatoprotective. An important herb in Ayurveda, where infusions of the bark, leaves, and root are used medicinally to treat conditions such as fever, diarrhea, and inflammations. The leaf is carminative and an astringent and is used as a tonic and an expectorant and for asthma, pain relief, gynecological complaints and as an anthelmintic; it is used as a treatment for hydrocele, pneumonia, and intestinal worms. In India, an infusion of the bark is used as a gargle for hoarseness. In Madagascar, the bark, triturated in oil, is used as an antirheumatic liniment. The bark is used as a tanning material.

DOI: 10.1201/9781003263500-111

G

Galinsoga parviflora (Cav.)

Synonyms: *Adventina parviflora* (Cav.) Raf.; *Galinsoga laciniata* Retz.; *Galinsoga quinqueradiata* Ruiz.; *Tridax parviflora*

Family: Asteraceae (Compositae)

Common names: English: Gallant soldier, Quickweed; **Hindi:** Peepali; **Malayalam:** Mukuthipoovu

Distribution: The plant is abundant in Central America, South America, Europe, West Indies, Mexico, Australia, Africa, and Asia.

Description: Erect, branched herbs, 30–40 cm high; stem smooth. Leaves opposite, 3–5 × 2–3 cm, ovate, acute, glabrous, serrate, 3-ribbed from base. Heads 4 × 3 mm, many, in axillary or terminal peduncled cymes; peduncles dichotomous, slender; bracts biseriate, 3 × 2 mm, ovate, acute, glabrous; receptacle paleate. Outer row of flowers ligulate, limb deeply 3-lobed, white. Inner flowers tubular, bisexual; corolla 1.5 mm long, 5-toothed; anthers obtuse at base. Achenes 2 mm long, obovoid, truncate, black, 3–4-angled, scabrid; pappus of 3–4 fimbriate hyaline (https://indiabiodiversity.org/species/show/229805).

Flowering and fruiting: December–April

Chemical constituents: Phytochemicals such as quercetin, beta-sitosterol, and gallic and hydroxybenzoic acid have been identified from *G. parviflora* (Mostafa et al., 2013)

Used in: Wound healing, blood coagulation issues, colds, and flu

Parts used: Leaves and flowers

Uses: It is commonly employed for the management of both infectious and noninfectious diseases. Its properties include antibacterial, antimalarial, antioxidant, antidiabetic, and anti-inflammatory (Mostafa et al., 2013; Govindarajan et al., 2018; Ripanda et al., 2023). The aerial portions of *Galinsoga* species are used in anti-inflammatory drug preparations in traditional medicine for the treatment of dermatological problems, such as eczema and lichens. This plant has been used as an anti-scurvy agent due. It is also used due to its high level of vitamin C, its ability to stop bleeding, and for treating cold sores, common cold, and flu. The flowers possess analgesic properties and are applied in making preparations for relief of toothache. It has been reported that aqueous extracts of aerial parts of the plant exert protective effects against damages caused by ultraviolet irradiation, and it is also used for wounds and eye diseases (Hamill et al., 2000; Pande et al., 2007; Surywanshi et al., 2015; Ali et al., 2017).

Methods of propagation: Seeds

DOI: 10.1201/9781003263500-113

Photo: C.T. Chandra Lekha

Galium aparine Linn

Synonyms: *Apanine vulgaris* Hill.; *Asterophyllum aparnine* (L.) Schimp. & Spenn.; *Crucianella purpurea* Wulff. ex Steud.; *Galium choroides* Rusby.; *Rubia aparnine* (L.) Baill.

Family: Rubiaceae

Common names: English: Clivers, Catchweed, Goosegrass, Sticky willy

Distribution: Europe, North Africa, Asia Minor, Siberia, Iran, Afghanistan, Pakistan, and India. In India, in the temperate Himalayas

Description: Cleavers are annuals with creeping straggling stems that branch and grow along the ground and over other plants. They attach themselves with the small, hooked hairs that grow out of the stems and leaves. The stems can reach up to 3 ft or longer and are angular or square-shaped. The leaves are simple, narrowly oblanceolate to linear, 5–40 × 4–6 mm and borne in whorls of 6–8 inflorescence axillary, 3-flowered; peduncle stout, 1.5–2 cm long. Flowers are tiny, star-shaped, white to greenish; are clustered in groups of 2 or 3; and are borne out of the leaf axils The corolla bears 4 petals. The fruit is globular 2–5 mm in diameter, covered with dense hooked hairs or setae, mericarps sometimes separated and free.

Flowering and fruiting: March–July

Chemical constituents: The plant contains iridoid monoterpenes (asperuloside), benzyl isoquinoline alkaloids (including protopine), beta-carboline alkaloids (including harmine), quinazoline alkaloids, and flavonoids (such as luteolin). The methanolic extract of the plant contains chlorogenic acid, scopoletin, and rutin. The plant is reported to contain several sterols. The root contains anthraquinones. It also contain iridoid glycosides, such as asperulosidic acid and 10-deacetylasperulosidic acid (Deliorman et al., 2001; Khare, 2007).

Used in: Folk

Parts used: Whole plant and leaves

Uses: Choleretic, stomachic, diuretic, refrigerant, lymphatic, alterative, antiscorbutic, detoxifier; nonsteroidal anti-inflammatory used for enlarged lymph nodes, especially cervical neck nodes, cystic and nodular changes in the glands, and modular goiter. Used as a cleansing drink for malignant conditions and skin disorders, including psoriasis (Khare, 2007). Poultices and washes made from cleavers were traditionally used to treat a variety of skin ailments, light wounds, and burns (Grieve, 1971). As a pulp, it has been used to relieve poisonous bites and stings (Jones, 1991).

Methods of cultivation: Seed

DOI: 10.1201/9781003263500-114

Garcinia indica Choisy.

Synonyms: Garcinia purpurea Roxb.

Family: Clusiaceae (Guttiferae)

Common names: English: Kokam butter tree, Mangosteen oil tree; **Gugrati:** Kokam; **Hindi:** Kokum; **Kannada:** Murginahuli, Muragal; **Malayalam:** Kaattampi, Kokkam; **Marathi:** Amsula, Bheranda, Kokambi, Ratamba; **Sanskrit:** Amlavrkshak, Vrkshaamla; **Tamil:** Murgal

Distribution: Evergreen forests of the Western Ghats from Konkan southward and Goa. Also cultivated in southern districts of cultivated in southern districts of Maharashtra and on lower slopes of Nilgiris (Khare, 2007)

Description: A large tree having elliptic, oblong or oblong. Lanceolate deep-green glossy leaves. Leaves 5.5–8 cm long and 2.5–3 cm broad. The flowers are fleshy, dark pink, solitary or in spreading clusters. The fruit is brown marbled with yellow and is crowned by 4 parted stalks, with few stigmas. There are 6–8 seeds per fruit.

Flowering and fruiting: November–February

Chemical constituents: The fruit rind contains a polyisoprenylated phenolic pigment, garcinol, and its isomer, isogarcinol, along with hydroxycitric acid, cyanidin-3-glucoside, and cyanidin-3-sambubioside. L-leucine and DNP-L-leucine hydrochloride have been reported from the leaves. Kokum butter contains fatty acids—palmitic 2%, stearic 57.5%, oleic 39.0%, linoleic 1.3%, and others 0.2% (Khare, 2007).

Used in: Ayurveda, Folk, Traditional, and Siddha

Parts used: Fruit and bark

Uses: Fruit: antiscorbutic, cholagogue, cooling, anti-bilious, emollient, and demulcent. Syrup from the fruit juice is given in bilious affections. Bark: astringent, oil, or kokum butter, used for dysentery and diarrhea with mucus. Applied externally to ulcerations, fissures of lips, chapped skin, and skin diseases. Fruit: 10–20-ml juice; root bark: 40–80-ml decoction (Khare, 2007). In particular, *G. indica* has been used in traditional medicine to treat inflammation, dermatitis, and diarrhea and promote digestion. It is also reported that *G. indica* has antioxidant, anti-obesity, anti-arthritic, anti-inflammatory, antibacterial, hepatoprotective, cardioprotective, antidepressant, and anxiolytic effects both in vitro and in vivo (Lim et al., 2021).

Methods of cultivation: Grafting

DOI: 10.1201/9781003263500-115

Gardenia gummifera L.

Synonyms: *Genipa arborea* (Roxb.) Baill.; *Genipa gummifera* (Lf.) Baill

Family: Rubiaceae

Common names: English: Gummy gardenia, Cambi gum tree; **Gujrati:** Dikamalli; **Hindi:** Dekamali; **Kannada:** KaduBikke, Dikkemalli; **Malayalam:** Kambimaram; **Marathi:** Dikamalli; **Sanskrit:** Hingupatrikaa, Hingushivaatikaa, Nadihingu, Venupatrikaa; **Tamil:** Dikamalapi, Sirukambi; **Telugu:** Chittamali, Telabampa

Distribution: Distributed in East Asia—central and southern India. Found in dry deciduous forests, dry scrub lands

Description: An evergreen shrub or small trees growing to 1–15 m tall. Leaves: simple, opposite, decussate; stipulate, stipules intra-petiolar, connate; petiole 1–4 mm, stout, glabrous; lamina 4–8 × 1.5–4 cm, ovate, obovate or obovate-oblong, base obtuse, round or subcordate, apex acute or obtuse. Leaf: margin entire, glabrous, shiny, coriaceous; lateral nerves 8–20 pairs, parallel, prominent, intercostal reticulate, prominent. Flowers are bisexual, axillary, solitary or in lax cymes; white in color, turns to yellow, calyx lobes 5, 4.5 mm, truncate, 1.5 mm, triangular, acute, corolla 4.5 cm across, cylindrical, 5-lobed, 2.5 × 1.5 cm, puberulous obtuse. Stamens 5, 1 cm long, included; ovary inferior 6 × 3.5 mm, 1-celled, pentagonous, ovules many; style 2.5 cm; stigma 5 angled, fusiform. Fruit: a berry, ellipsoid or oblong, 4 × 3 cm across. Seeds: rugose.

Flowering and fruiting: June–August

Chemical constituents: The leaves of *G. gummifera*, indicated the presence of alkaloid, cardiac glycosides, coumarins, flavonoids, phenols, phytosteroids, quinines, saponins, steroids, tannins, terpenoids, fats, gum, and mucilage. The resin is acrid, bitter, thermogenic, cardiotonic, carminative, antispasmodic, stimulant, diaphoretic, anthelmintic, antiseptic, and expectorant. It is traditionally used in conditions of cardiac debility, obesity, lipolytic disorders, bronchitis, neuropathy, and splenomegaly and is given to children for nervous disorders and diarrhea due to dentition. Used to cure diabetes, high blood pressure, insomnia, anxiety, and many other conditions.

Used in: Ayurveda

Parts used: Leaves, resin, and gum

Uses: The leaf extract of *G. gummifera* is used against kapha, skin disease, indigestion, and worm infestation. It is useful in treating digestive problems, including dyspepsia and diarrhea, and used as an astringent and expectorant for nervous conditions and spasms. It is said to be carminative, anthelmintic, and diaphoretic and is also claimed to be helpful in stomach upset, flatulence for cleaning foul ulcers and wounds (Jangam et al., 2022).

Methods of propagation: Seeds

DOI: 10.1201/9781003263500-116

Gentiana kurroo Royle.

Synonyms: *Gentianodes kurroo* (Royle) S. Omar, S.I. Ali, & M. Qaiser; *Pneumonanthe kurroo* (Royle) G. Don.; *Tretohiza kurroo* (Royle) J. Sojak.

Family: Gentianaceae

Common names: English: Himalayan gentian, Indian gentian root; **Hindi:** Karu; **Sanskrit:** Traayamaana

Distribution: Native to northern temperate areas of the world. The species is distributed in the Himalayan region across India, Pakistan, and Nepal. It is endemic to the northwestern Himalayas and commonly grows in Kashmir, Himachal Pradesh, and adjoining hills of the northwestern Himalayas. They are listed as critically endangered by the International Union for Conservation of Nature (IUCN).

Description: A rosette forming annual or perennial herbs. The shoot system is represented by flowering branches only with cauline leaves. Flowers: complete, actinomorphic, hermaphrodite, bracteates, pedicellate, hypogynous and pentamerous. Petals: 5 gamopetalous and infundibuliform, deep blue, dotted white at the throat, and white at the base from inside. Stamens: 5 epipetalous. Gynoecium: bicarpellary and syncarpous. Ovary: unilocular containing numerous ovules. Fruit: an achene (Raina et al., 2003).

Flowering and fruiting: August–November

Chemical constituents: The rhizome and roots of *G. Kurroo* contain iridoid glycoside; major component was identified as 6-cinnamoylcatalpol. The most important constituents of the drug are secoiridoid bitter compounds, amarogentin, and gentiopicroside, together with traces of swertiamarin and sweroside. The roots also contain alkaloids, gentianine and gentioflavine, xanthones, bitter oligosaccharides, gentiobiose, and gentianose. Amarogentin, gentiopicrin, swertiamarin, and sioeroside (iridoid monoterpenes) are toxic constituents (Khare, 2007).

Used in: Ayurveda and Unani

Uses: Sialagogue, digestant, stimulant, antispasmodic, anti-inflammatory, emmenagogue. Used for alkalosis, feeble digestion in older adults from gastric acid deficiency (increases gastric juices without altering pH); also used for jaundice, nausea, vomiting, travel sickness, diarrhea, malaria and nervous exhaustion, hepatitis, urinary tract infections, pruritus, and eczema (Khare, 2007).

Methods of cultivation: Rhizome cuttings and seeds

DOI: 10.1201/9781003263500-117

Gmelina asiatica L.

Synonyms: *Gmelina attenuata* H. R. Fletcher; *Gmelina lobata* Gaertn.; *Gmelina paniculata* H. R. Fletcher; *Gmelina parvifolia* Roxb.; *Premna parvifolia* Roth

Family: Lamiaceae

Common names: English: Asian bush beech; **Hindi:** Badhara, Nag phool; **Kannada**: Guludu mara, Kalshivani; **Malayalam:** Cherkumizhi, Kumilamaram, Mulkumizhu; **Marathi:** Kalishivan, Lahanashivan; **Sanskrit:** Gopabhadra, Gopachandra, Vikrani; **Tamil:** Kadambal, Mulkumizh, Nilakummala, Pedanalli; **Telugu:** Chirugumudu, Gumadi, Kavagummudu, Peddanelli

Distribution: Bangladesh, Cambodia, China, India, Indonesia, Malaysia, Myanmar, Sri Lanka, Thailand, and Vietnam. In India found in Andaman and Nicobar Islands, Andhra Pradesh, Bihar, Gujarat, Karnataka, Kerala, Maharashtra, Odisha, Rajasthan, Tamil Nadu, Uttar Pradesh, and West Bengal. Found in deciduous forests and scrub jungles from plains to 500 m

Description: Bushy shrub erect or straggling, armed with spines 12–15 ft tall. Bark light brown, corky with lenticular tubercles, branches and branchlets obtusely quadrangular, rigid, tomentose, spines axillary about 1–3 cm long. Leaves: simple, opposite, rhomboid-obovate or ovate, 0.7–5 × 0.5–3 cm across, margins entire or rarely lobed, apex acute rarely acuminate, petiole robust, slender, canaliculated, pubescent about 1–2.5 cm long, exstipulate. Inflorescence: racemes axillary or terminal, 2–16 flowered, slightly curved or erect, about 8 cm long, bracts linear lanceolate pubescent. Flowers: bisexual, zygomorphic, pubescent, yellow, pedicel about 5 mm long, calyx cupular, 4-toothed, teeth triangular about 1 cm long, corolla infundibular, 4-lobed, lobes pubescent, yellow, upper lobe and lower lobe opposite to each other, ovate, apex acute, lateral lobes oblong, corolla tube narrow, curvate, pubescent. Stamens: 4, didynamous, 2 much exserted, 2 near the throat, rarely underdeveloped 5th stamen present, filaments filiform, slender, shining yellow, anther oblong, yellowish brown, basifixed, Ovary ovoid, style slender, yellow stigma bi-lobed. Fruit a drupe, subglobose, abut 2–3 cm in diameter, green when young later yellow and black when completely matured with a bony endocarp.

Flowering and fruiting: Almost throughout the year.

Chemical constituents: Alkaloids, carbohydrates, glycosides, coumarins, quinones, saponins, steroids, terpenoids, proteins, phytosterols, tannins, and flavonoids. 1, 2-Benzenedicarboxylic acid, diisooctyl ester (31.22%) was the prevailing compound in chloroform extract, and monolinoleoylglycerol trimethylsilyl ether (38.51%) was the major constituent of ethanolic extract.

Used in: Folk

Parts used: Leaves, fruits, and roots

Uses: The plant is widely esteemed as a demulcent and alterative. The root is used as a blood purifier and in the treatment of rheumatism, incontinence, gonorrhea, and syphilis, as well as for treating infections of the bladder and to promote micturition. A decoction of the roots is used externally as a treatment against rheumatism, yaws, and nervous diseases. The macerated leaves are used in urogenital affections. Soaking the leaves in water renders a mucilaginous mixture, which is credited with antibacterial properties. An infusion of the shoots is given internally as a treatment against rheumatism, yaws, and nervous diseases.

DOI: 10.1201/9781003263500-118

The juice of the fruit is applied to portions of the feet affected by "alipunga", a kind of eczema. The juice is also considered an "anti-limatik" (a species of leech of the genus *Haemadipsa*). The fruit, pounded with lime, is applied as a poultice to the throat as a treatment for cough. Combined with lime and garlic, the fruit is vigorously applied to the body in treating cases of dropsy. The fruit juice is a traditional treatment for dandruff—it is applied to the scalp once every 3 days.

Methods of cultivation: Seeds

Gnidia glauca (Fresen.) Gilg.

Synonyms: *Gnidia eriocephala* Meisn.; *Lasiosiphon glaucus* Fresen.; *Lasiosiphon eriocephalus* Decne

Family: Thymelaeacae

Common names: Kannada: Akkigida, Mukkude; **Malayalam:** Nanjinar, Nanju; **Marathi:** Datpadi; **Tamil:** Nacchi-nar, Vandakollipatti

Distribution: India, Sri Lanka, and Africa; in India throughout; in the Western Ghats, along margin of evergreen forests between 500 m and 2000 m

Description: Small trees up to 7 m tall. Bark brownish, irregularly scaly when mature; blaze yellowish. Young branchlets angular to terete, pubescent when young, later glabrous. Leaves: simple, alternate, spiral, clustered at twig ends; petiole 0.3 cm long, slender, planoconvex in cross section, pubescent when young; lamina 4–8.5 × 1–2 cm, narrow elliptic to elliptic-obovate, apex acute with mucronate tip, base obtuse or attenuate, margin entire, pubescent beneath; midrib flat or slightly canaliculate above; secondary nerves up to 10 pairs, not prominent; tertiary nerves broadly reticulate or obscure. Flowers: flower hermaphrodite, spathe cymbiform, green, spadix yellow, tricarpellary ovary, stigma lobulate, fruits are berries yellow, in terminal heads, with involucrate silky bracts; pedicel 0.2 cm long. Fruit: indehiscent, dry, small, 1-seeded.

Flowering and fruiting: January–May

Chemical constituents: The aqueous extract of flower contains phytochemicals like alkaloids, steroids, saponin, coumarin, protein, flavonoids, diterpenes, phytosterol, cardiac glycosides, and carbohydrates; the ethanolic extract shown presence of alkaloids, steroids, saponin, tannin, coumarin, flavonoids, diterpenes, phytosterol, phenols and cardiac glycosides while petroleum ether extract found with alkaloids, steroids, phytosterol, and cardiac glycosides only. The aqueous extract of leaf contains secondary metabolites like alkaloids, steroids, saponin, emodins, protein, and phytosterol; the ethanolic extract shown presence of alkaloids, saponin, steroids, tannin, diterpenes, phenols, and cardiac glycosides while petroleum ether extract observed with alkaloids, saponin, steroids, phytosterol, and cardiac glycosides (Godghate et al., 2015).

Parts used: Leaves and roots

Uses: Leaf powder is used for smallpox, and root is bruised and fried in oil for application on abscesses (Vinayaka et al., 2009) as it is used for treatment of cancers, burns, wounds, abdominal pain, snakebites, and sore throat. Similarly, leaves are applied to treat backache, joint aches, contusions, and swellings. It is considered as a powerful vesicant (Ghosh et al., 2011, 2015). It possesses remarkable antineoplastic activity. It is used as an antiviral agent against rabies (Teklehaymanot and Giday, 2007).

Methods of propagation: Seeds

DOI: 10.1201/9781003263500-119

Gynandropsis gynandra (Linn.) Briq.

Synonyms: *Gy nandropsis pentaphylla* L. DC; *Cleome pentaphylla* Linn.; *Gymnogonia pentaphylla* (L.) R. Br. ex Steud.; *Pedicellaria pentaphylla* (L.) Schrank.

Family: Capparidaceae

Common names: Bengali: Hurhuria, Shulte; **English:** African cabbage, Spider wisp, Cat's whiskers; **Gugrati:** Talvani, Dhelitalavan; **Hindi:** Hulhul, Hurhur, Kavalia; **Kannada:** Naram bele Soppu, Nayeetulasi; **Malayalam:** Atunarivela; **Marathi:** Tilvan, Bhatvan, Mabli, Tilavana, Tilvant; **Sanskrit:** Tilaparni Ajagandhaa, Pashugandhaa, Suuryaavartta; **Tamil**: Nalvelai; **Telugu:** Vaminta, Vayinta

Distribution: Throughout warmer parts of India

Description: Herb, about 2 ft high, flowers white to purplish; weed of waste places and fields. Leaves: mostly 5 foliate pinnately compound; leaf stalk is 20–50 mm long with glandular hairs. Inflorescence: corymbose-racemes, flowers mostly white stamens 6, gynandrophore present. Fruit: a capsule 4–8 cm; seeds dark brown, 1.5 mm in diameter.

Flowering and fruiting: Late spring–late summer

Chemical constituents: The seeds are reported to contain cleomin, hexacosanol, free beta-sitosterol, and kaempferol; also glucosinolates. The medicinal properties of the seeds are attributed to the presence of cleomin. Seeds also contain 1% tannins (Khare, 2007).

Used in: Ayurveda

Parts used: Leaves, seeds, and roots

Uses: Leaves and seeds in the same way as mustard. Bruised leaves rubefacient and vesicant, used as counterirritant in headache, neuralgia, and rheumatic affections. Roots decoction, febrifuge. Seeds, anthelmintic; externally counterirritant. Applied as poultice to sores with maggots. An infusion is given for coughs. The *Ayurvedic Pharmacopoeia of India* recommended the use of the seed in chlorosis, chronic obstructive jaundice and enlarged prostate. A decoction or mixture of bubbled leaves and/or roots is regulated to encourage labor pain in pregnant ladies. It is used to treat stomach-throb and constipation, conjunctivitis, and serious thread worm disease, as well as to relieve burning chest pains. The leaves have anti-inflammatory activities and are utilized to treat joint inflammation (Adhikari and Paul, 2018).

Methods of cultivation: Seeds

DOI: 10.1201/9781003263500-120

Photo: Dr. Srinath Rao

H

Helleborus niger Linn.

Synonyms: *Helleborus altifolius* Steud.; *Helleborus grandiflorus* salisb

Family: Ranunculaceae

Common names: Bengali: Kala kutki **English:** Black hellebore, Christmas rose; **Gugrati:** Koddu; **Hindi:** Khorasani-kutki, Kalikatuki; **Kannada**: Katukarohini; **Malayalam:** Kadagaruganie, Katukarohini; **Marathl:** Balakadu **Sanskrlt:** Kaṭuracīni, Kaṭurohini, Kaṭuki Vakrāgra, Krsnabhedi; **Tamil:** Kadagaruganie, Katurohini; **Telugu:** Katukarogani, Katukarohini

Distribution: Native to sub-alpine woods in Southern and Eastern Europe. Grown in Indian gardens

Description: A perennial herb with a cylindrical, brownish black, knotted, brittle, fleshly, subterranean, bracteated, definite rhizome, with its numerous branches much interlaced, and giving off many stout, fibrous, straight brown roots. Leaves: from the extremities of the rhizome with red petioles, pedate, the lateral divisions deeply divided into 2–4 nearly separate lobes successively smaller toward the petiole, coriaceous, nearly evergreen, smooth, shining, dark green above, paler and reticulated beneath, lobes coarsely serrate above. Flower: stalks terminating the rhizome, surrounded at the base with a loose entire bract, shorter than the leaves, cylindrical, smooth, tapering, mottled with pink below, 1- (rarely 2-) flowered, with 2 or 3 large, ovate, acute, concave bracts a little below the flower. Fruits: follicular, sessile; pericarp leathery, dehiscing along the ventral suture. Seeds: several, in two rows, oval, black, shining; embryo small, at the base of plentiful horny endosperm (Bentley and Trimen, 2006).

Flowering: December–February

Chemical constituents: The rhizome contains cardiac glycosides; helleborin, helleborein, hellebrin, and others based on helleborigenin. Helleborin has a burning, acrid taste and is a narcotic. Helleborein has a sweetish taste and is a highly active cardiac poison. Helleborin and veratrin (steroidal saponins), hellebrin or helleborein, (steroid glycoside) are main constituents of the root and leaves. The plant irritates mucous membranes. A related species, *Helleborus virdis* Linn. (bear's foot, green hellebore) is known as Kaali Kutaki and Krishna_bhedi. The plant contains magnoflorine and corytuberine. The roots and rhizomes gave hellebrin, desgluco-hellebrin, hellebrigenin, bufatetraenolide, beta-ecdysterone, and β-hydroxyecdysterone (Khare, 2007; Prajapati et al., 2007; Kishor Kumar and Lalitha, 2017).

Used in: Ayurveda, Homeopathy, Siddha, and Unani

Parts used: Roots

Uses: Digitalis-like action in cardiac disorders, drastic purgative, abortifacient, diuretic, local anesthetic, narcotic. In homeopathy, it is used for treating various health disorders related to mind, head, mouth, stomach, skin, and joints. Urinary disorders such as scanty urine and frequent urge to urinate can also be cured using this remedy. It provides relief from chronic headache. It reduces excessive salivation and bad breath. It helps in treating skin disorders such as itching and edema and in reducing depression and melancholic weeping. It relieves pain in arms and legs.

Traditionally, the *H. niger* root is used as a hydragogue cathartic and anthelmintic and in epilepsy, dropsy, hypochondriasis, mania melancholia, chronic skin affections, and worms. The powdered roots are given in dyspepsia and as a purgative (Board, 2005; Nadkarni, 2010).

Methods of cultivation: Seed

DOI: 10.1201/9781003263500-122

Hibiscus sabdariffa (L.)

Synonyms: *Abelmoschus cruentus* (Bertol.) Walp.; *Fucaria sabdariffa* (L.) Ulb.; *Hibiscus cruentus* Bertol; *Hibiscus fraternus* L.; *Hibiscus gossypifolius* Mill.

Family: Malvaceae

Common names: Bengali: Chukar, Lalmista, Patwa; **English:** Roselle, Jamaican sorrel, Natal sorrel, Red sorrel; **Hindi:** Lalambori, Patwa; **Kannada:** Pulachikai, Pundi beeja; **Marathi:** Lal ambadi, **Tamil:** Pulichai, **Telugu:** Yerra gongura

Distribution: Native to the West Indies; Found throughout India in waste places near villages and cultivated throughout India

Description: Erect shrub about 5–6 ft high, stem green or reddish, sparsely pubescent. Leaves: polymorphic, entire or palmately 3–5 lobed, 4–11 × 1–1.5 cm, cunate at base, margins serrate acute at apex, mid-nerve with a nectary at base, petiole long about 4–8 cm reddish purple. Flowers: axillary or in racemose panicles, pink or yellow, with a dark-red spot at the base of each petal, epicalyx 8–12 lobed adnate to calyx at base, persistent. Calyx: 5-lobed, distinctly nerved, campanulate, base 2–4 cm long, purple with a nectary on mid-nerve, persistent. Corolla: 5-lobed, 4–5 cm across, with purple center, blue and dark-colored at the base, campanulate or cylindric. Staminal column usually shorter or almost as long petals. Filaments are the slender stalks that support the anthers. In *Hibiscus sabdariffa*, they join to form a tube. Ovary: superior, 5 locular, rarely up to 10 locular, axile placentation, ovules 3 or more per locule, style 1 or 5 branched distally, stigma discoid, capitate. Fruit: capsule, globose-cylindrical, apex apiculate, acute or acuminate, dehiscing loculicidally, glabrous or hairy. Seed many, reniform or subglobose, smooth, glabrous, or hairy.

Flowering: April–July; **Fruiting:** August–November

Chemical constituents: The plants are rich in anthocyanins, as well as protocatechuic acid. The dried calyces contain the flavonoids gossypetin, hibiscetine, and sabdaretine. The major pigment, formerly reported as hibiscin, has been identified as daphniphylline. Small amounts of myrtillin (delphinidin 3-monoglucoside), Chrysanthenin (cyanidin 3-monoglucoside), and delphinidin are also present. The red calyxes are the most used are characterized by their concentration anthocyanin. Delphinidin 3-sambubioside and cyanidin 3-sambubioside are the major anthocyanin (Cissé et al., 2009). Roselle: The seeds are a good source of lipid-soluble antioxidants, particularly gamma-tocopherol. A number of compounds have also been isolated and characterized from roselle, including flavonoids, anthocyanidins, triterpenoids, steroids, and alkaloids (Mishra, 1999).

Parts used: Leaves

Uses: Roselle is used in many folk medicines. It is valued for its mild laxative effect, ability to increase urination, relief during hot weather, and treatment for cracks in the feet, bilious, sores and wounds (Yadong et al., 2005). Leaves are diuretic, choretic, febrifuge, and hypotensive. In African folk medicine, roselle leaves are used for their antimicrobial, emollient, antipyretic, diuretic, anthelmintic, sedative properties, and as a soothing cough remedy, whereas in India, leaves are poultice on abscesses (Neuwinger, 2000). Tablets of roselle extract (1 g), given 3 times a day (for a total of 3 g/day), significantly lowered serum cholesterol. Aqueous extract of roselle was as effective as captopril in treating mild to moderate hypertension (Herrera-Arellano et al., 2004). Preparation of the plant is used to treat carbuncles, conjunctivitis, inflammation of the skin, and scalding. Roselle has been reported to possess a lactogenic activity. It was also reported that roselle is considered as a possible

DOI: 10.1201/9781003263500-123

anti-obesity agent. Extracts from roselle are also known to have effect on inflammatory disease and cancer (Chewonarin et al., 1999). The fleshy calyx is used in the preparation of a refreshing and cooling beverage. Tender leaves and stalks are eaten as salad and as a pot herb and are used for seasoning curries. Seeds have been used as an aphrodisiac coffee substitute. Fruits are edible and are useful in arteriosclerosis and as an intestinal antiseptic.

Methods of cultivation: Seeds

Photo: Dr. Ramakrishna Akula

Hippophae rhamnoides Linn.

Synonyms: *Argussiera rhamnoides* (L.) Bubani; *Elaeagnus rhamnoides* (L.) A. Nelson.; *Hippophaes rhamnoideum* (L.) St-Lag.; *Rhamnoides hippophae* Moench.; Osyris rhamnoides (L.) Scop

Family: Elaeagnaceae

Common names: English: Seabuckthorn, Sand thorn, Himalaya berry; **Hindi:** Chharma, Tarwaa

Distribution: Distributed in more than 20 countries of Europe and Asia. In India, found in Northwest Himalayas at an altitude of 2305–5000 m

Description: A spiny deciduous shrub, its height reaches up to 10–13 ft. Leaves grow on alternate sides of the stem, narrow and lanceolate, having a silvery-green upper side and a darker lower side. It is dioecious; that is, male and female flowers grow on separate plants. The male inflorescence is built up of 4–6 apetalous flowers, while the female inflorescence normally consists of only 1 apetalous flower and contains 1 ovary and 1 ovule. oval or lightly roundish fruit that grows in compact clusters, similar to grape bunches. Their color ranges from pale yellow to dark orange.

Flowering and fruiting: July–September

Chemical constituents: The berries contain polyphenols 3,4-dihydroxy benzoic acid and p-Coumaric acid. They are an important source of vitamins for people living in cold, long-winter regions; contain high concentration of vitamins A, C, and E; carotenoids, flavonoids, and health-beneficial fatty acids, as well as higher amounts of vitamin B12 than other fruits (carotene 30–40 mg) and vitamins B1, B2, B6, and C, 50–600 mg, and E, 160mg (Khare, 2007; Zakynthinos and Varzakas, 2015; Sharma and Kalkal, 2018). Leaves, fruits, and seeds possess an impressive array of antioxidant compounds. Moreover, apart from being a particularly rich source of vitamin C, the berries are packed with vitamin E, carotenoids, flavonoids, sterols, stanols, superoxides, and polar lipids. Leaves are just as rich in physiologically active beneficial compounds. The list includes beta carotene, vitamin E, flavonoids, and ellagic, ferulic, and folic acid and important concentrations of minerals such as calcium, magnesium, and potassium.

Used in: Folk

Parts used: Leaves and fruit

Uses: Fruit: astringent, antidiarrheal, stomachic, antitussive, antihemorrhagic. Sea buckthorn preparations are used internally for stomach ulcer, duodenal ulcer, and other illnesses of the alimentary organs; externally in cases of burns, bedsores, and other skin complications induced by the treatment with X-rays and other radiations (Khare, 2007). Fruits are used against infections, cold, and flu, if administered internally as tea, juice, syrup. Remedies derived from the fruit, seeds, leaves and bark of *H. rhamnoides* have shown promising results in treating mucous membranes injuries such as ulcers, gastrointestinal disorders, and vaginal problems.

Additionally, recent studies have revealed that the essential oils and juices extracted from this plant have beneficial effects on the cardiac system while also considerably lowering the levels of harmful cholesterol. These oils are very effective in the treatment of a wide range of skin conditions (dermatitis, eczema, rosacea, acne, psoriasis) and skin injuries (burns, bedsores, sun damage). Remedies made from its fruit, leaves, and oils to support the immune system, to improve eyesight and enhance muscle nourishment; compounds that can be found in the bark and the leaves of this shrub have antitumor and

DOI: 10.1201/9781003263500-124

anticancer action. Homeopathic products made from this plant help decrease stress, have a hepatoprotective and cytoprotective action, and destroy microbes and harmful bacteria. The sea buckthorn has unique anti-aging qualities, and this is why it is extensively used in thousands of skin care products (Guliyev et al., 2004; Bal et al., 2011).

Methods of cultivation: Seeds; vegetative propagation. Preferred propagation is from stem cuttings or suckers.

Hiptage benghalensis (L.) Kurz.

Synonyms: *Banisteria benghalensis* L.; *Hiptage madablota* Gaertn.;

Family: Malpighiaceae

Common names: Bengali: Madhabilata; **English:** Clustered hiptage, Helicopter flower; **Gugrati:** Madhavi; **Hindi:** Madhavi lata; **Kannada**: Adirganchi, Madhvi, Vasant duti; **Malayalam:** Chittilakody, Njarmbodal, Seethambu; **Sanskrit:** Atimuktaka, Maadhavi, Madhu madhavi Vaasanti, Pundrika, Mandaka, Vimukta; **Tamil:** Madhavi, Karipakkukodi, Vasandagala-malligai; **Telugu:** Madhavi tige

Distribution: *H. benghalensis* (L) Kurz is native to subtropical and tropical Asia parts of southern China, Taiwan, India, Nepal, Indonesia, Malaysia, Myanmar, the Philippines, Sri Lanka, and Thailand. It is found throughout India, predominantly in Western Ghats, Konkan, Deccan, Kumaon, Assam and Andaman island, chiefly in damp places up to an altitude of 1500 m (Meena et al., 2014).

Description: Large woody climbers. Leaves: opposite 7–12 × 2.5–5 cm, elliptic-oblong to lanceolate, base attenuate or obtuse, apex acuminate, with a pair of glands on petioles at the base of lamina. Flowers: 5-merous, bisexual, in axillary or terminal racemes; peduncles articulated with pedicels. Sepals: 5, 6–7 mm long, oblong, one larger with a large gland at the base outside. Petals: 5, white, 1.5–2 × 1.2–1.5 cm, clawed, reflexed, margin crisped. Stamens: 10, unequal, 0.8–1.5 cm long; filaments united at base. Ovary: 3-celled, 3-lobed, 4–6 mm long; style approximately 1.8 cm long; stigma capitate. Fruit: a 3-winged samara, 1 wing larger, about 4 × 2 cm. Seeds: 0.8–1 cm across, globose.

Flowering and fruiting: January–May

Chemical constituents: The stem and its bark contain friedelin, epifriedelinol, octacosanol, α-amyrin, beta-sitosterol, and its beta-D-glucoside. The root bark gave a nitrogenous glucoside and hiptagin, identical with endecaphyllin and a glucosyl xanthone, mangiferin (Khare, 2004, 2007).

Used in: Ayurveda, Siddha, and Traditional

Parts used: Fruit and seeds roots

Uses: The kernel of seeds is prescribed for reducing abdominal girth (obesity). Leaves: used in chronic rheumatism, asthma, and skin diseases. Bark: used in bronchial asthma. Fruit, seed, and root: powder 3–5 g; paste 5–10 g. The bark leaves and flowers are aromatic, bitter, acrid, refrigerant, vulnerary, expectorant, cardiotonic, anti-inflammatory, and insecticidal. They are useful in burning sensations, wounds, ulcers, cough, and asthma. Whole-plant powder is used as an analgesic in babies suffering from high fever. A water-boiled plant decoction is administrated orally in case of weak mothers after childbirth. The leaves are used as a temporary cure in asthmatic problems (Dey et al., 2011). In the *Bhavaprakasha*, the seed kernel of Madhavilata taken with honey checks the growth of abdomen due to obesity. The bark of *H. benghalensis* is used in case of rheumatoid arthritis and dyspnea. *H. benghalensis* flowers are used as a good coolant taken with drinking water. Madhavi root, with buttermilk, was also given for reducing the waist girth. Sushruta prescribed heart wood in internally in obesity, jaundice, urethral discharges and externally in chronic skin diseases, Seed oil also given for internally for flatulence and biliousness and for increasing slimy secretions of the body (Khare, 2004, Meena et al., 2014).

Methods of cultivation: Seed

DOI: 10.1201/9781003263500-125

Holoptelea intigrefolia Roxb.

Synonyms: *Ulmus integrifolia*
Family: Ulmaceae
Common names: Bengali: Nata karanja, Kalmi; **English:** Indian elm; **Gujarati:** Kanjo; **Hindi:** Chilbi; **Kannada:** Nilavahi; **Malayalam:** Aval, Nettaval; **Marathi:** Ainasadada; **Sanskrit:** Chirivilva; **Tamil:** Avilthol; **Telugu:** Nali
Distribution: The native distribution of the plant can be seen in Asia's tropical region, including India, Nepal, Sri Lanka, Indochina, Cambodia, Laos, Myanmar, Vietnam, Burma, and China.
Description: It is a medium-sized to large glabrous deciduous tree, 25 m high. Bark: 6–8 mm thick, whitish or yellowish gray, exfoliating in irregular flakes and with an offensive smell when freshly cut. Branchlets: pubescent. Leaves: simple, elliptic, ovate, acuminate, base rounded or subcordate in shape. Apex: acuminate, margin entire, distantly serrate when young, coriaceous, glabrous above, appressed pubescent punctate below; lateral nerves 6–9 pairs, pinnate, prominent, intercostal reticulate, prominent. Flowers: greenish yellow, polygamous, and found in short racemes or fascicles, polygamous, appear before leaves, 5–8 mm across, greenish purple, tepals 4 or 5, free, anthers pubescent; female flowers with longer pedicels; ovary superior, compressed, long stipitate, 2-winged, 1-celled, ovule 1; style bifid. Fruit: a samara 3 cm across, suborbicular with membranous wing, 1 seeded.
Flowering: January to February; **Fruiting:** April to May
Chemical constituents: Terpenoids, sterols, saponins, tannins, proteins, carbohydrates, flavonoids, and alkaloids, 1, 4-naphthalenedione have been isolated from leaves, whereas Holoptelin-A and B have been isolated from stem bark. Friedelin, epifriedelinol, β-D-glucose, and 2-aminnaphthaquinone are reported in stem bark. Hexacosanol, octacosanol, β-amyrin, and β-sitosterol are some other elements isolated from leaves.
Parts used: Bark and leaves
Uses: Various parts of the tree have been found to be useful in the treatment of bronchitis and obesity Traditionally, the bark and leaves are used as bitter, astringent, and anthelmintic and helps in treatment of inflammation, acid gastritis, dyspepsia, colic, intestinal worms, vomiting, edema, piles, wound healing, leprosy, diabetes, hemorrhoids, dysmenorrhea, diarrhea, rheumatism, polyuria, helminthiasis, tuberculosis, fistula, flatulence, vitiligo, filariasis, and skin diseases. Bark and leaves are applied externally on the white patches in leukoderma.
Methods of cultivation: Seeds

DOI: 10.1201/9781003263500-126

Humulus lupulus Linn.

Synonyms: *Humulus cordifolius* Miq.; *Humulus vulgaris* Gilib.; *Lupulus communis* Gaerten.; *Lupulus humulus* Mill.

Family: Cannabaceae

Common names: English: Hops

Distribution: Native to Europe and Asia. Conditions for its successful cultivation are reported to exist in Kashmir and parts of Himachal Pradesh.

Description: *Humulus lupulus* is a perennial herbaceous plant up to 10 m (33 ft) tall. It has simple leaves with 3–5 deep lobes that can be opposite or alternate. The plant is dioecious, with male and female flowers on separate plants. The fragrant flowers are wind-pollinated. The staminate (male) flowers do not have petals, while the pistillate (female) flowers have petals enveloping the fruit. The female flower cones (or strobili) are known as hops. The fruit is an achene, meaning that it is dry and does not split open at maturity. The achene is surrounded by tepals, and lupulin-secreting glands are concentrated on the fruit.

Flowering and fruiting: July–August

Chemical constituents: Hops contain bitter principles: lupulin, containing humulone, lupulone, and valerianic acid; volatile oil (0.3%–1%), including humulene; flavonoids, including xanthohumol; and polyphenolic tannins, asparagine, and estrogenic substances. Bitter principles stimulate the digestive system. Valerianic acid is sedative. The resin components, lupulone and humulone, are antiseptic against Gram-positive bacteria. Asparagine is diuretic. Hop mash or extract is used in the preparation of toothpaste for inhibiting Gram-positive bacteria and in hair preparations for preventing dandruff formation. It is also used in skin-lightening creams (Khare, 1974).

Used in: Unani

Parts used: Flowers

Uses: Flowers: sedative, hypnotic, nervine tonic, diuretic, spasmolytic on smooth muscle, analgesic, astringent. Used for nervous diseases, intestinal cramps, menopause, insomnia, neuralgia, and nervous diarrhea. It is also used as a tonic in stomach and liver affections. As a blood cleanser, the root is used like sarsaparilla. Application: in mood disturbances, such as restlessness and anxiety, sleep disturbances.as sedative, soporific, spasmolytic, and aromatic bitter and indicates its use for excitability, restlessness, disorders of sleep, and lack of appetite. Hop cones consist of the whole, dried female inflorescences of *H. lupulus* (Khare, 1974).

Methods of cultivation: Seeds

DOI: 10.1201/9781003263500-127

Hydnocarpus kurzii (King) Warb.

Synonyms: *Hydnocarpus heterophylla* auct. non-Bl.; *Taractogenos Kurzii* King.
Family: Achariaceae
Common names: Bengali: Chaulamugra, Dalmungri, Garo; **English:** Chalmogra; **Hindi:** Mograa; **Sanskrit:** Tuvaraka; **Tamil:** Niradi-muttu
Distribution: Bangladesh, Myanmar, Thailand, Laos, and Vietnam. In India, found in Assam and Tripura
Description: Trees: 10–20 m tall, with narrow crown and hanging branches; bark: plain gray, brown or almost black outside, often with white patches. Leaves: bifarious, variable, lanceolate, oblong-lanceolate, elliptic-oblong, obovate-oblong or even elliptic, usually cuneate, and somewhat inequilateral at base, 9–25 × 3–7 cm, membranous and minutely appressed pubescent when young, soon glabrescent; petioles swollen and slightly geniculate at upper end, 1–3 cm long. Male flowers: 8–12 mm across, pale yellow, 3–5 each in two compact fascicles atop a common peduncle 7–15 mm long; pedicels 7–10 mm long, all strigose, fulvous, and tomentose. Sepals: ovate-orbicular, concave, about 5 × 4 mm, fulvous, tomentose. Petals broadly ovate to ovate-orbicular, cuneate at base, ciliate along margins, approximately 4 mm long; epipetalous scales obovate, densely white pilose at apex. Stamens 20–30; filaments approximately 4 mm long, densely pilose. Female flowers 2–5 in a compact fascicle atop a ca 5 mm long peduncle, rarely solitary by abortion; pedicels ca 12 mm long, strigose, fulvous, tomentose. Sepals and petals are often fewer than in male flowers. Staminodes: 10–16. Ovary: ovoid, obscurely 6–7-ribbed densely fulvous tomentose, with 4 parietal placentae; stigmas 4. Fruit: a berry. Berries: globose somewhat protruded at apex, 8–10 cm across; pericarp reddish brown, velvety tomentose, afterward scurfy. Seeds 12–18, tightly packed in fruit, angular-ovoid, 2.2–3.2 × 1.6–2 cm.
Flowering and fruiting: April (November), probably throughout the year in flashes; fruiting afterward
Chemical constituents: The plant is a source of chaulmoogra oil (*Oleum Chaulmoogra*), which contains hydnocarpic, chaulmoogric, gorlic, oleic, palmitic acids, and lower homologues of hydnocarpic acid (Khare, 2007). The lignans hydnowightin, hydnocarpin, and neohydnocarpin have been isolated in the plant. The oil contains chaulmoogric and hydnocarpic fatty acids.
Used in: Ayurveda, Siddha, and Unani
Parts used: Seeds
Uses: Action: antileprotic, dermatic, febrifuge, and sedative. Used parenterally for leprosy; also for psoriasis, eczema and dermatitis. The oil mixed with neem oil or oil of *Psoralea corylifolia* is used in leprosy (Khare, 2007), applied externally, and by direct infiltration of the lesions, it has been shown to be effective in decreasing the size of nodules, numb patches, and skin lesions. The oil has also been recommended as a topical application to treat rheumatism, sprains and bruises, sciatica, and chest complaints and for dressing wounds. The oil is known for its blood-purifying properties. A decoction of the bark is drunk as a general tonic to improve health and in the treatment of internal disorders and skin diseases. It has cytotoxic potential against various forms of cancer. Fruit: eaten as a remedy for leprous sores, boils, and vomiting (Jain and DeFilipps, 1991).
Methods of cultivation: Seeds and root

DOI: 10.1201/9781003263500-128

Hygrophila auriculata (Schumach.) Heine

Synonyms: *Asteracantha auriculata* Nees.; *Asteracantha indaviana* De Wild & T. Durand; *Asteracantha longifolia* (L.) Nees; *Hygrophila longifolia* (L.) Kurz.; *Hygrophila schulli* M.r. Almeida & S. M. Almeida; *Hygrophila spinosa* T. Anderson; *Ruellia longifolia* (L.) Roxb

Family: Acanthaceae

Common names: Bengali: Shulamrdana; **English:** Marsh barbel; **Gujarati:** Ekharo, Kanta sheriyo; **Hindi:** Bhankari, Kshura, Tal makhana; **Kannada:** Gokantaka, Kolavalike; **Malayalam:** Vaya schulli; **Marathi:** Ekhara; **Sanskrit:** Kokilaksha; **Tamil:** Neermulli; **Telugu:** Enugu palleru, Kokilaksi

Distribution: Around the world found in Sri Lanka, Myanmar, Indonesia, Malaysia, and throughout the plains of India, in moist habitats such as marshy margins of canals

Description: It is a stout herb 3–6 ft high. Branches are erect, unbranched, nodes swollen and hairy. Leaves: dorsiventral, smooth and even with a prominent midrib, stalk less, densely hairy 10–15 cm long, occurs in whorls of 6 at the node. Straight yellow about 4 cm long; spines are present in the axil of each leaf. Flowers occur in 4 pairs at each node, are about 3 cm long, purple blue in color, and 2-lipped, upper lip 2-lobed and lower lip 3-lobed. Calyx: 4 lobed, lobes unequal, lanceolate acute; outer one longest, 2–2.5 cm long. Stamens: 4 anthers about 2–2.5mm long. Ovary: oblong, ovules 8, style filiform. Fruit: a capsule, 4–8-seeded, seeds orbicular about 3 mm on long curved retinacula.

Flowering: August–March; **Fruiting:** April–October

Chemical constituents: The important constituents of this plant are flavonoids, terpenoids, lupeol, betulin, and fatty acids. Phytochemically, the whole plant contains phytosterols, tannins, carbohydrates, flavonoids, terpenoids, and sterols. Seed oil contains uronic, palmitic, stearic, oleic, and linoleic acids. Apigenin-7-O-glucuronide and apigenin-7-oglucoside were isolated from the flowers. Lupeol, betulin, and stigmasterol were isolated from the plant (Godbole et al., 1941; Balraj and Nagarajan, 1982; Rastogi and Mehrotra, 1993; Usha et al., 2007).

Used in: Ayurveda, Unani, and Traditional

Parts used: Leaves, seeds, and root

Uses: It is an aphrodisiac, tastant, and restorative mucilaginous. Useful in inflammation calculus, hyperactive, thirst, diarrhea, poisoning, pain, anemia, abdominal disorders, flatulence, urine retention, burning sensation, rheumatoid arthritis, Prameha, vision disorders, and blood disorders. Seeds are used as a tonic and for antidiarrheal and conception promoter. The leaf of this plant is used to treat for pain, poison, pretentiousness, abdominal disorders, anemia, constipation, and urinary disorders, and the roots have medicinal properties, such as coolant, diuretic, aphrodisiac, tonic, and anti-inflammatory and are useful in dropsy, hyper thirst, calculus, strangury, flatulence, and Vata-associated disorders (Rao, 2009). Its leaf is useful for cough and anal fistula; its seed is useful in blood disorders. The intake of a root decoction is useful in jaundice, and topical application of its leaf paste is useful in Prameha. Its root and a whole-part decoction are useful for rheumatoid arthritis (Gaur, 2007). Topical application of a leaf paste is useful in lumbago and arthralgia. The intake of ash, along with cow urine or water, is useful for inflammation (Tripathi, 2009). The intake of the decoction and other medicinal herbs alleviates insomnia (Sharma and

DOI: 10.1201/9781003263500-129

Cakradatt, 2007). The intake of root decoction is useful in anasarca. The intake of powder prepared from talmakhana fruit powder and sugar, along with milk, acts as an aphrodisiac (Tripathi, 2005; Sarvananda and Premarathna, 2018).

Methods of propagation: Seeds

Photo: Dr. Srinath Rao

Hyptis suaveolens (L.) Poit.

Synonyms: *Ballota suaveolens* L.; *Bystropogon graveolens* Blume; *Bystropogon suaveolens* (L.) L'Hér.; *Gnoteris cordata* Raf.; *Gnoteris villosa* Raf.; *Hyptis congesta* Leonard; *Hyptis ebracteata* R. Br.; *Hyptis plumieri* Poit.

Family: Lamiaceae (Labiatae)

Common names: Bengali: Bilatti Tulas; **English:** Pignut; **Hindi:** Vilaiti Tulsi; **Malayalam:** Nattapoochedi; **Marathi:** Bhustrena, Jungli Tulas; **Sanskrit:** Bhustrna; **Telugu:** Gabbu rotta, Mahabeera, Sirna Tulasi

Distribution: Throughout India on wastelands

Description: Shrubs, to 1.5 m high; stem obtusely 4-angular, thinly hairy. Leaves: ovate, acute, hispid below, glabrate above; petiole to 5 cm long. Flowers: in clusters of 1–12; calyx tube 8 mm long, tubular, 10-ribbed, glandular hairy, teeth spinulose, 4 mm long; corolla 5 mm long, lobes short, glabrous inside, blue. Nutlets: 4 × 2.5 mm, compressed, with a ridge on dorsal surface, pubescent, deep brown, mucilaginous when wet.

Flowering and fruiting: Almost throughout the year

Chemical constituents: β-sitosterol, β-sitosterol glycoside, rosamarinic acid, methyl rosmarinate, α-pinene, oleanolic acid, oleanic acid, ursolic acid, α-phellandrene (Gayathri et al., 2021)

Used in: Traditional

Parts used: Entire plant and leaves

Uses: The entire plant is used for headache, convulsions, migraine, catarrh, and skin complaints; essential oil is used as a remedy for toothache and is applied on forehead for headache. A root extract is given for hematuria. The bark for diarrhea and dysentery. Stems, with leaves, are used in a decoction for coughs, asthma, and respiratory infections; skin diseases; and colds; the smoke from the dry sticks, along with fresh leaves, are used in honey collection so the honeybees do not bite when the honey is collected. A decoction of the leaves is used for stomachache or indigestion; leaves are crushed and applied for sprain and swellings; in convulsions, a child bathed with a mixture of mother's urine and crushed leaves; leaf extract or juice is applied on measles; infusions stimulant, depurative, carminative, febrifuge, for colds, flu, fever, yellow fever, malaria, and constipation; squeezed leaves for boils, eczema, cuts and wounds, as antiseptic healer; the leaves' juice act as an insect repellent against mosquitoes; leaves spread on a cot ward off bedbugs. Seeds soaked in water are applied on wounds, allowing pus to flow out; the blackish seeds soaked in a glass of milk and taken for spermatorrhea. Dried aromatic leaves are used in smoking cattle sheds to repel flies and mosquitoes. It has medico-religious beliefs. In veterinary medicine and insecticide, plants are kept on eggs being sat on for incubation, and plants kept in cages to protect poultry from insect and fungal infestation; leaves for sucking insect from livestock body (Quattrocchi, 2012). The leaves of the plant and rhizome of *Zingiber officinale* are crushed together into paste. The paste (half a teaspoon) is taken twice daily to cure dysentery (*Herbal Cures: Traditional Approach*).

It is a carminative, antispasmodic, anti-soporific, antirheumatic, anti-cephalalgia, and lactagogue. It is used in catarrhal and uterine affections, parasitical cutaneous diseases, epistaxis (Gayathri et al., 2021).

Uses: In China, it is used to combat fever and to invigorate health. In Malaysia, a decoction of 15 g of dried herb is used to combat fever, assuage headache, and promote digestion; the plant is used externally to soothe skin rashes and eczema and resolve swellings. In the

DOI: 10.1201/9781003263500-130

Philippines, the plant is used to stop spasms, treat rheumatism, and promote menses, as well as to invigorate health. A decoction of the plant is used to assuage stomachache and treat dysentery. In Taiwan, the aerial parts are used to assuage pain and skin discomfort, whereas in Vietnam, the plant is used to increase milk secretion.

Methods of propagation: Seeds

Photo: Dr. Srinath Rao

Hyssopus officinalis Linn.

Synonyms: *Hyssopus alopecuroides* Fisch ex Benth.; *Hyssopus altissimus* Mill.; *Hyssopus angustifolius* M. Bieb.; *Hyssopus aristatus* Godr.; *Hyssopus beugesiacus* Jord & Fourr.

Family: Lamiaceae (Labiatae)

Common names: English: Hyssop; **Hindi:** Diyaanku; **Sanskrit:** Dayaa-kunji

Distribution: Native to Europe and temperate Asia. Occurs in West Himalayas from Kashmir to Kumaon

Description: Hyssop is a brightly colored shrub or subshrub that ranges from 30 to 60 cm (12–24 in.) in height. The stem is woody at the base, from which grows a number of upright branches. Its leaves are lanceolate, dark green, and 2–2.5 cm long (Pérez Maté, 2002). Flowers: pink, blue, or, more rarely, white. Fruits: oblong shaped tetra-achenes.

Flowering and fruiting: Summer

Chemical constituents: The main constituents of the oil were myrtenyl acetate (74.08%), camphor (6.76%), germacrene (3.39%), spathulenol (2.14%), caryophyllene oxide (2.13%) and β-caryophyllene (2.10%), with lesser amounts of cis-sabinol (1.75%), β-bourbonene (1.47%), and bornyl acetate (1.42%; Fathiazad et al., 2011). Garg et al. (1999) found that the main volatile constituents of *H. officinalis* essential oil from a variety of locations included β-pinene, limonene, β-phellandrene, 1,8-cineole, pinocamphone, isopinocamphone, pinocarvone, germacrene-D, and methyleugenol.

Used in: Ayurveda, Folk, and Unani

Parts used: Oil

Uses: Stimulant, carminative, sedative, antispasmodic, diuretic, pectoral. Used for bronchitis, coughs, and colds. Induces heavy sweating in fevers, increases blood pressure. Emmenagogue. Used externally for bruises, discolored contusions, and cuts.

Used as expectorant. Hyssop has been used in folk medicine for centuries for stimulation of the circulation and for treatment of a variety of ailments, such as upper respiratory illness, asthma, cough, sore throat, intestinal infections, gastrointestinal upset, gallbladder disease, poor appetite, urinary tract infections, and dysmenorrhea. Hyssop is also used topically in gargles, medicinal baths, and creams for skin irritation, burns, and frostbite.

Methods of cultivation: Seed

DOI: 10.1201/9781003263500-131

Ibris amara L.

Synonyms: *Biauricula amara* (L.) Bubani.; *Ibris affinis* Jord.; *Ibris crenata* Lam.; *Ibris decipiens* Jord.

Family: Cruciferae; Brassicaceae

Common names: English: Rocket candytuft, Clown's mustard

Distribution: Native to Europe; cultivated in gardens. Reported to occur in Chamba.

Description: *I. amara* is a branched erect annual typically 10–30 cm tall, sometimes reaching 40 cm. Its stems are more or less hairy below and glabrous above, with leaves scattered along their lengths. Leaf blades are simple, with those lower on the stem spathulate and reaching about 8 cm long and 1.2 cm wide, and those higher up are lanceolate or oblong-cuneate and as small as 3 cm long and 0.4 cm wide. All are more or less sparsely dentate toward their apices.

The inflorescence is a lax racemose corymb, bearing 10–30 flowers, and elongating to 10 cm when in fruit. Individual flowers are about 1 cm across, borne on a filiform pedicel, which is either spreading or ascending, and about 1 cm long. There are 4 petals, which in wild-type individuals are white or pale violet, with cultivated varieties available in deeper shades of pink, violet, or fuchsia. The petals are prominently unequal, with the outer two nearly twice as long as the sepals, with larger petals, which are 5–8 mm long, obovate-oblong, and cuneate beneath. The fruit is a siliculae typically 5 mm, reaching 7 mm in diameter, with deeply notched apices, and wings that widen somewhat toward the top. The styles are as long as or slightly exceed the apical notch, with valves that have reticulate veins. The reddish-brown seeds are about 3 mm long, broadly ovate, and weakly winged beneath.

Flowering and fruiting: March–May

Chemical constituents: The seeds contain a mustard oil and a glycoside, glucoiberin. The plant contains sulfur-containing glucosinolates; also contains bitter and toxic tetracyclic terpenoids, and cucurbitacin E and I.

Parts used: Seeds

Uses: Used for gout, rheumatism, also for bronchitis and asthma; as a tonic in enlargement of heart to allay excited action of the heart. The seed extract exhibited cytotoxicity against renal and brain tumors and melanoma cell lines. The activity may be attributed to the presence of cucurbitacin E and I.

Methods of cultivation: Seeds

DOI: 10.1201/9781003263500-133

Ichnocarpus frutescens R. Br.

Synonyms: *Apocynum frutescens* Linnaeus; *Echites frutescens* Wallich ex Roxburgh; *Gardenia volubilis* Loureiro; *Ichnocarpus ovatifolius* A. de Candolle; *I. volubilis* (Loureiro) Merrill; *Micrechites sinensis* Markgraf.

Family: Apocynaceae

Common names: Bengali: Dudhi, Syamalota; **English:** Black creeper; **Hindi:** Kalidudhi, Siamalata; **Kannada:** Karehambu; **Malayalam:** Paalvally; **Marathi:** Krishnasarwa, Kantebhouri; **Sanskrit:** Gopavalli, Krishna Saarivaa, Krishna-muuli, Syamlata, Sariva; **Tamil:** Karunannari, Makalikilanzhu, Manipilaan kodi, Udar kodi; **Telugu:** Illukatte, Nalateage

Distribution: Uttar Pradesh, Madhya Pradesh, Bihar, Assam, and the Sundarbans

Description: It is a woody shrub with lianas sprawling to 33 ft in maximum length and 6 cm in diameter. The bark produces a creamy white sap. The leaves are simple opposite, up to 11 cm long and 4.5 cm wide, oblong or elliptic-ovate, apex, obtuse-apiculate, base acute-attenuate, margin entire. Inflorescence is a head of several flowers. Flowers: 3.5 mm long, in terminal or axillary panicled cymes; pedicels 3 mm long; calyx 1 mm long, lobes ovate, acute, pubescent; corolla white, salver-shaped, contracted at mouth, tube 2 mm long, hairy inside, throat densely white villous, lobes twisted, oblong, acuminate; stamens included, anthers deeply sagittate at base, apiculate at apex; carpels free, stigma columnar. Mericarps follicular, to 15 cm long, slender, rusty puberulous; seeds many, compressed, crowned with long silky coma.

Flowering: May–August; **Fruiting:** August–December

Chemical constituents: Leaves contain flavones, namely, apigenin and luteolin; glycoflavones, that is, vitexin and isovitexin; proanthocyanidin and phenolic acids, vanillic, syringic, and sinapic acids; and protocatechuic acid (Daniel and Sabnis, 1978). Ursolic acid acetate, kaempferol, kaemferol-3-galactoside (trifolin), and mannitol were also identified from leaves (Khan et al., 1995). The roots contain 2-hydroxy-4-methoxybenzaldehyde. And β-sitosterol (Khastir and Sengupta, 1960). Flowers contain quercetin and quercetin-3-O-β-D-glucopyranoside (Singh and Singh, 1987). The whole plant contain n-butyl sorboside, kaempferol, and its glucoside (Khare, 2007).

Used in: Ayurveda, Folk, and Siddha

Parts used: Roots

Uses: Root: demulcent, diuretic, alterative, diaphoretic; used in fevers, dyspepsia and cutaneous affections. The roots of the plant are used as a substitute for Indian sarsaparilla and are often mixed with the roots of *Hemidesmus indicus* (Khare, 2007). Formulations made with this plant are given orally to treat dysentery, glossitis, hematuria, measles, bleeding gums, convulsions, cough, and delirium. It is generally called as a blood purifier. The roots are reported to possess demulcent, tonic, diaphoretic, and diuretic properties used in fever, dyspepsia, and skin troubles. The whole plant is used as tribal medicine in atrophy, bleeding gums, convulsions, cough, delirium, dysentery, glossitis, hematuria, measles, and night blindness and to relieve pain due to insect bites, splenomegaly, and tuberculosis. Plant is also used in abdominal and glandular tumors (Chaudhary et al., 2012).

Methods of cultivation: Seeds

DOI: 10.1201/9781003263500-134

Ipomoea aquatica Forssk

Synonyms: *Convolvulus adansonii* Desr.; *Convolvulus repens* Vahl.; *Convolvulus rostratus* Zipp.ex Span.; *Ipomoea incurva* G. Don.; *Ipomoea natans* Dinter & Susseng.

Family: Convolvulaceae

Common names: Bengali: Kulmi saag; **English:** Water morning glory, Water spinach; **Gujarati:** Nalini bhaji; **Hindi:** Kalmi saag; **Kannada:** Neeru billi gadde, Neeru hambu; **Malayalam:** Ballel, Kozhuppa; **Marathi:** Nalichi bhaji; **Sanskrit:** Kalambi, Karemu, Naalikah; **Tamil:** Sarkaraivalli; **Telugu:** Tutikura

Distribution: Distributed in all tropical regions of the world: Central and South America, tropical Africa, southwest islands of the Indian Ocean, India, Bangladesh, China, Pakistan, Southeast Asia, Indonesia, and Pacific islands

Description: Creeping or floating aquatic herbs, stems hollow, rooting at the nodes. Leaves: alternate, varies in form, usually oblong-lanceolate or narrowly triangular, 5–10 × 2–6 cm, base hastate, apex acute; petiole 6–10 cm long. Flowers: purplish white, solitary or few in cymes. Sepals: subequal, 6–8 mm long, oblong-lanceolate, membranous, glabrous. Corolla: funnel-form, about 5 cm long, pale purple to nearly white, tube to 2 cm long, lobes obscure. Stamens: included, filaments unequal, hairy at the base. Ovary: glabrous. Capsule: globose; seeds 2 or 4, minutely pubescent (https://indiabiodiversity.org/species/show/230).

Flowering and fruiting: November–March, in some cases throughout the year.

Chemical constituents: It contains vitamins, flavonoids, amino acids, alkaloids, lipids, steroids, saponin, phenols, reducing sugars, tannins, beta-carotene, glycosides, and minerals.

Used in: Ayurveda, Traditional, and Unani

Parts used: Leaves and young shoots

Uses: It is used against piles, and nosebleeds, as an anthelmintic, and to treat high blood pressure. In Ayurveda, leaf extracts are used against jaundice and nervous debility. In indigenous medicine in Sri Lanka, water spinach is supposed to have insulin-like properties (Nagendra Prasad et al., 2008). It is traditionally used to treat diabetes, constipation, opium poisoning (juice is used to induce vomiting), heavy metal toxicity, piles, and ringworm (a bud paste is applied). In the Unani system, it is used as carminative as a remedy to treat inflammation, fever, jaundice, bronchitis, biliousness, and liver malfunction. Plant juice is considered as purgative in nature (https://www.easyayurveda.com/2020/01/23/water-spinach/).

Methods of propagation: Seeds

DOI: 10.1201/9781003263500-135

Photo: C.T. Chandra Lekha

Ipomoea digitata Linn.

Synonyms: *Ipomea paniculata* R. Br. Burm.; *Ipomoea mauritiana* Jacq.

Family: Convolvulaceae

Common names: **Bengali:** Bhuh Kumdaa, Bhooi Kumhdaa, Bhumi kumar; **English:** Milky yam; **Gugrati:** Vidaaree Kand; **Hindi:** Vidaaree Kanda, Bhuh Kumdaa, Bhui Kumbhadaa; **Kannada**: Kandapalaasha, Bhuumi kuushmaanda, Nelkumbal, Naadakumbala; **Malayalam:** Paalmutakku; **Marathi:** Bhui Kohalaa; **Sanskrit:** Bhuumi kuushmaanda. Kshira-vidaari, Kshir valli, Payasvini, Swaadukandaa, Vidari Ikshugandha; **Tamil:** Nilappuchani, Paalmudamgi; **Telugu:** Paalagummudu, Nelagummudu

Distribution: Tropical India in moist regions. *I. digitata* is found growing in hotter and moist regions of India, namely, Bihar, Orissa, West Bengal, Assam, Deccan, Konkan, and Kerala.

Description: An extensive perennial climber with large, ovoid, and tuberous roots. The leaves are large, palmately 5-7-lobed ovate, lanceolate, acute or acuminate, glabrous, and with prominent nerves beneath. Flowers are widely campanulate and few to many in the axillary corymbose cymes. Corolla is purple and campanulate-infundibuliform. Ovary 4-celled. Capsules are small and ovoid. Four black and woolly seeds in each fruit.

Flowering: August–September

Chemical constituents: Rhizomes have taraxerol acetate and beta-sitosterol. Fresh leaves contain 6.3 mg/100 g of carotene (Khare, 2007). The root contains pterocarpan tuberosin; pterocarpanone-hydroxytuberosone; two pterocarpenes, anhydrotuberosin and 3-O-methylanhydrotuberosin; and a coumestan tuberostan. An isoflavone-puerarone and a coumestan-puerarostan.

Used in: Ayurvedic, Folk, and Siddha

Parts used: Rhizome and tuberous roots

Uses: Cholagogue, galactagogue, alterative, demulcent, and purgative. Resin from root: uses similar to jalap. Flour of raw rhizome is given for enlargement of liver and spleen, as well as for menorrhagia, debility, and fat accumulation (Khare, 2007). It is also used to treat gynecology ailments, such as abnormally long and heavy menstrual periods. The plant is used as an aphrodisiac and a tonic and for increasing breast milk. For the medicinal purpose, the tuberous roots of the plant are used. The dried roots are boiled in milk, added to ghee and sugar, and given to improve weight and moderate menstrual discharge. The root powder is mild purgative and useful in constipation. The root powder is given in emaciation, debility, poor digestion. In spermatorrhea (involuntary discharge of semen without orgasm), the juice of fresh root is given with cumin/jeera and sugar.

Methods of cultivation: Tuberous roots and rhizomes

DOI: 10.1201/9781003263500-136

Ipomoea muricata (Linn.) Jacq.

Synonyms: *Bonanox muricata* (L.) Raf.; *Ipomea turbinata* Lag.; *Convolvulus muricatus* Linn.

Family: Convolvulaceae

Common names: English: Clove bean, Traveler's midnight lilies; **Malayalam:** Nithya vazhutana; **Sanskrit:** Krishnabija; **Tamil:** Kattu Talai

Distribution: The Himalayas, West Bengal, Bihar, Orissa, Maharashtra, and South India

Description: *I. muricata* is a fast annual climber, with funnel-shaped and white, reddish, or lilac flowers. The stems are twining and muricate. Th leaf texture is glabrous, and its shape is ovate or orbicular flowers lavender to purplish, tube narrow and cylindric, 3–6 cm long; limb approximately 5 cm across. Sepals: subequal, ovate to elliptic lanceolate; the 2 outer ones 6–7 mm long, attenuated into a fleshy caudate awn 4–6 mm long; the inner 7–8 mm long, obtuse or slightly emarginate, the awn somewhat shorter; bracts about 4 mm long, leafy, oblong or obovate-oblong, acute, caducous.

Flowering: April–May

Chemical constituents: The seeds contain resin glycosides, which are laxative. Lysergol is also present in the seeds. It exhibits hypotensive, psychotropic, analgesic, and uterus and intestine-stimulating properties. The presence of indole alkaloids is reported in the seeds (Khare, 2007).

Used in: Ayurveda, Folk, Siddha, and Unani

Parts used: Seeds, plant, and juice

Uses: Purgative, febrifuge. Seeds: cardiac depressant, spasmolytic, hypotensive, antibacterial, and antifungal. The plant's juice destroys bedbugs (Khare, 2007). The seeds, stems, and leaves are said to be effective in treating several types of skin ailments, such as chronic and gangrenous wounds, cuts, and blisters due to burns (Ysrael, 2003). The seeds are used as a vulnerary and are considered a very efficacious antidotal remedy for poisoning, and the plant sap is used as an insecticide.

Methods of cultivation: Seed

DOI: 10.1201/9781003263500-137

Photo: C.T. Chandralekha

Ipomoea pes-caprae (Linn.) Sweet.

Synonyms: *Ipomea biloba* Forsk.

Family: Convolvulaceae

Common names: English: Goat's foot creeper; **Sanskrit:** Chhagalaantri, Maryaada-valli; **Tamil:** Adambu, Attukkal, Musattalai

Distribution: Near the sea, especially on the west coast

Description: A prostrate herb. The stem of this plant is succulent, running along the ground and rooting at the nodes, green in color, herbaceous, prostrate, cylindrical, up to 30 cm in height, bearing numerous branches and 0.3–0.4 cm in thickness. The root of *I. pes-caprae* is large and thick, about 3 m long, and 5 cm in diameter. The leaves are simple, alternately arranged, dark green, and glabrous. The leaf shape can be variable but is typically ovate, orbicular, or oblong. The leaf base is truncate to shallowly cordate, and the apex is usually notched to deeply cleft but is sometimes rounded or truncated, goat footprint–like leaf shape. The petioles vary in length, ranging from 2 to 15 cm. On young leaves, the petioles are commonly reddish, becoming yellowish-green as they age. There is a pair of nectar-producing glands on the underside of each leaf blade at its juncture with the petiole. These nectaries are red on new leaves, turning black with age, attracting ants, which defend the plant against herbivorous insects. Flowers: pinkish lavender, funnel-shaped (6.4–7.6 cm wide) that bloom throughout the summer and fall. The fruits are ovoid to flattened-globose, and capsules are dehiscent, usually measuring 1.3–1.9 cm long and wide. The seed is rounded to trigonous (three-sided); covered with dense, velvety hairs; and is 1.5–2.5 cm long (Brown and Frank, 2020).

Flowering and fruiting: Throughout summer

Chemical constituents: Antispasmodic isoprenoids, beta-Damascenone, and E-phytol have been isolated from the extract. The antispasmodic activity was found to be in the same range as that of papaverine. The leaves and seeds contain indole alkaloid. Plant also contains a steroid; an amide, pentatriacontane, triacontane volatile oil; and behenic, melissic, butyric, and myristic acid (Khare, 2007).

Used in: Ayurveda and Siddha

Uses: Astringent, stomachic, laxative, antidiarrheal, antiemetic, analgesic. Leaf: diuretic, anti-inflammatory. Used in colic, prolapsus ani; externally in rheumatism. Essential oil of leaves: antagonistic to histamine. Leaf extract is used for different types of inflammation, including injuries caused by poisonous jellyfish. Clinical trials have proved that an extract ipomoea pes-caprae extract (IPA) inhibited the action of jellyfish toxins. Its topical application inhibited carrageenan-induced paw and ear edema induced by arachidonic acid or ethyl phenylpropionate in rats. The crude extract of leaves also shows inhibitory effect on prostaglandin synthesis in vitro. Crude extract (IPA) of the leaves has also been shown to antagonize smooth muscle contraction induced by several agonists via nonspecific mechanism (Khare, 2007).

Methods of cultivation: Seeds

DOI: 10.1201/9781003263500-138

Ixora coccinea Linn.

Synonyms: *Ixora grandiflora* Bot.; *Ixora bandhuca* Roxbg.
Family: Rubiaceae
Common names: Bengali: Rangan; **English:** Jungleflame ixora; **Hindi:** Rangan; **Kannada:** Kisukare; **Sanskrit:** Bandhuka, Paaranti; **Tamil:** Vetchi, Thechii
Distribution: Southwestern Peninsular India; cultivated throughout India
Description: A small shrub, leathery leaves, ranging from 3 to 6 in. in length, and produces large clusters of tiny flowers
Flowering and fruiting: Summer
Chemical constituents: The plant contains important phytochemicals such as lupeol, ursolic acid, oleanolic acid, sitosterol, rutin, leucocyanidin, anthocyanins, proanthocyanidins, and glycosides of kaempferol and quercetin (Baliga and Kurian, 2012).
Used in: Ayurveda, Folk, and Siddha
Parts used: Leaves, flowers, and roots
Uses: Herb: astringent, antiseptic, blood-purifier, sedative, anti-leukorrheic, antidiarrheal, anti-catarrhal. Used in dysmenorrhea, hemoptysis, bronchitis. Root: astringent, antiseptic (used against scabies and other skin diseases). Flowers: prescribed in dysentery and dysmenorrhea. The saponifiable fraction of the petroleum ether extract of roots exhibited anti-inflammatory activity in carrageenan-induced paw edema in albino rats. The plant substrate removes heavy metals, such as lead, cadmium, and mercury, from polluted water (Khare, 2007). The plant possesses antioxidative, antibacterial, gastroprotective, hepatoprotective, antidiarrheal, antinociceptive, antimutagenic, antineoplastic, and chemo-preventive effects (Baliga and Kurian, 2012).
Methods of cultivation: Stem cuttings

DOI: 10.1201/9781003263500-139

J

Jasminum grandiflorum (L.)

Synonyms: *Jasminum aureum* D. Don.; *Jasminum officinale* Linn. var. grandiflorum (L.) Kobuski.; *Jasminum floribundum* R. Br. ex Fresen.

Family: Oleaceae

Common names: English: Royal jasmine, Catalonian jasmine; **Hindi:** Chameli; **Kannada:** Jati Maliga, Sanna Jati Mallige; **Malayalam:** Pichi; **Sanskrit:** Baliharsa, Chetaki, Hridyagandha, Malati, Rajaputrika; **Tamil:** Anankam, Pichi, Jatimalli; **Telugu:** Adavimalle, Jati, Sannajati

Distribution: Native of Asia, Afghanistan, and Persia. Widely cultivated in warm temperate, subtropical, and tropical climates all over the world for its scented flowers. Cultivated throughout India

Description: Large scrambling sub-erect twining evergreen shrub; leaves opposite 3–7, ovate to elliptic in shape, 3 paired foliates ending with a single leaf at the tip; leaflets elongate-lanceolate, acute, 7–11 terminal leaflet. Flowers: terminal and axillary, cymes, white in color, fragrant, petals 5 forming a basal tube 15–25 mm long. Calyx: lobes long and linear, sepals 5. The fruit is a black berry, elliptic to globose in shape.

Flowering and fruiting: July–October

Chemical constituents: Presence of alkaloids, glycoside, flavonoid, terpenes, tannin, resin, and salicylic acid

Used in: Ayurveda and Folk

Parts used: Whole plant

Uses: For medicinal purpose various parts of plant such as leaves and flowers are used. The plant parts exhibit astringent/constricts tissues, acrid, thermogenic/heat-generating, aphrodisiac, antiseptic, pain-relieving, purifying, increase flow of blood in pelvic region, emollient, urine stimulating, antiparasitic, deobstruent, teeth-cleaning, pus-discharging, and tonic properties.

The leaves are used in treatment of swollen spongy gums, ulcers, and loose teeth; toothache; skin diseases; pain in ear; pus in ear; ulcer; painful periods; psoriasis; and wound. Flowers are used in eye diseases, ulcers, skin diseases, itching, diseases of the teeth, and the like. Jasmine oil has antidepressant, anti-inflammatory, antiseptic, aphrodisiac, and sedative properties and used in treatment of depression, nervous exhaustion, stress-related conditions, and in childbirth. It is used during labor to strengthen the contraction. In the case of paralysis, a root paste is applied. To increase milk production in lactating mothers, a poultice of flowers is applied to breasts (Chopra et al., 1986).

Methods of propagation: By cuttings, layering. Layering in the field is done with 1-year-old shoots; a slanting cut is made approximately halfway through the shoot some 50 cm from the end; the cut is buried about 10–15 cm deep with the top remaining aboveground. After about 4–6 months, the rooted layers can be separated from the parent plant and transplanted.

DOI: 10.1201/9781003263500-141

Photo: Dr. Srinath Rao

Jatropha glandulifera Roxb.

Family: Euphorbiaceae

Common names: English: Glandular jatropa; **Hindi:** Bagharenda, Jangali-erandi; **Kannada:** Kari thuraka haralu; **Malayalam:** Atala, Ktalavanakku; **Marathi:** Jangali erandi; **Sanskrit:** Akhukarnika, Dravanthi, Vyaaghrairanda; **Tamil:** Adalai, Eliyamanakku, Vellakattukottai; **Telugu:** Adavi amadam, Dundaga aaku

Distribution: South India and Bengal

Description: It is an evergreen shrub or treelet 3–5 m tall, with sticky transparent juice; trunk is stout branches with glandular leaves, leaves simple, alternate, 5–15 cm long and 2.5–5.5 cm wide, base rounded or flat, leaves with dense glandular hairs, entire 3-lobed lobes triangular, pointed, sometimes toothed, leaf stalks 2–9 cm long, glandular hairy. Flowers born in cymes, greenish yellow. Fruit a capsule, 3-lobed, glandular; seeds 3.

Flowering and fruiting: Throughout the year

Chemical constituents: The plant contains alkannins (iso-hexenylnaphthazarin). The presence of alkannins in this plant (a member of the Euphorbiaceae family) should be considered as an exception. The root gave jatropholone A, fraxetin, and a coumarinolignan (Khare, 2007).

Used in: Ayurveda, Folk, and Siddha

Parts used: Seeds, seed oil, and roots

Uses: Root and oil from seed: purgative. Oil: antirheumatic, anti-paralytic. Used externally on ringworm and chronic ulcers. Root: used for glandular swellings Latex: applied to warts and tumors. Dosage: seed—50–100 mg powder (Khare, 2007). Also used in treating skin diseases, rheumatism, piles, snakebites, and wounds (https://ayurwiki.org/Ayurwiki/Jatropha_glandulifera_-_Dravant). It is reported for various therapeutic actions like-analgesic properties, inflammation, in asthma, and for bronchitis. It extensively reported as emmenagogue, as analgesic, in inflammation, asthma, bronchitis and used as antidote to scorpion stinge. The roots were reported to cure piles. The viscid sap (latex) and leaf is used for warts, tumors, and wound healing (Bhagat and Kulkarni, 2003).

Methods of cultivation: Seeds

DOI: 10.1201/9781003263500-142

Juglans regia Linn.

Synonyms: *Juglans duclouxiana* Dode

Family: Juglandaceae

Common names: English: Walnut tree, Carpathian walnut, Madeira walnut; **Hindi:** Akhrot; **Kannada:** Akrotu; **Sanskrit:** Akshoda, Akshoda-ka, Akshota, Shailbhava, Pilu, Karparaal, Vrantphala; **Tamil:** Akrottu

Distribution: Native to Iran; now cultivated in Kashmir, Himachal Pradesh., Khasi Hills and the hills of Uttar Pradesh.

Description: *J. regia* is a large deciduous tree, attaining heights of 80–120 ft and a trunk up to 6 ft in diameter, commonly with a short trunk and a broad crown. The bark is smooth, olive brown when young and silvery gray on older branches and features scattered broad fissures with a rougher texture. The leaves are alternately arranged, 25–40 cm long, odd-pinnate with 5–9 leaflets, paired alternately with 1 terminal leaflet. The largest leaflets are the three at the apex, 10–18 cm long and 6–8 cm broad; the basal pair of leaflets are much smaller, 5–8 cm long, with the margins of the leaflet entire. Male flowers: in drooping catkins 5–10 cm long, tepals 4, ovate, approximately 2 mm long; stamens 10–20, subsessile; anthers 2 mm long. Female flowers: 1–3, terminal on short spikes; involucral tube of fused bract and bracteoles, about 3.5 mm long, tomentose, glandular, obscurely 4-toothed and irregular at the margin; tepals 4, linear, 2.5–4 mm long, alternating with the teeth; margin sparsely pubescent; ovary: about 3.5 mm long, ovoid, inferior; style about 2 mm long; stigmas 2, recurved, plumose to fimbriate, exserted ripening in the autumn into a fruit with a green, semi-fleshy husk and a brown, corrugated nut. Fruit is a drupaceous nut, not a true nut; up to 5 cm long, ovoid to subglobose; epicarp green, glandular; endocarp 2-valved; seed: 2- to 4-lobed at the base. Following full ripening of fruit, the removal of the husk reveals the wrinkly walnut.

Flowering and fruiting: February–April

Chemical constituents: From the volatile oil of the leaves terpenoid substances (monoterpenes, sesquiterpenes, diterpene and triterpene derivatives) and eugenol have been isolated. Fatty acids, including geranic acid; alpha-and beta-pinene, 1,8, cincole, limonene, beta-eudesmol, and juglone are also important constituents of the volatile oil. The leaves contain many chemicals, the most prominent of which are aesculin, taxifolin (Pantocid), quercetin glucuronide, kaempferol-rhamnoside, syringetin-O-hexoside, myricetin-3-O-glucoside, myricetin-3-O-pantocid, and epicatechin. The unripe fruit husk also gave napthoquinones. The kernels of Indian walnuts contain 15.6% protein, 11% carbohydrates, 1.8% mineral matter (sodium, potassium, calcium, magnesium, iron, copper, phosphorus, sulfur, chorine). Iodine (2.8 mcg/100 g), arsenic, zinc, cobalt, and manganese are also reported. Kernels are also rich in vitamins of the B group, vitamin A 30IU/100 g, and ascorbic acid (3 mg/100 g). The walnut fruits are valuable and tasty. The juice of unripe fruits showed significant thyroid hormone–enhancing activity (prolonged use of such extract may cause serious side effect (Khare, 2007).

Used in: Ayurveda, Siddha, Traditional, and Unani

Parts used: Leaves, bark, and dried cotyledons

Uses: Leaves and bark: alterative, astra-laxative, antiseptic, mild hypoglycemic, anti-inflammatory, anti-scrofula, and detergent. An infusion of leaves and bark is used for herpes, eczema, and other cutaneous affections; externally to skin eruptions and ulcers. Volatile oil: antifungal, antimicrobial. Key application of leaf is for mild, superficial inflammation

DOI: 10.1201/9781003263500-143

of the skin and excessive perspiration of hands, and walnut-hull preparations are used for skin diseases and abscesses. Walnut is eaten as a dry fruit. Because of its resemblance to the brain, it was thought, according to the "doctrine of signatures", to be a good brain tonic. Walnuts are also eaten to lower cholesterol levels (Khare, 2007; Almedia et al., 2008; Carvalho et al., 2010; Gupta et al., 2019). The bark, branches, and exocarp of the green immature fruit is used to treat, gastric, liver, and lung cancer in China (Baytop, 1999; Liu et al., 2004).

Methods of cultivation: Seed

Justicia gendarussa Burm. f.

Family: Acanthaceae

Common names: English: Willow-leaved justicia; **Marathi:** Bakas, Kala adulasa; **Sanskrit:** Krishna Vaasaa, Nila-nirgundi, Krishna nirgundi, Nila-manjari. kasanah, vaidyasinha; **Tamil:** Karunochhi, Vadaikkuthi

Distribution: Throughout the greater part of India and Andaman Islands

Description: An erect, branched shrub that can reach heights up to 1.5 m. Its leaves are lanceolate, 4–14 cm long and 1–2.5 cm wide. The color of the leaves can be white, green, or gray and hairy on both sides. They are bitter, acrid, and thermogenic. The flowers are small, grow as 4–12-cm-long spikes at the end of branches or in leaf axils. The color of the flowers is white or pink with purple spots. The flowers are hermaphrodite, and the bark is a dark purple. The capsule is about 1.2 cm long and smooth.

Flowering: January–April

Chemical constituents: The leaves contain beta-sitosterol, an alkaloid, lupeol, friedelin, and aromatic amine (Khare, 2007).

Used in: Ayurveda and Siddha

Uses: Febrifuge, diaphoretic, emetic, emmenagogue. Infusion of leaves: given internally in cephalalgia, hemiplegia, and facial paralysis. Fresh leaves: used topically in edema and rheumatism. Bark: emetic (Khare, 2007). Leaf extract was proven to potentially become a male nonhormonal contraceptive method due to its competitive and reversible inhibition of the spermatozoan hyaluronidase enzyme. The plant is already used as traditional contraceptive method in Indonesia. The plant compound patentiflorin A contained in *J. gendarussa* has shown to have a positive activity against several HIV strains. The juice of the leaves can be drizzled into the ear for earache. To treat external edema, an oil made from the leaves is used (Zhang, Hong-Jie et al., 2017; Ratih et al., 2019). It is said to be useful for the treatment of asthma, rheumatism, and colic in children. Used as a treatment for skin problems such as eczema.

Methods of cultivation: Seeds

DOI: 10.1201/9781003263500-144

Justicia procumbens Linn.

Synonyms: *Rostellularia procumbens* (L.) Nees

Family: Acanthaceae

Common names: English: Water willow; **Gugrati:** Pitpapado; **Hindi:** Karambal; **Kannada:** Hucchu nela bevu; **Malayalam:** Cheriyaoridalthamara, Cherutharthaval, pittakkoti; **Marathi:** Karambal, Pitpapada, Kalmashi; **Sanskrit:** Parpata, renu, varatikta; **Tamil:** Knteyu, Kotakacalai, Kukkurm

Distribution: Western Ghats, west coast from Konkan to Kerala; abundant in the rainy season

Description: *J. procumbens* is a slender, often tufted, prostrate or ascending, branched perennial herb. The stems are 10–40 cm. The leaves are elliptic to oblong-ovate or ovate, 7–20 mm long and 5–20 mm wide, obtuse at both ends, and wholly or slightly crenate as to margin, glabrous or sparsely hairy, 1.5–3-cm-long petioles. The flowers are pink, 6–7 mm long and borne 1–5 cm long and about 5 mm in diameter in terminal, very thick cylindrical spikes.

The sepals and bracts are green, linear-lanceolate and hairy; calyx 3–4 mm long, 4-partite deep; uneven, circular, subacute, or acute segments, 4–6 mm thick, outdoor hairy, 2-lipped corolla; stamens 2; tied to the upper part of the corolla; unequally spaced anthers, ovary tip and style base hairy. The fruit (capsule) is slightly hairy and about 4 mm long, glabrous; orbicular seeds 1 × 1 mm, brown, striate drogues (Amritha et al., 2020).

Flowering and fruiting: June–December

Chemical constituents: The plant contains naphthofuranones, including justicidin A, B, C, D, G, and H; and diphyllin, The flowers contain peonidin glucoside (Khare, 2007).

Used in: Ayurveda, Folk, and Traditional

Parts used: Leaves and root

Uses: Used for the treatment of osteoporosis. Essential oil is antifungal. Its medicinal properties, including anti-inflammatory, anti-arthritic, antiarrhythmic, anti-asthmatic, anti-proliferative, anti-allergic, antibacterial, cytotoxic, and anti-HIV activities activity have been reported for this plant. Conventionally, it is used for the treatment of fever, inflammation, cold, cough, asthma, cancer, snakebites. A decoction of the leaves and root decoction has been used against typhoid fever. For treating asthma, cough, backache, flatulence, and many skin conditions, a decoction or an infusion is used. A decoction of the leaves is used in curvature care and bone diseases. Root juice is used in cough treatment where blood is coughed (Biswas et al., 2017a; Joshi et al., 2000; Amritha et al., 2020).

Methods of cultivation: Seeds

DOI: 10.1201/9781003263500-145

Justicia repens (L.)

Synonyms: *Adathoda spicata* Burm. ex Steud.; *Diclipta repens* L. Juss.; *Rungia repens* Nees.
Family: Acanthaceae
Common names: English: Creeping rungia; **Gujarati:** Parapata; **Hindi:** Khamor; **Marathi:** Ghati pitpapra; **Sanskrit:** Parpatha; **Tamil:** Kodaga saleh
Distribution: Found throughout the plains and lower parts of hills in India. They usually grow as a weed in moist habitats such as riverbanks, ditches, channels, and bunds of paddy fields.
Description: A annual or perennial herb growing to 30–40 cm. Leaves 3–5 × 1.5 cm, elliptic-lanceolate, acute at both ends, puberulous; petiole to 0.5 mm. Spike to 6 mm, terminal; bracts 4-farious, 2 rows barren, similar to 7 × 3 mm, elliptic-lanceolate, thinly pubescent, margin scarious; bracteoles about 4 × 0.7 mm, lanceolate, acuminate, ciliate; Flowers bi-lipped, calyx 3.5 × 5 mm, lobes 5, subequal, lanceolate; corolla to 5 × 2 mm, bluish purple with pinkish white striations; stamens 2, on the throat of the corolla tube, filaments puberulous, anthers cells unequal, appendaged; ovary globose, 1 mm, puberulous, style hairy, stigma hispid. Capsule to 5 mm, elliptic-oblong, pubescent; seeds suborbicular.
Flowering: June–July; **Fruiting:** July–August
Chemical constituents: Contain flavonoids notably, chrysoeriol 4'-mono and 7,4'-diglucosides, luteolin 7-glucoside, apigenin 7-glucoside, kaempferol 3-O-α-L-rhamnopyranosyl (1 3) β-D-glucopyrano-side, delphinidin 3,5-diglucoside, lutein, and kaempferol (Modi and Shah, 2017).
Used in: Ayurveda
Parts used: Stems, roots, and whole plant
Uses: The herb is antioxidant, anti-inflammatory, antimicrobial, diuretic, febrifuge and vermifuge. The dried, pulverized herb is used to treat cough and fever. The fresh, bruised leaves are mixed with castor oil and applied to the scalp to cure scaly fungous infection that usually occurs in children (*Tinea capitis*). The herb is also used to treat infections, coma, and snakebites.
Methods of propagation: Seeds

DOI: 10.1201/9781003263500-146

K

Kaempferia galanga Linn.

Synonyms: *Alpinia sessilis* J. Konig; *Kaempferia marginata* Carey ex Roscoe.; *Kaempferia plantaginifolia* Salisb.; *Kaempferia procumbens* Noronha.

Family: Zingiberaceae

Common names: Bengali: Ekangi, Bhuichampa; **English:** Aromatic ginger, Resurrection lily, Maraba; **Hindi:** Chandramula, Sidhoul; **Kannada:** Kachchra; **Malayalam:** Karchura; **Marathi:** Kapur kachari; **Sanskrit:** Sugandha-vachaa, Chandramuula, Chandramuulikaa; **Tamil:** Kachoram

Distribution: Throughout the plains of India, cultivated in gardens

Description: It is a stemless herb arising from tuberous rootstocks with fibrous cylindrical roots. The rhizome has dark reddish-brown skin, and the soft interior is nearly white. The leaves usually 2–3, spread horizontally, dark green, broadly elliptical to slightly flat with circular outline, measuring 8–15 cm wide. The blade is often lying flat to the soil the top surface is smooth while the bottom surface is cobweb-hairy. The inflorescence is sessile, emerging from between the leaves. It is 4–12 flowered. The sepal is 2–3 cm long. The petal is white, with tube 2.5–5 cm long and lobes 1.5–3 cm long. Its lip is broadly reversed egg-shaped, divided to about halfway or more, white or pale purple with violet to purple spots at the base. Each lateral lobe is about 2–2.5 cm × 1.5–2 cm. Another abortive stamen has an imperfect anther that is oblong-reversed egg-shaped to oblong-lance-shaped, 1.5–3 cm long and white. Their fertile stamen is 10–13 mm long, with two lobes deeply connective with abruptly bent lobes (Ibrahim, 1999).

Flowering and fruiting: Summer

Chemical constituents: The essential oil from rhizomes contain n-pentadecane, ethyl-p-methoxy cinnamate, ethyl cinnamate, carene, camphene, borneol, p-methoxystyrene, p-methoxy cinnamate, p-methoxy-trans-cinnamic acid and cinnamaldehyde. Insecticidal activity of the oil is attributed to ethyl cinnamates. Ethyl-p-methoxy-cinnamate shows monoamine oxidase inhibitor and a cytotoxic principle Ethyl-p-methoxy-trans cinnamate is the main compound in the root (Khare, 2007).

Used in: Ayurveda and Siddha

Parts used: Leaves, flowers, rhizome, and tubers

Uses: The tuber is stimulant, carminative, expectorant, and diuretic and is used for respiratory ailments like cough, bronchitis, and asthma. The rhizomes exhibit cytotoxic activity. The leaves and flowers exhibit antiphlogistic and vitamin P activity (Khare, 2007) As folk medicine, the rhizome is employed as an antibacterial; for the treatment of hypertension, asthma, rheumatism, indigestion, cold, and headache; and for relieving abdominal pain and toothache (Kanjanapothi et al., 2004). The rhizome is also used to treat wind and phlegm, restore digestive heat, and help circulate the blood (Gyatso and Hakim, 2010). The powdered rhizome mixed with honey is an expectorant used to treat productive cough and pectoral affection. In addition, oil prepared from the rhizome is applied over the nasal region to relieve nasal congestion (Khare, 2004). The preparation also can be used to treat wounds and applied to rheumatic region (Peter, 2004). Roasted rhizomes are applied as hot poultice in rheumatism (Khare, 2004).

Methods of cultivation: Rhizome and tuber

DOI: 10.1201/9781003263500-148

Kaempferia parviflora Wall. ex Baker

Synonyms: *Kaempferia rubromarginata*
Family: Zingiberaceae
Common names: English: Black ginger, Thai ginseng
Distribution: Found in tropical areas such as Malaysia, Sumatra, Borneo Island, and Thailand. Bangladesh. In India mostly in Mizoram
Description: It is a stem less rhizome forming herb about 20 cm tall. Rhizomes are small, having deep purple color, hence the name black ginger. The leaves are thin, about 8–16 cm long, rounded at the base. Green in color. The stalk of the leaf is short and channeled. Flowers are few in a stalk, less central cluster. The bracts are lance-shaped, green, and about 2.5 cm I length. The calyx is longer than the bract.
Chemical constituents: Contains 2 major constituents: 5,7-dimethoxyflavone and 5,7,4′-tri-methoxyflavone. In addition, it contains flavonoids and methoxyflavone substances (Sutthanut et al., 2007; Saokaew, 2017).
Used in: Folk and Traditional
Uses: Its rhizome has been long used as folk medicine. It is widely believed to reduce perceived effort, improve physical work capacity, and prolong hill trekking. It shows following properties: antiallergenic, anti-inflammatory antimutagenic, antidepressive, anticholinesterase, antimicrobial, anticancer, anti-peptic ulcer, cardioprotective, anti-obesity activity, and aphrodisia. Flavonoids contained in Krachaidum rhizome extract were reported to possess antioxidant activity, neuroprotective effects, and cognition-enhancing effects. It enhanced the response to sexual erotic stimuli (Sutthanut et al., 2007; Saokaew, 2017).
Methods of propagation: Rhizome

DOI: 10.1201/9781003263500-149

Kigelia pinnata Jacq. DC

Synonyms: *Bignonia africana* Lam.; *Crescentia pinnata* Jacq.; *Kigelia africana* (lam.) benth; *Kigelia pinnata* DC; *Tecoma africana* (Lam.) G. Don.

Family: Bignoniaceae

Common names: English: Sausage tree; **Hindi:** Balam khira, Jhar fanoos; **Kannada:** Aanethoradu kaayi, Mara sowthe; **Tamil:** Yanai pudukku; **Telugu:** Enuga thondamu, Kijili, Naagamalle

Distribution: Native to tropical Africa; found in India also

Description: Medium to large tree. Bark: gray, smooth, usually flaking in larger specimens. The tree is evergreen in regions where rainfall occurs throughout the year but deciduous where there is a long dry season. Leaves: opposite, crowded near the ends of branches, imparipinnate with 3–5 pairs of leaflets and a terminal leaflet. Leaflets: oblong, leathery with rough hairs on both surfaces; margins entire or with some small teeth. Leaves: 30–50 cm long with 6–10 oval leaflets up to 20 cm long and 5.7 cm in diameter. The terminal leaflet can be either present or absent. Flowers: large, dark maroon with yellow veining, bell-shaped in pendulous sprays up to 12-flowered, orange to maroon or purplish green, and about 10 cm wide. The flowers hang down from branches on long flexible stems 7–20. ft long. Fruit: a huge, woody berry about 30–90 cm long and up to 18 cm in diameter, sausage-shaped, and weighing up to 7 kg.

Flowering and fruiting: August–October

Chemical constituents: The roots, wood, and leaves have been found to contain naphthoquinones, dihydroisocoumarins, flavonoids, and aldehydic iridoids. Among the naphthoquinones, kigelinole, isokigelinole, pinnatal, and isopinnatal have been isolated. From the root and its bark, the usual plant substances stigmasterol, β-sitosterol, ferulic acid, the naphthoquinones lapachol, 6-methoxymellein, and two new phenolic compounds have been isolated. Kigelin is the main component of the plant.

Used in: Folk

Parts used: Leaves, stem, bark, and fruits

Uses: The plant is used includes digestive problems, fainting, anemia, sickle-cell anemia, epilepsy, respiratory ailments, hepatic and cardiac disorders, and nutritional illnesses, such as kwashiorkor, rickets, wasting, weakness, and skin problems. The plant is used particularly to remedy a wide range of skin problems. Powders and infusions of the bark, leaves, stems, twigs, or fruits are all used to clean and dress flesh wounds and open sores. The bark, stems, twigs, leaves, and fruits are infused and taken orally, or applied locally, to relieve rheumatism, sprains, hematoma, and bruising; snakebite antidotes are made with an infusion of the fruits, stems, leaves, twigs, or bark, taken orally or rubbed onto the bite. Infectious diseases, including leprosy, impetigo, and worm infestations in the blood, are all treated with this plant; dermal complaints and infections, such as whitlows, cysts, acne and boils, are treated with traditional medicines containing the fruit and, less frequently, the bark; sore eyes are treated with drops made from flower sap mixed with water.

Methods of propagation: Seeds

DOI: 10.1201/9781003263500-150

L

Lactuca sativa Linn

Synonyms: *Lactusa scariola* Linn. var. sativa C. B. Clarke.

Family: Asteraceae (Compositae)

Common names: Bengali: Kahu, Salad; **English:** Garden lettuce; **Gugrati:** Loni, Ghol; **Hindi:** Kahu, Salaad Pattaa; **Kannada:** Saladdi soppu; **Tamil:** Salattu virai; **Telugu:** Kavu, Shallattu

Distribution: Native to southern and West Asia. Cultivated throughout India as a cold-weather crop

Description: It is a rosette herb, annual or biennial, 30–120 cm high, with a thin taproot. Flowering stems erect, slightly angular, green to yellow to whitish, glabrous, robust, densely leafy, branched in upper half. Leaves: rosette, oblong to obovate or elliptic or ovate, 5–20 cm long and 4–20 cm wide, runcinate to entire, undulate and dentate or entire, rounded or obtuse, petiole-like attenuate or cuneate, glabrous. Cauline leaves many, ovate or broadly elliptic, entire, usually undulate or denticulate, acute or rounded, rounded, amplexicaul or cuneate, glabrous. Synflorescent paniculiform, with up to 400 heads. Peduncle glabrous, covered with bracts. Peduncle bracts obtuse, amplexicaul. Heads with 7–15(–35) flowers. Involucral bracts 12.0 mm long, 1.0–1.7 mm wide, yellowish green; outer involucral bracts glabrous; inner involucral bracts ovate, obtuse. Receptacle flat, pitted, without scales. Corolla ligulate, 12.0–14.0 mm long, yellow; ligule sometimes streaked violet. Achenes obovoid in outline, compressed, 3.0–4.0(–5.0) mm long, often muricate at apex, brownish gray or whitish, with 5–7 ribs on each face, with a filiform beak; beak half as long as the corpus, whitish. Pappus white, 3.0–4.0 mm long, flexible, of simple bristles.

Flowering and fruiting: Midsummer

Chemical constituents: The leaves contain calcium, phosphorus, iron, thiamine, riboflavin, carotene, iodine, and fluorine. Aqueous extract of roots give a guanine-type sesquiterpene glycoside and, lactoside C, along with known glycosides, lactoside A and macro-cliniside A (Khare, 2007). It also contain glycosides, steroids, phenolics, tannins, resin, alkaloid (lactucarium, a mixture of lactocin and three bitter principles: lectucin, lectopicrin, and lactucic acid), and organic acid (oxalic acid, malic acid, citric acid). *L. sativa* contains antioxidants flavanol, quercetin, caffeic acid, and ascorbic acid; inorganic compounds: sodium, potassium, magnesium, iron, copper, chloride, sulfur, and phosphorus; and vitamins: vitamin A, vitamin B1, vitamin B2, nicotinic acid, vitamin C, vitamin E, vitamin G, vitamin K, and folic acid (Anonymous, 1987, 1997; Nadkarni, 2009).

Used in: Unani

Parts used: Leaves

Uses: Plant: used in painful ulcers and burns. A dietary allowance of 10 g of lettuce is adequate to meet the vitamin K requirement of the body (Khare, 2007). It stimulates the secretion of digestive enzymes, increases appetite, promotes the peristalsis of the human intestinal wall, and prevents constipation.

Methods of cultivation: Shoot and leaf cuttings

DOI: 10.1201/9781003263500-152

Launaea sarmentosa (Willd.) Kuntze

Synonyms: *Launaea pinnatifida* Cass.; *Microrhynchus sarmentosus* (Willd.) DC.; *Microrhynchus bellidifolius* (Cass.) DC

Family: Asteraceae (Compositae)

Common names: English: Beach launea; **Gujrati:** Alamiro, Bhopathari; **Hindi:** Pathri; **Kannada:** Bonapatre; **Malayalam:** Kadalkozhuppa, Kozhikkaalan veru; **Marathi:** Sagar pathari; **Tamil:** Chentam, Eluttani, Kattari

Distribution: Bangladesh, India, coastal Madagascar, Mauritius, Sri Lanka, South Africa, and Seychelles. Abundantly growing species of sandy area. Distributed throughout India

Description: Glabrous, procumbent, perennial herb with prostrate creeping stem. Stem: 30–90 cm long, striate, rooting at the node, and bearing rosette leaves. Leaves: crowded near the base and in clusters at the nodes, rosulate, attenuate at base, oblanceolate, to spathulate, 10–12 cm long, obtuse to rounded at the apex, margins denticulate, teeth indurated. Flowers: in capitula 1.0–1.5 cm long, solitary, borne on small bract-covered stalk 0.5–1.2 cm long, arising from the nodes; bracts 3–4-seriate, often purplish, margins white, membranous, 0.3–0.5–1.2 × 0.1–0.2 cm; those of outer series oblong, obtuse; inner ones linear-lanceolate; florets 20 in number; pappus unimorphic, 0.6–1.0 cm long, white, deciduous; corolla yellow, tube 0.4–0.5 cm long, ligules 0.6–1.0 cm long yellow in color, anthers yellow in color. Fruits: a cypsela black, 4–5 ribbed, 0.3–0.4 cm long, glabrous, minutely wrinkled, truncate, or narrowed.

Flowering and fruiting: October–May

Chemical constituents: Leaves contain taraxasterol. Roots contain taraxerol acetate. Preliminary phytochemical screening of shows the presence of alkaloids, tannin, and steroids (Yusriyya et al., 2013).

Used in: Ayurveda and Tribal

Parts used: Leaves

Uses: The plant is antirheumatic, diuretic, and galactagogue. The juice of the plant is used as a soporific for children. It is applied externally in the treatment of rheumatic affections, either combined with the oil of *Pongamia glabra* or with the juice of the leaves of *Vitex leucoxylon*. Crushed roots mixed with sugar is taken for treating inflammation of urethra. The roots are exploited for the cure against jaundice. It has also been used as a coolant, diuretic, and demulcent against allergic infections (Anjaria et al., 2002). Root is used mainly in a lehya preparation known as hilibeys, which is used by mothers after childbirth. It is also used for various abdominal disorders, especially due to Vai (Vata), and in urinary infections. *L. sarmentosa* is also reported to possess tonic, soporific, diuretic, and aperient properties and is used as a substitute for *Taraxacum* (*T. officinale*). Latex from *L. sarmentosa* is also commonly used by fishermen to heal skin injuries caused by fish spines while fishing. Other uses in mainland India include the use of the whole plant in gout and the leaf in rheumatism. Among the islanders of the Indian Ocean, the whole plant is used as a bath decoction to treat skin diseases and fish stings (Jain and Srivastava, 2005; Pullaiah, 2006; Yusriyya et al., 2013).

Methods of propagation: Seeds and cuttings

DOI: 10.1201/9781003263500-153

Photo: Dr. Srinath Rao

Lawsonia inermis Linn.

Synonyms: *Alcana spinosa* (L.) Gaertn.; *Lawsonia speciosa* L.; *Lawsonia spinosa* L.; Rotantha combretoides Baker.

Family: Lythraceae

Common names: Bengali: Mehadi; **English:** Henna; **Gugrati:** Mehandi **Hindi:** Hinaa, Mehndi; **Kannada:** Korate, Mehndi; **Malayalam:** Mayilanji, Pontalasi; **Marathi:** Mehandi; **Sanskrit:** Madayanti, Madayantikaa, Mendika, Ranjak; **Tamil:** Marithondi, Marudum; **Telugu:** Mehendi akulu, Gorintaku

Distribution: Native to Arabia and Persia; now cultivated mainly in Haryana and Gujarat; to a small extent in Madhya Pradesh and Rajasthan

Description: Henna is a tall shrub or small tree, standing 1.8–7.6 m tall (6–25 ft). It is glabrous and multi-branched, with spine-tipped branchlets. The leaves grow opposite each other on the stem. They are glabrous, sub-sessile, elliptical, and lanceolate (long and wider in the middle; average dimensions are 1.5–5.0 × 0.5–2 cm, acuminate (tapering to a long point), and have depressed veins on the dorsal surface. Henna flowers have 4 sepals and a 2 mm (0.079 in.) calyx tube, with 3-mm spread lobes. Its petals are ovate, with white or red stamens found in pairs on the rim of the calyx tube. The ovary is 4-celled, 5 mm long, and erect. Henna fruits are small, brownish capsules, 4–8 mm in diameter, with 32–49 seeds per fruit, and open irregularly into 4 splits.

Flowering and fruiting: Late spring or early summer

Chemical constituents: The leaves contain naphthoquinones, in particular lawsone; coumarins (laxanthone I, II and III); flavonoids, luteolin and its 7-O-glucoside, acacetin-3-O-glucoside; all parts contain tannins. Chloroform and ethanol extracts of leaves exhibit promising antibacterial activity against *Shigella* and *Vibrio cholerae*. Leaf extract shows antifungal activity against several pathogenic bacteria and fungi. Henna paint is used as a medicament for treatment of hands and feet for mycosis. The antimycotic activity is due to lawsone, a naphthoquinone. Extract of the stem bark shows hepatoprotective activity (Khare, 2007).

Used in: Ayurveda and Folk

Parts used: Leaves, stem, and roots

Uses: Leaves: astringent, antihemorrhagic, antispasmodic, oxytocic, antifertility, antifungal, antibacterial. Used externally to treat skin infections (tinea); also as a hair conditioner. the use of the leaves in dysuria, jaundice, bleeding disorders, ulcers, prurigo, and other obstinate skin diseases. The leaf is also recommended in giddiness and vertigo. Stem bark and root, probably due to the presence of isoplumbagin and lawsaritol, exhibit anti-inflammatory activity experimentally. Evidence shows that the henna leaf might be able to decrease the formation of sickled cells in individuals with sickle cell anemia (Khare, 2007).

Methods of cultivation: Seeds and stem cuttings

DOI: 10.1201/9781003263500-154

Leea aequata Linn.

Synonyms: *Leea ancolana* Miq.; *Leea hirta* Roxb. ex Hornem.; *Leea hirsuta* Bl.; *Lee hispida* Gagnep.; *Lee humilis* Hassk.

Family: Vitaceae

Common names: English: Leea; **Sanskrit:** Kaakajanghaa, Nadikaantaa, **Telugu:** Surapadi

Distribution: *L. aequata* is usually a shrub, less often a small tree, distributed in Bangladesh, Bhutan, Cambodia, China, India, Malaysia, Myanmar, Nepal, Philippines, Thailand, and Vietnam. In India, found in all states

Description: It is a large shrub, young branches villous. Leaves: alternate, (1-)2(-3)-pinnate; rachises 7–20 cm long, angled, pubescent; petioles 6–15 cm long; stipules oblong-obovate, 3–8 × 3–6 cm, hairy, caducous; leaflets 5-many, oblong-lanceolate or lanceolate, cuneate to truncate, sometimes subcordate or rounded at base, sharply serrate at margin, acuminate to caudate at apex, membranous, hispid with scattered gray hairs above, rough with scattered rounded brown peltate glands beneath; petiolules of lateral leaflets 3–12 mm long, those of terminal ones 2.5–3.5 cm long, pubescent. Inflorescences: subterminal, 6–10 cm long, more or less glandular-hairy; peduncles 1–4 cm long; bracts conspicuous, ovate to oblong, 6–8 mm × about 5 mm. Flowers: in corymbs, 5-merous, greenish white. Calyx: lobed halfway down, glabrous to densely pubescent, covered with pearl glands outside. Corolla: lobes 2–3 × 1–1.5 mm. Staminal tube 1.5–2.5 mm long; lobes deeply cleft; filaments 1–1.2 mm long; anthers 1–1.2 mm long. Ovary: 4–7 with locules; style 1–2 mm long. Fruits: globose-depressed, 6–7 mm, orange-red, black when ripe; seeds: 3–6.

Flowering and fruiting: July–December

Chemical constituents: A new neolignan, (7*S*,8*R*)-9′-*O*-acetylcedrusin has been identified from the arial parts of *Leea aequata* Linn. and a new lactam, (3*S*,4*S*)-4-chloro-3-hydroxypiperidin-2-one, along with 21 known compounds, including 5 lignans, 4 flavonoid glycosides, and others, were isolated from the ethanoic extract of the aerial parts of *L. aequata* (Tun et al., 2019).

Used in: Traditional

Parts used: Leaves, stem, and root

Uses: Stem and root: astringent, anthelmintic. Used for indigestion, jaundice, chronic fever, and malaria. Essential oil: inhibits the growth of *Mycobacterium tuberculosis*; also inhibits the growth of *Micrococcus pyogenes* var. aureus and *Pasteurella pestis*. Root, tuber, and stem: mucilaginous, astringent. Leaves and twigs: antiseptic; used for poulticing wounds (Khare, 2007).

Methods of cultivation: Seeds

DOI: 10.1201/9781003263500-155

Leonotis nepetaefolia (L.) R. Br.

Synonyms: *Leonotis nepetifolius* (L.) Mill.; *Phomis nepetaefolia* L.

Family: Lamiaceae (Labiatae)

Common names: English: Chrismas candlestick, Lion's ear; **Hindi:** Bada guma, Lal guma; **Kannada:** Deepa shoole, Goagadde, Kaduthumbe; **Marathi:** Deepamaal; **Sanskrit:** Granthiparni, Kaakapuchha

Distribution: Throughout the warmer parts of India

Description: It is an erect annual herb that grows 2.5 m high, occasionally reaching up to 3 m. The stem is unbranched at the base and loosely branched toward the apex with strongly angled stems having appressed retrose hairs that are longer at the nodes. The leaves are smooth with a toothed margin; they are large, ovate, lobed, acute, and winged in the upper part. The inflorescence comprises axillary dense, globose multi-flowered verticillasters. Flowers are orange and borne in spiny clusters. Floral leaves are lanceolate and deflexed; bracts are linear, highly spinous pointed, and deflexed. Calyxes are tubular, incurved, and hairy with 8–9 sharp, pointed teeth. The plant typically has a bilabiate, orange-scarlet and hairy corolla with the upper lip densely woody, and the lower lip made up of 3 lobes. Stamens are 4 and didynamous; the ovary has 4 lobes, and the fruit contains 4 ovoid nutlets (Tiwari, 2019).

Flowering and fruiting: November–December

Chemical constituents: The root contains n-octacosanol; n-octacosanoic acid; quercetin, 4,6,7-trimethoxy-5-methylchromene-2one; campesterol; and beta-sitosterol-beta D-glucopyranoside. The plant contains 4,6,7-trimethoxy-5-methyl-chromene-2-one. The leaves contain neptaefolin, neptaefuran, neptaefuranol, neptaefolinol, leonitin, neptaefolinin, and (–)-55, 6-octadecadienoic acid. The seed oil contains oleic, linoleic, palmitic, and stearic acids. The fatty oil, extracted from the seeds, is similar to olive oil (Khare, 2007).

Used in: Ayurveda, Folk, and Traditional

Parts used: Leaves and seeds

Uses: Leaves: spasmolytic. Ash of flower head: applied to burns and scalds, in ringworm and other skin diseases. The *Ayurvedic Pharmacopoeia of India* recommends the root in cough, bronchitis, and dyspnea. The seeds possess feeble antimalarial activity. The seed extract showed 100% toxicity against *Alternaria alternata* and marked toxicity against *Aspergillus niger.* Root: 5–10 g powder (Khare, 2007). Leaves of *L. nepetifolia* are brewed as a tea for fever, coughs, womb prolapse, malaria, and suggested to be beneficial to bone and lung health. The roots of *L. nepetifolia* are considered to be the botanical sources of granthiparna, an Ayurvedic herb. In many traditional health practices. *L. nepetifolia* is used to treat bronchial asthma, diarrhea, fever, influenza, malaria, cough, epilepsy, womb prolapse, skin ailments, and rheumatism (da Silva Almeida et al., 2018; Oliveira et al., 2015).

Methods of cultivation: Seeds

DOI: 10.1201/9781003263500-156

Leonurus cardiaca Linn.

Synonyms: *Leonurusglabra*; *Cardiaca stachys*
Family: Lamiaceae (Labiatae)
Common names: English: Common motherwort, Lion's tai
Distribution: Native to Europe; also distributed in Himalayas from Kashmir to Kumaon
Description: Perennial, herbaceous plant. The plant grows to about 60–100 cm in height. It has a squarish stem, which is clad in short hairs and is often purplish, especially near the nodes. The leaves are opposite, have serrated margins, and are palmately lobed with long petioles; basal leaves are wedge-shaped with 3 points while the upper leaves have 3–5. They are slightly hairy above and grayish beneath. Flowers appear in leaf axils on the upper part of the plant and have 3-lobed bracts. The calyx of each flower is bell-shaped and has 5 lobes. The corolla is irregular, 8–12 mm (0.3–0.5 in.) long, fused, long-tubed with two lips. The upper lip is convex and covered with white hairs, and the lower lip is 3-lobed, downward-curving, and spotted with red. The flowers are pink to lilac in color often with furry lower lips. There are 4 protruding stamens, 2 short and 2 longer, and the fruit is a 4-chambered schizocarp.
Flowering and fruiting: Late summer
Chemical constituents: The plant contains diterpene bitter principles, iridoid monoterpenes, flavonoids including rutin and quercitrin, leonurine, betaine, caffeic acid derivatives, tannins, and traces of a volatile oil. The herb contains several components with sedative effect: alpha-pinene, benzaldehyde, caryophyllene, limonene and oleanolic acid (Khare, 2007). The herb contains the alkaloid leonurine (Kuhn and David, 2000). (Among other chemical constituents, it also contains stachydrine; Kuchta et al., 2013) bitter iridoid glycosides (leonuride), diterpenoids, flavonoids (including rutin and quercetin), tannins, volatile oils, and vitamin A.
Used in: Folk and Unani
Parts used: Leaves, stem, and flowers
Uses: Stomachic, laxative, antispasmodic, diaphoretic, emmenagogue, used in absent or painful menstruation, premenstrual tension, menopausal flushes. Hypnotic, sedative. Used as a cardiac tonic. (Studies in China have shown that motherwort extracts show antiplatelet aggregation actions and decrease the levels of blood lipids.) Key application: In nervous cardiac disorders and as adjuvant for thyroid hyperfunction (Blumenthal, 1998) and as antispasmodic (*The British Herbal Pharmacopoeia*, 1993). The British Herbal Compendium indicated its use for patients who have neuropathic cardiac disorders and cardiac complaints of nervous origin The herb is a slow acting adjuvant in functional and neurogenic heart diseases. Its sedative and spasmolytic properties combine well with *Valeriana officinalis* or other cardioactive substances (Khare, 2007). Stachydrine is extracted from the leaves of motherwort and has demonstrated various bioactivities for the treatment of fibrosis, cardiovascular diseases, cancers, uterine diseases, brain injuries, and inflammation (Chenget al., 2020).
Methods of cultivation: Seeds

DOI: 10.1201/9781003263500-157

Lepidagathis cristata Willd.

Synonyms: *Acanthus cristatus* Rottler ex Nees; *Lepidagathis rupestris* Bedd.; *Lepidagathis shuteri* T. Anderson.; *Lepidagathis subarmata* C. B. Clarke

Family: Acanthaceae

Common names: English: Crested lepidagathis; **Hindi:** Bukhar Jadi; **Kannada:** Gantu kalaugedde; **Marathi:** Bhui gend, Bhu terada; **Telugu:** Mullabanthi

Distribution: All over India

Description: A perennial herb mostly found in dry places and wastelands. Plant is a stiff herb with almost no stem. Branches are about 20 cm long, procumbently rising out from a hard central globose rootstock head on the ground and spread out. Leaves: 3–6 cm long and 0.5–1 cm broad, are stalk less, alternate, elliptic, serrate and usually lineolate, velvet-hairy, pointed at both ends, margin entire. Flowers are sessile, capitates, pale pink, 2-lipped; the upper lip is notched, and the lower lip is divided into 3 lobes. The heads terminal or axillary densely crowed at the base of the stem, bracts elliptic, spinescent, bracteolate. Calyx has 5 lobes, hairy, and corolla is white with a brown color or purple spots. Anther: didynamous, 2-celled and exerted stamens 4. Fruit: a capsule oblong with seeds (Ahirwar and Ahirwar, 2021).

Flowering and fruiting: January–March

Chemical constituents: Extract of whole plant material contains alkaloids, flavonoids, phenolic compounds, saponins, tannins, and terpenoids. Cristatin A (6-hydroxyluteolin, 6-hydroxyluteolin-7-apioside), a tryptophan-derived alkaloid, and cycloartenol, stigmasta-5, 11(12)-dien-3b-ol were found in whole-plant extract. Inflorescence of *L. cristata*, contains oleic acid, 3-(octadecyloxy) propyl ester. Leaf contains heptadecane, 9-hexyl and ethyl iso-allocholate. Root contains heptadecane, 9-hexyl, octadecane, 3-ethyl-5–(2-ethylbutyl) (Ranganatham and Nagarajan, 1980; Maghdu and Palaniyappan, 2014; Kamaladevi, 2017; Ahirwar and Ahirwar, 2021).

Used in: Folk, Siddha, and Traditional

Parts used: Leaves, roots, and inflorescence

DOI: 10.1201/9781003263500-158

Leucas aspera Spreng

Synonyms: *Leucas minahassae* Koord., nom. nud.; *Leucas obliqua* Buch.-Ham. ex Dillwyn; *Leucas plukenetii* (Roth) Spreng.; *Phlomis aspera* Willd.

Family: Labiatae; Laminceae

Common names: Bengali: Darunaphula, Hulkasha; **English:** White dead nettle; **Gugrati:** Kulnnphul; **Hindi:** Goma madhupati; **Marathi:** Bahuphul, Tumba; **Sanskrit:** Dronpushpi, Phalepush-paa, Kutambak; **Tamil:** Thumbai; **Telugu:** Tunni

Distribution: Throughout India in cultivated fields, wastelands, and along the roadside

Description: *L. aspera* is an annual, branched, herb growing to a height of 15–60 cm, with a stout and hispid acutely quadrangular stem and branches. Leaves are opposite decussate, sub-sessile or shortly petiolate, linear or linearly lanceolate, obtuse, pubescent up to 8.0 cm long and 1.25 cm broad, with entire or crenate margin; petiole 2.5–6 mm long; flowers white, sessile small, in dense terminal or axillary whorls; bracts 6 mm long, linear, acute, bristle-tipped, ciliate with long slender hairs; calyx variable, tubular, 8–13 mm long; tube curved, contracted above the nutlets, the lower half usually glabrous and membranous, the upper half ribbed and hispid; mouth small, very oblique, not villous, the upper part produced forward; teeth small, triangular, bristle-tipped, ciliate, the upper tooth being the largest. Corolla: 1 cm long; tube 5 mm long and pubescent above, annulate in the middle; upper lip 3 mm long, densely white, woolly; lower lip about twice as long, the middle lobe obviate, rounded, the lateral lobes small, subacute. Fruit: nutlets, 2.5 mm long, oblong, brown, smooth, inner face angular and outer face rounded (Prajapati et al., 2010).

Flowering and fruiting: November–February

Chemical constituents: The plant gave oleanolic acid, ursolic acid, and beta-sitosterol. The root contains a triterpenoid, leucolactone, and the sterols sitosterol, stigmasterol and campesterol (Khare, 2007). Furthermore, studies reveal the presence of various phytochemical constituents mainly triterpenoids, oleanolic acid, ursolic acid and β-sitosterol, nicotine, sterols, glucoside, diterpenes, phenolic compounds (4–(24-hydroxy-1-oxo-5-n-propyltetracosanyl)-phenol; Prajapati et al., 2010).

Used in: Ayurveda, Folk, and Siddha

Parts used: Flowers and leaves

Uses: Carminative, antihistaminic, antipyretic, febrifuge, antiseptic. Used in jaundice, anorexia, dyspepsia, fever, helminthic manifestation, respiratory, and skin diseases. Flowers: given with honey for coughs and colds to children. Leaves: juice is used as an external application for psoriasis, chronic skin eruptions and painful swellings. An alcoholic extract of leaves shows antibacterial activity (Khare, 2007). Medicinally, it has been proven to possess various pharmacological activities like antifungal, antioxidant, antimicrobial, antinociceptive, and cytotoxic activity (Prajapati et al., 2010). Bruised leaves are applied locally in snakebites (Shirazi, 1947; Rai et al., 2005).

Methods of cultivation: Seeds

DOI: 10.1201/9781003263500-159

Lippia nodiflora Rich.

Synonyms: *Phyla nodiflora* (Linn.) Greene.

Family: Verbenaceae

Common names: English: Wild sage; **Hindi:** Jalpapli, Jal buti; **Kannada:** Nelahippali; **Malayalam:** Kattuthippali, Nirthippali; **Marathi:** Adali, jalapimpali; **Sanskrit:** Jalapippali; **Tamil:** Paduthalai; **Telugu:** Bokkena

Distribution: Throughout India, in wet places and riverbanks.

Description: A small perennial creeping/spreading herb. Stem squarish hairy, creeping and rooting at the nodes. Leaves are small, opposite, spathulate, base cuneate, apex rounded and deeply serrate. Inflorescence is an axillary globose head later becomes spicate. Flowers: 5-merous, pink or white, sessile, 1–2.5 cm long when mature and 6–9 mm across. Calyx: deeply bilabiate, compressed, hairy. Corolla: white to pale blue, bilabiate, upper lip erect, bifid, the lower 3-lobed, the central lobe the largest. Stamens: 4, didynamous, included. Ovary: 2-celled, ovules 2, stigma oblique. Fruits: drupe dry, splitting into 2, 1-seeded pyrenes.

Flowering and fruiting: Throughout the year

Chemical constituents: Aerial parts are reported to contain flavonoids, flavone aglycones, and flavone sulfates (Khare, 2007).

Used in: Ayurveda, Siddha, and Unani

Parts used: Whole plant and leaves

Uses: Plant: cooling, febrifuge, diuretic. A poultice is used as maturant for boils. Leaves: an infusion is given to women after delivery. The leaf juice enters into hair oils for alopecia areata. Whole plant juice: 10–20 ml is used (Khare, 2007). Furthermore, it is found to possess analgesic, anti-inflammatory, antioxidant, antinociceptive, antimicrobial, antipyretic, antitumor, lipid peroxide scavenging, and free radical scavenging activities (Amir et al., 2011). The plant is effective against bronchitis, respiratory diseases, arthritis, fever, dyspepsia, hookworm, gonorrhea, ulcers, stomachic, wounds, burning sensation, asthma, thirst, loss of consciousness, diaper rash, erysipelas, neuralgia, sores, spasms, and vertigo (Kirtikar and Basu, 1918, 1991; Anonymous, 1969; Amir et al., 2011).

Methods of cultivation: Propagated both vegetatively and by seeds. Plants break up during flooding and can quickly reestablish in moist soil as floodwater subsides.

DOI: 10.1201/9781003263500-160

Luffa acutangula (Linn.) Roxb.

Synonyms: *Cucumis acutangulus, Cucurbita acutangula, Luffa foetida, Luffa drastica, Cucumis operculatus* Roxb., *Luffa gosa* Ham.

Family: Cucurbitaceae

Common names: Bengali: Titotorai, Titojhinga, **English:** Ribbed or ridged gourd; **Hindi:** Turai; **Kannada**: Heerikai; **Malayalam:** Athanga; **Marathi:** Dodaka; **Sanskrit:** Dhamargava, Katukoshataki, Tiktakoshtaki; **Tamil:** Itukari, Itukarikkoti; **Telugu:** Beerakaya

Distribution: Throughout India; cultivated throughout greater parts of India

Description: It is an annual climber. Five-angled, glabrous stem is brownish yellow in color along with tendrils up to 6-fid. Leaves are simple, alternate and orbicular in outline with 15–20 cm long, palmately 5–7-angled, triangular to broadly rounded lobes and pale green in color. Veins and vein islets are prominent. Flowers are regular and unisexual and consist of yellow petals. Female flowers are yellow-colored, solitary, 2–15 cm long on pedicels, with inferior, longitudinally ridged ovary and 3-lobed stigma while male flowers are light greenish in color, consist of three free stamens with yellow corolla inserted into the receptacle tube. Fruits are cylindrical, pale yellowish brown in color, bitter in taste, tapered toward the base and are covered with 8–10 prominent ribs. Inner part of the fruit is three chambered, fibrous and easily detachable from the outer part. Seeds are elliptical and black-colored (Shendge and and Belemkar, 2018).

Flowering and fruiting: August–September

Chemical constituents: The plant contains the oleanane-type triterpene saponins. It is reported to exhibit antitumor activity (Khare, 2007).

Used in: Ayurveda, Folk, and Traditional.

Parts used: Whole plant, leaves, and seeds

Uses: Plant: purgative, diuretic. Used for edema, splenic enlargement, cough, and asthma. Seeds: emetic, expectorant. It is reported to exhibit antitumor activity. The fruit juice is used as a homeopathic drug in hepatic congestion, irritation, and inflammation of gastric mucosa. Leaf, fruit, root: 10–20-ml juice (Khare, 2007). Crude extract of plant and its isolated compounds possess broad pharmacological activities such as antidiabetic, hepatoprotective, antiulcer, anticancer, immunomodulatory, antihyperlipidemic, antioxidant, antimicrobial, central nervous system depressant, analgesic, and anti-inflammatory (Shendge and Belemkar, 2018). The fruit widely used for diabetes treatment (Kanaka et al., 2013). Apart from this, the plant is also used by the tribes of western Maharashtra on insect bite. Fruit powder is applied topically to treat swollen hemorrhoids. The kernel of the seed is used as an efficient remedy for dysentery while the juice of the fruit is applied to cure a headache (Dandge et al., 2010). Oral administration of seed powder is extensively used for the treatment of urinary bladder stone in Rajasthan (Katewa et al., 2004). Local application of pulverized leaves is reported to be useful in splenitis, hemorrhoids, ringworm infection, and leprosy while the juice of the leaves is administered into the eye for treatment of granular conjunctivitis in children (Mahbubar, 2013). In addition, the fruit possesses demulcent and diuretic properties while the seeds have purgative, emetic and anthelmintic properties. The dried fruit powder is useful in preventing premature graying of hair. The root of the plant is laxative and used in dropsy (Nadkarni, 1996).

Methods of cultivation: Seeds

DOI: 10.1201/9781003263500-161

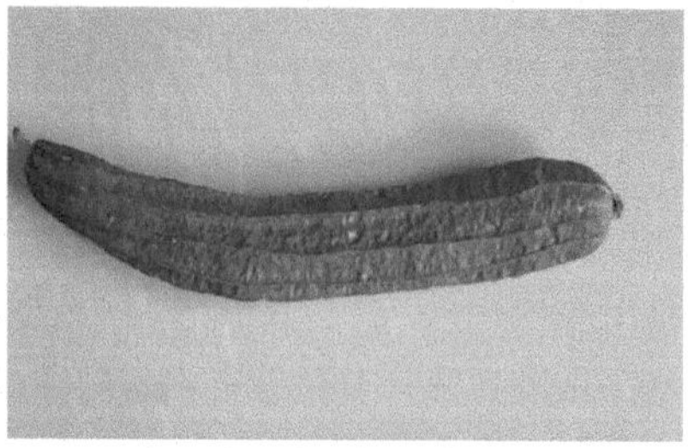

Photo: Dr. Srinath Rao

Luffa cylindrica (Linn.) M. J. Roem.

Synonyms: *Luffa aegyptiaca* Mill.; *Luffa pentandra* Roxb; *Cucumis acutangulus* L.; *Cucurbita acutangula* (L.) Blume; *Luffa foetida* Cav.; *Luffa plukenetiana* Ser.; *Luffa hermaphrodita* Singh & Bandhari; *Momordica luffa*Vell.

Family: Cucurbitaceae

Common names: English: Smooth luffa, Spongegourd, Sponge cucumber; **Kannada:** Tuppada hirekai; **Sanskrit:** Dhaamaargava, Mahaakoshtaki, Mahaajaalini, Raajakoshataki; **Tamil:** Mozhukupeerankai, Pikku

Distribution: Cultivated throughout greater parts of India

Description: Annual, or trailer. Tendrils: slightly pubescent, 3–6-fid. Stem: 5-angled, finely hairy to glabrous. Leaves: palmately 5-lobed, dark green, orbicular-cordate, 8–25 cm across, lobes triangular, lanceolate, acute apiculate, entire or sinuate, scabrous. Petiole: 5–15 cm long. Flowers: bright yellow, pedicellate, 5–6 cm across; male racemose, racemes axillary, 12–25 cm long, 15–20-flowered, female flowers in the same axil as males. Probrract fleshy, ovate, with 3–7 glistening glands on the upper surface. Calyx: tube short, broadly campanulate, slightly pubescent; lobes triangular-lanceolate, longer than tube. Petals: obovate-cuneiform, 2.5–3.5 cm long, 1– 2.5 cm broad, obtuse. Stamens: 3–5, filaments 6–8 mm long. Ovary: cylindrical, finely appressed hairy. Fruit: cylindrical and fusiform, 20–50 cm long, 6–10 cm across, smooth. Seeds: dull black, elliptic-ovoid, about 10–12 mm long, 6–8 mm broad, with about 1-mm-wide margin.

Flowering and fruiting: September–October

Chemical constituents: Triterpenoid saponins like lucyoside O have been isolated from the leaves. The roots of the mature plants contain an acidic pentacyclic triterpene, bryonolic acid (Khare, 2007). Further the phytochemical screening of *L. cylindrica* revealed that the plant contained anthocyanins, glycosides, flavonoids, triterpenoid, cardiac glycosides, saponins, carbohydrates, proteins, alkaloids, and tannins (Al-Snafi, 2019a).

Used in: Ayurveda and Siddha

Parts used: Leaf, flower, fruit, seeds, seed oil, and stem juice

Uses: Plant: used against pharyngitis, rhinitis, mastitis, edema, swellings, and burns. Leaves: used for chronic bronchitis. Leaf juice is given for amenorrhea. Flowers: used for treating migraine. Effective in controlling obesity, as well as the side effects of steroids. It also finds application in topical medication for skin disorders and hemorrhoids, as well as used as antitussive, antiallergic, and anti-inflammatory activity. An aqueous extract of seeds showed strong fibrinolytic activity. It also showed anticancer activity in transplanted tumors. Leaf, flower, fruit: 10–20-ml juice (Khare, 2007). The leaves of the plant are usually prepared in decoction and consumed to treat malaria. Various parts of the plant are also used in other African folk medicine traditions to treat ailments ranging from fever, jaundice, tumors, leprosy, and wounds to bleeding from bowels or bladders and alleviate pain and inflammation (Partap et al., 2012; Azeez et al., 2013). The plant is reported to have anti-tubercular and antiseptic properties. Seeds of the old fruits are used in traditional medicine as anthelmintic, stomachic, and antipyretic. Its seeds are used in the treatment of

DOI: 10.1201/9781003263500-162

asthma, sinusitis, and fever. Its seed oil is reported to possess antioxidant, antimicrobial, anti-inflammatory, and bronchodilator properties. Fruit is used in the treatment of hyperglycemia, ascites, hepatotoxicity, biliary, and intestinal colitis and in enlarged spleen. The juice of stem is used in the treatment of several respiratory disorders (Sharma et al., 2014).

Methods of cultivation: Seeds

Luffa echinata Roxb.

Synonyms: *Luffa echinocarpa* Fenzl. ex Hook. Fil.; *Luffa erinocarpa* Fenzel ex Steud.; *Luffa longystyla* Edgew.; *Luffa bondel* Buch.-Ham. ex Steud.; *Luffa bindaal* Roxb.

Family: Cucurbitaceae

Common names: Bengali: Deyatada; **English:** Bitter sponge gourd, Bristly luffa; **Hindi:** Bidali, Dutar toru; **Kannada:** Daevadaali, Daevadangari; **Marathi:** Devadali, Devadangari, Kukadvel; **Sanskrit:** Akhuvishaka, Devadaali, Devataadaka; **Tamil:** Kocaratacitta, Kukaitali, Panibira; **Telugu:** Paanibeera

Distribution: Throughout India

Description: It is a slender climber, with bifid bristly or smooth tendrils. Stem is sulcate, slightly hairy. Leaves: kidney-shaped, round 5-lobed, tip rounded. Leaves: simple alternate, bristly on both surfaces; leaf stalk is stout up to 12 cm long. Flowers are white, about 2.5 cm across, male flowers born on a 15-cm raceme; 5–12 flowers per raceme, sepal tube about 5–6 mm long, hairy sepals lance-shaped; petals ovate 1–1.2 cm long, blunt, hairy at the base. Stamens: 3 with filaments united, 3–5 mm long. Ovary is ovoid; fruit is ashy oblong ovoid, 2–5 cm long, densely covered with 4–7 mm long bristles. Seeds: minute, ovoid, black.

Flowering and fruiting: August–September

Chemical constituents: The fruit contains chrysoeriol and its glycosides as principal flavonoids. Seeds contain cucurbitacin B, triterpene alcohols, and a saponin with oleanolic acid as sapogenin (Khare, 2007).

Used in: Ayurveda, Folk, and Siddha

Parts used: Fruit

Uses: Fruit: purgative (intensely bitter and fibrous). An infusion is given in biliary and intestinal colic; also in nephritis and chronic bronchitis. The aqueous extract of the fruit is beneficial in jaundice. Fruit extract of the plant exhibited Hypoglycemic activity. Dosage: fruit—1–3 g powder (Khare, 2007). It is also used in Ayurveda to treat hepato-splenic affections, skin diseases, inflammation, sinusitis, intestinal colic, hemorrhoids, edema, cough, and urinary disorders. The plant also possesses antioxidant, analgesic, anti-inflammatory, antidepressant, anxiolytic, antiepileptic, hepatoprotective, antibacterial, antifungal, antiulcer, and anticancer activity (Modi and Kumar, 2014).

Methods of cultivation: Seeds

DOI: 10.1201/9781003263500-163

Lycopus europaeus Linn.

Synonyms: *Lycopus albus*; *Lycopus aquaticus*

Family: Lamiaceae (Labiatae)

Common names: English: Gipsywort, Bugleweed; **Hindi:** Jalneem

Distribution: Western Himalayas in Jammu and Kashmir and Himachal Pradesh

Description: Perennial herb, spreading by branching stolons. Stems: 40–120 cm high, erect, quadrangular, pubescent. Leaves: opposite, short-petioled, 3–12 cm long, 1–5 cm broad, lanceolate to ovate or elliptical, pubescent particularly along veins and glandular-punctate; base cuneate; apex acute to acuminate; margins sinuate or coarsely serrate with deepest lobes near base, or base pinnatifid in lower leaves. Inflorescences: axillary, verticillate, bracteate; bracts lanceolate, 3–5 mm long, with narrow pointed apex. Calyx: campanulate, 3.5–4.5 mm long, 13-nerved, pubescent; teeth 5, equal, longer than tube, awned. Corolla: white with small purple spots, about 4 mm long, nearly radially symmetrical, tubular below, gradually widening above, 4-lobed; upper lobe slightly wider and crenate on outer margin. Stamens: 2, protruding from corolla; anthers 2-loculed; staminodes 2, minute. Style gynobasic. Fruit: 4 nutlets, 1.6–1.9 mm long, 1.0–1.3 mm broad, with a corky crest (Al-Snafi, 2019b).

Flowering: June–September; **Fruiting:** October

Chemical constituents: *L. europaeus* contained alkaloids, coumarin, tannic acid, phenolic compounds, flavonoids, minerals, essential and volatile oils and many other chemical constituents (Al-Snafi, 2019b). The leaves contain lithospermic acid. The plant contains luteolin-7-glucoside, ursolic acid, caffeic acid, chlorogenic acid, sinapic acid, ellagic acid, and other derivatives of phenolic acid. The antioxidant activity of the plant is partially attributed to rosmarinic acid. Antigonadotropic activity of the leaf extract is attributed to phenolic precursors (Khare, 2007).

Uses: Cardioactive, diuretic, peripheral vasoconstrictor, sedative, narcotic, antihemorrhagic, antitussive, and thyrostat. In mild thyroid hyperfunction, enlargement of thyroid (Khare, 2007). *L. europaeus* is used traditionally to cure coughs, respiratory disorders, and as a natural remedy for sleeplessness. It is also used as a natural treatment for hyperthyroidism and for some symptoms of Grave's disease, such as palpitations. It is also used in vascular excitement; hemorrhage, in small amounts, resulting from determination of blood to the lungs, kidneys, or gastrointestinal organs; albuminuria, with frequent pulse; cough, with copious expectoration of mucus or mucopus, especially debilitating chronic cough; and wakefulness and morbid vigilance, with inordinately active circulation (Al-Snafi, 2019b).

Methods of cultivation: Seeds

 DOI: 10.1201/9781003263500-164

M

Machilus macrantha Nees

Synonyms: *Persea macrantha* (Nees) Kosterm

Family: Lauraceae

Common names: English: Large flowered bay tree, Indian persea; **Hindi:** Kurma; **Kannada:** Gulmavu; **Malayalam:** Kurilamavu, kurma Malamavu; **Marathi:** Gulumb; **Tamil:** Annai kuru. Kolar mavu; **Telugu:** Nara

Distribution: It is found in Sri Lanka Deccan and Peninsular region of India in the Western Ghats, and South, Central and South Maharashtra Sahyadri, up to an altitude of 2100 m

Description: Evergreen trees, to 30 m high, bark 20–25 mm thick, surface pale brown, mottled with dark blotches, scurfy and thinly scaly, rough, exfoliations small, brittle; blaze pinkish; branchlets glabrous. Leaves simple, alternate, clustered at the tip of branchlets, estipulate; petiole 15–40 mm long, stout, grooved above, glabrous; lamina 6.5–20 long 3.7–10 cm broad, oblong, elliptic-oblong, base oblique or acute, apex obtuse or obtusely acute, margin entire, glabrous, glaucous beneath, coriaceous; lateral nerves 6–12 pairs, pinnate, prominent, intercostal reticulate. Flowers: bisexual, 10–12 mm across, pale yellow, in panicles from upper axils and terminal; perianth tube very short, tepals 6, subequal, in 2 series, 4–5 mm, obovate, puberulous; persistent, spreading or reflexed in fruit; stamens 9 perfect, those of first and second row opposite the perianth lobes, introrse, with long filaments, those of third row opposite the first row, extrorse with slender filaments and a pair of stipitate glands at their base, filaments pubescent 2.5 mm; anthers 4-celled; staminodes 3, in row 4 and opposite the row 2, 4 mm long, stalked, arrow-shaped, pubescent; ovary half inferior, sessile, 1-celled; ovule 1, pendulous; style, slender, 2 mm; stigma discoid. Fruit: a berry 15–18 mm across, globose, green with white specks, aromatic, with a basal persistent rim of perianth; epicarp red when ripe; seed 1, globose.

Flowering and fruiting: December–April

Chemical constituents: Phytochemical studies have shown the presence of different phytochemical groups like alkaloids, carbohydrates, proteins and amino acids, sterols, phenols, tannins, and mucilage. The major alkaloid reported is machiline from root; macranthine, norligans, and B-sitosterol in leaf and bark; and an uncrystallized base, machiline hydrochloride, and arabinoxylan from the bark (Niranjan et al., 2018)

Parts used: Leaves and bark

Uses: Traditionally, the bark of the plant is used in the treatment of asthma, rheumatism, tuberculosis, and consumption while the leaves are used externally on ulcers. newer pharmacological activities like anti-inflammatory, antioxidant, central nervous system depressant, hypotensive effects were also reported. The leaf as well as bark are used extensively by folklore practitioners in treating various ailments such as rheumatism, asthma, ulcer, bruise, mental upset, fractures, swellings, weakness, and debility (Niranjan et al., 2018). Commercially, this plant is widely used for manufacturing wood, furniture, and incense sticks, as well as for forestation to conserve nature.

Methods of cultivation: Seeds

Further reading: http://www.biotik.org/india/species/p/persmacr/persmacr_en.html

DOI: 10.1201/9781003263500-166

Madhuca indica J. F. Gmel.

Synonyms: *Madhuca longifolia* (Koen.) Macb. var. latifolia (Roxb.) Cheval.; *Bassia latifolia* Roxb

Family: Sapotaceae

Common names: Bengali: Mahwa, Maul, Mahwla; **English:** Mahua tree, Butter nut tree; **Hindi:** Mahuaa; **Kannada:** Tuppe; **Marathi:** Mahwa, Mohwra; **Sanskrit:** Madhuuka, Gundapushpa; **Tamil:** Ieluppai; **Telugu:** Ippa

Distribution: A large evergreen tree found in South India, and evergreen forests of the Western Ghats from Konkan southward. The tree is planted in most parts of India.

Description: *M. indica* is an evergreen or semi-evergreen tree. It attains height up to 70 ft. The tree matures, starts bearing fruit at 8–15 years and fruits up to 60 years. Leaves are 10–30 cm long. Leaves are thick and leathery; most of leaves pointed at the tip. Flowers are small and fleshy. Flowers are dull or pale white in color. Calyx: coriaceous, densely clothed rusty tomentum. Corolla: yellowish-white, tube, fleshy. Stamens: 20–30, usually 24 or 26, anthers hispid at the back with stiff hairs. Fruits: 2–6 cm long, fleshy and greenish (Chaudhary et al., 2015).

Flowering and fruiting: March–May

Chemical constituents: The fruit pulp yielded a number of triterpenoids (including alpha- and beta-amyrin acetate), as well as n-hexacosanol, beta-D-glucoside of beta-sitosterol, and free sitosterol. The nutshell gave beta-sitosterol glucoside, quercetin, and dihydroquercetin. The corollas are rich source of sugars, vitamins, phosphorus, calcium, and iron; magnesium and copper are also present. The sugars identified are sucrose, maltose, glucose, fructose, arabinose, and rhamnose. The seeds yielded saponins—2,3-di-O-glucopyranoside of bassic acid (saponin A and saponin B). A mixture of saponins from seeds exhibits spermicidal activity (Awasthi and Mitra, 1968; Hariharan et al., 1972; Khare, 2007; Chaudhary et al., 2015; Badukale et al., 2021).

Used in: Ayurveda, Folk, Siddha, and Unani

Parts used: Flowers, seed oil, and bark

Uses: Flowers: stimulant, demulcent, laxative, anthelmintic, bechic. Seed oil: galactogenic, anticephalgic, emetic. The flowers are further used as a cooling agent, aphrodisiac, astringent, demulcent, tonsillitis, helminths, pharyngitis, bronchitis, impotency, inflammation, and eczema. Flowers are used in the cure of eye diseases. The flower juice is beneficial for the treatment of skin disease. Used for pneumonia, skin diseases, piles. Bark: astringent, emollient. Used for tonsilitis, gum troubles, diabetes, ulcers. Bark, seed oil, and gum: antirheumatic. In Ayurveda, the flower without stalk or calyx in asthma and phthisis. The bark is a good remedy for itching, swellings, fractures, and snakebites, as well as diabetes mellitus. Mahua oil is used for the treatment of skin diseases, rheumatism, and headache and as a laxative. Fruits are astringent and largely employed as a lotion for chronic ulcers, in acute and chronic tonsillitis, and in pharyngitis (Khare, 2007; Khare et al., 2018).

Methods of cultivation: Seeds

DOI: 10.1201/9781003263500-167

Majorana hortensis Moench

Synonyms: *Origanum majorana* L.; *Thymus majorana*
Family: Lamiaceae
Common name: English: Sweet marjoram; **Hindi:** Murwa; **Kannada:** Maruga; **Marathi:** Marwa; **Sanskrit:** Ajanmasurabhi; **Tamil:** Marikozhundhu, Kuvalamayam; **Telugu:** Maruvamu
Distribution: This herb is native to Mediterranean region and cultivated in many countries of Asia, North Africa, and Europe.
Description: Marjoram is a perennial evergreen herb growing to a height of about 40 cm. It has a square, red-brown stem. Leaves are smooth, simple, petiolated, ovate to oblong-ovate, 0.5–1.5 cm long, 0.2–0.8 cm wide, with obtuse apex, entire margin, symmetrical but tapering base, and reticulate venation. The texture of the leaf is extremely smooth due to the presence of numerous hairs. During summer, the plant produces tiny white to pink flowers.
Flowering and fruiting: June–September
Parts used: Whole plant
Used in: Folk and Traditional
Chemical constituents: Caffeic and rosmarinic acid, carvacrol flavonoids, linalool, terpenes, triterpenoids, sabinene, sabinene hydrate. Leaf extract contains vanillic acid, gallic acid, ferulic acid, caffeic acid, syringic acid, *p*- and *m*-Hydroxybenzoic acid, coumaric acid, neochlorogenic acid, protocatechuic acid, chlorogenic acid, cryptochlorogenic acid, and caftaric acid are phenolic acids (Bina and Rahimi, 2017).
Uses: Sweet marjoram has traditionally been used to relieve muscle spasms, insomnia, nausea, and headaches. In aromatherapy, marjoram is used to purportedly help calm individuals who have feelings of emotional instability or who are prone to hysteria or irritability. It is also used in relieving symptoms of asthma, eczema, and skin irritation, Alzheimer's disease, anorexia, antidote to poisons, antifungal, antimicrobial, antioxidant, antiseptic, antiviral, aromatherapy, arthritis, bleeding (urogenital), blood circulation, blood cleanser, cancer, cancer preventative, clear sinuses, colds, colic, decongestant, depression, diabetes, digestion, diuretic, dizziness, emotional distress, emotional instability, flavoring, flu, fumigant, gallstones, gas, gastrointestinal disorders, headache, heart disease, high blood pressure, inflammation of ear, insomnia, kidney disease prevention, laryngitis, laxative, liver protection, lower back pain, menstrual disorders, menstrual flow stimulant, migraines, muscle relaxant, muscle spasms, nasal inflammation, nausea, neurasthenia, painful menstruation, paralysis, rheumatism, rhinitis, sedative, sneezing, stiff joints, thrombosis, and vaginal discharge (Bina and Rahimi, 2017).
Methods of propagation: The plant is usually propagated by stem cuttings; survival efficiency is not very significant because of poor rooting.

DOI: 10.1201/9781003263500-168

Malva sylvestris Linn.

Synonyms: *Malva ambigua* Guss.; *Malva maruitiana* L.; *Malva erecta* C. Presl.; *Malva gymnocarpa* Pomel.

Family: Malvaceae

Common names: English: Cheese flower, Common mallow, Blue mallow, High mallow; **Hindi:** Gurchanti, Socholi; **Kannada:** Sanna bindige gida, Seeme bende; **Sanskrit:** Suvarchalaa

Distribution: Temperate Himalayas from Punjab to Kumaon, up to 2400 m; Maharashtra, Karnataka, and Tamil Nadu

Description: *M. sylvestris* Linn is an herbaceous perennial with an erect or decumbent branched stem up to 1 m (sometimes 1.5 m) long. The minutely ridged stems are covered with fine soft hairs, sometimes with a slightly bulbous base. The leaves are alternate, with a petiole up to 20 cm long, simple but palmate, up to 7 cm long by 10 cm wide, with stellate hairs (i.e., several strands radiating from a common center) and prominent veins on the underside The flowers are reddish-purple with dark stripes and occur in axillary clusters of 2–4 along the main stem with the flowers at the base opening first. There are 5 petals, each being up to 20 mm long, and 5 sepals, 3–6 mm long, which are fused in the lower half and have broadly triangular lobes. There is also an epicalyx with oblong segments, two-thirds as long as calyx lobes (2–3 mm long × 1.5 mm wide).

Flowering and fruiting: August–March

Chemical constituents: The herb contains sulfated flavanol glycosides, mucilage, and tannins. Flowers contain malvin (an anthocyanin), malvidin diglucoside, tannins, carotene, and ascorbic acid (Khare, 2007). Some reports revealed the presence of malvone A, a naphthoquinone and different known monoterpenes, aromatic compounds, and a tetrahydroxylated acyclic diterpene (Veshkorova et al., 2010).

Used in: Ayurveda and Unani

Parts used: Shoots, flowers, and root

Uses: Mucilaginous, emollient, laxative, antitussive, pectoral, antibacterial. Infusion is used for coughs and colds and irritation of the bronchi. Phagocyte stimulant. Key application: In irritation of the mucosa and throat and dry, irritative cough. The plant flowers are used as a remedy for cut wounds, eczema, dermal infected wounds, bronchitis, digestive problems, and inflammations. Traditionally, the medicinal plant has been used to treat several infections and diseases, such as cold, burn, cough, tonsillitis, bronchitis, digestive problems, eczema, and cut wounds. Roots and shoots are used for toothache, genital tract infections, and dermatitis. Shoots are employed to treat toothache, genital tract infections, hemorrhoids, and constipation. Leafy flowered stems are employed for cold, cough, throat pain, tonsils, and bladder problems. Seeds/mericarps are used to treat inflamed or injured skin and as a demulcent, emollient, and diuretic. Young leaves are used for burns, skin injuries, diarrhea, pectoral, and rheumatism. Flowers are used for acne, skin problems, eyes, throat pain, and cough. In tisanes, the flowers and leaves are used to help heal internal wounds, lesions of the mucous membranes and stomach ulcers. The flowers and immature fruits are efficacious to cure whooping cough (Khare, 2007; Pirbalouti et al., 2009, 2010).

DOI: 10.1201/9781003263500-169

Memecylon umbellatum Burm.f.

Synonym: *Memecylon edule* Roxb. var. rubro-caeulea (Thw) Clarke; *Memecylon umbellatum*; *Memecylon edule*

Family: *Melastomataceae*

Common names: English: Ironwood tree; **Hindi:** Anjan; **Kannada:** Arachare; **Malayalam:** Kannaav; **Marathi:** Anjan; **Sanskrit:** Anjan; **Tamil:** Pande-kaya; **Telugu:** Alli, Lakhonde

Distribution: Peninsular India and Sri Lanka; in the Western Ghats, throughout, evergreen to semi-evergreen forests, up to 1200 m

Description: Shrub to small tree up to 8 m tall. Bark pale brown, fissured, thinly flaky, when old blaze yellowish brown. Leaves simple, opposite, decussate; petiole approximately 1 cm long, canaliculate, glabrous; lamina 2.5–5 × 1.5–3 cm, elliptic to elliptic-ovate, apex acuminate, sometimes acute, base acute to subattenuate, margin entire, coriaceous, glabrous, drying brown; midrib canaliculate; secondary nerves obscure; tertiary nerves obscure. Inflorescence clusters of distinctly peduncled dense cymose umbels, axillary or on lateral tubercles 1.5–3 cm across, sessile or very short peduncled, many-flowered; flowers subsessile. Calyx: 2–3 mm wide; flowers deep blue. Fruit: a berry, globose, yellow, 0.5 cm across, mature fruit bluish black. Seeds: 1.

Flowering and fruiting: February–March

Chemical constituents: The leaves contain a yellow glycosidal substance umbelatone (4-hydroxy methyl-3-methly- but-2-ene-4,1-oxide) β-amyrin, sitosterol, glucoside, oleanolic and ursolic acid, tartaric acid and malic acid (1.38%), a resin (6%), and calcium oxalate (1.44%; Asolkar et al., 1956; Anonymous, 1998).

Used in: Traditional

Parts used: Bark, leaves, and fruits

Uses: Leaves are used as astringent; their lotions used for eye defects, gonorrhea, leucorrhea treatment of bone fracture, herpes, wounds, and snakebite and yield a yellow dye. Fruits are edible and astringent. The bark can be applied to leg for bruises and the root in a decotion is useful in excessive menstrual discharge. The leaves are said to have antidiarrheal properties.

Methods of cultivation: Vegetative, with softwood cuttings

DOI: 10.1201/9781003263500-170

Mimusops elengi (L.)

Synonyms: *Mimusops parviflora* R. Br.; *Mimusops lucida* Poir.; *Mimusops javensis* Burk.; *Kaukenia javensis.* (Burck) Kuntez.; *Kaukenia elengi* (L.) Kuntze

Family: Sapotaceae

Common names: Bengali: Bakul, Bohl; **English:** Spanish cherry, West Indian medlar, Bullet wood; **Gujarati:** Babhuli, Bolsari, Varsoli, Vovoli; **Hindi:** Bakul, Bolsari, Maulsarau; **Kannada:** Pagademara; **Malayalam:** Bakulam, Elengi, Ilanni; **Marathi:** Bakhor, Bakula, Barsoli, Ovalli; **Tamil:** Alagu, Ilangi, Kesaram, Magilam; **Telugu:** Pogada

Distribution: Native of India. It is cultivated in North and Peninsular India and Andaman Islands.

Description: Large evergreen tree with dark gray fissured bark and densely spreading crown. Leaves: simple, alternate, oblong, glabrous and leathery with wavy margins, petiolate, petiole 2–2.5 cm, lamina 10–15 cm long 4–8 cm broad, apex shortly acuminate, base round. Flowers: white, fragrant, pedicle deflexed. Calyx: hairy, segments 8, the outer 4, ovate lanceolate, the 4 inner narrower than outer. Corolla: longer than calyx lobes 24, in 2-series, the inner series of 8, the outer of 16, linear, oblong, acute. Stamens: 8, filaments short, anthers slightly twisted, staminodes 8, alternate with the stamens, lanceolate, hairs white. Ovary: hairy, 6–8 celled, style grooved, 5 mm; stigma minutely fimbricate. Fruit: a berry, ovoid, yellow, 2.5–3 cm long, and 1.5–2 cm broad, fleshy, epicarp thin. Seed: single, oblong-ellipsoid, laterally compressed.

Flowering and fruiting: January–April

Chemical constituents: Leaves: contain sterols, reducing sugars, and tannins, Roots: a steroidal saponin. Stem bark: spinasterol and taraxerol, flowers, D-mannitol, βsitosterol and β-sitosterol-D-glycoside; Seeds contain pentacyclic triterpene acids, mimusopic and mimusopsic acids.

Used in: Ayurveda

Parts used: Bark, flower, fruit, and seeds

Uses: For medicinal purposes, its leaves, fruit, bark, and flowers are used. The flowers bark and fruits are astringent in action. Bark is used as a cardio-tonic, astringent, antiulcer, and anticonvulsant and to reduce fever. The leaves are used as analgesic, antidiabetic, antihypertensive, and antidote to snakebite. The pulp of the fruit is used in curing chronic dysentery. A dried flower powder is used as a brain tonic. The seeds of tree are purgative. The aerial parts are diuretic in action. For improving fertility in women fruit pulp is eaten. Ripe fruit is given to promote delivery.

Methods of propagation: Seeds and cutting

DOI: 10.1201/9781003263500-171

Mitragyna parvifolia (Roxb) Korth.

Synonyms: *Nauclea parvifolia* Roxb.; *Stephegyne parvifolia* (Roxb) Korth.
Family: Rubiaceae
Common names: Bengali: Dharakadam, Gulikadam; **English:** Leichhard tree, Cheesewoood; Water kadamba; **Gujrati:** Kalam; **Hindi:** Guri, Kaim; **Kannada:** Nira kadamba, Sanna kadamba; **Malayam:** Kathamamarm Nirkkatampu, Puucckkatamp; **Marathi:** Kalam; **Sanskrit:** Vitanah, Giri kadamba, Kadamba; **Tamil:** Niculam, Nirkkatampu, Chinna kadambu; **Telgu:** Niru kadamba, Ajaghnamu, Ambusaramu, Nirakadambu
Distribution: Throughout the temperate part of the world. It is found throughout the greater parts of India up to an altitude of 1200 m. Tree is found scattered in deciduous forests. In India, mostly found in Thar Desert in Rajasthan.
Description: *M. parvifolia* (Roxb) Korth. is a medium to large deciduous tree with rounded crown mainly arboreal in nature up to 25 m tall. Leaves: simple, opposite, decussate; stipules 1 cm long, interpetiolar, ovate-oblong or obovate, foliaceous, membranous, caducous; petiole 10–40 mm long, stout, grooved above, glabrous; lamina 5–16 cm × 2.5–10 cm, variable, ovate, orbicular, elliptic or ovate-oblong, base obtuse, attenuate, acute or subcordate, apex obtuse or round; margin entire, glabrous, coriaceous; lateral nerves 6–10 pairs, pinnate, prominent beneath; intercostal scalariform, slender; domatia present. Flowers: bisexual, creamy white, 10–12 mm long, in terminal heads; peduncle supported by a pair of bract-like oblong leaves; bracteoles small, subulate; calyx tube short, truncate, rim even; corolla tube funnel-shaped, 8 mm long, villous inside, lobes 4–5 oblong, reflexed; stamens 5, attached toward the apex of corolla tube; anthers apiculate; ovary 2-celled, inferior, ovules many; style filiform; stigma mitriform, hollow at base. Fruit capsules in globose heads, 2–3 mm long, wavy, separating into 2 cocci, brown; seeds many, small and 10-ribbed. Bark: 20–25 mm thick, gray black, smooth exfoliations thin, unequal and fibrous blaze pink, traversed by whitish rays. Branchlets: young branchlets angular to subterete, glabrous (Choudhary and Jain, 2016).
Flowering: May–June; **Fruiting:** August–October
Chemical constituents: Mitraphylline, isomitraphylline, pteropodine, isopteropodine, speciophylline, and uncarine F are the major oxindole alkaloids isolated from *M. parvifolia* leaves (Choudhary and Jain, 2016).
Used in: Ayurveda
Parts used: Leaves, stem, and bark
Uses: *M. parvifolia* (Roxb.) Korth. is recommended in Ayurvedic texts for prevention and treatment of anti-arthritic, antipyretic, anticonvulsant, anthelmintic, antimicrobial, anti-inflammatory, antinociceptive, antiproliferative, and antioxidant activity. The bark is used for muscular pain; the bark and root, antispasmodic and febrifuge (Choudhary and Jain, 2016). Fresh leaves sap is used by tribes for treating jaundice in the Chenchus, Yerukulas, Yenadis, and Sugalis of Guntur district, Andhra Pradesh. Its leaves alleviate pain and swelling and are used for better healing of wounds and ulcers. The stem bark is used in the treatment of biliousness and muscular pains by the local inhabitant of Tumkur district.

DOI: 10.1201/9781003263500-172

The bark and roots used to treat fever, colic, muscular pain, burning sensation, poisoning, gynecological disorders, cough, and edema and as an aphrodisiac. The fruit juice augments the quantities of breast milk in lactating mothers and works as a lactodepurant (Rao and Pullaiah, 2001; Choudhary and Jain, 2016).

Methods of cultivation: Seeds

Morinda citrifolia L.

Family: Rubiaceae

Common names: English: Indian mulberry, Noni; **Hindi:** Noni; **Kannada:** Noni; **Marathi:** Noni

Distribution: Native to Southeast Asia and Australia. Throughout India on seashores

Description: A small tree or large ever green shrub about 8–10 m in height and 15–20 cm in stem diameter. Twigs are 4-angled. Leaves: simple opposite glossy membranous, glabrous, elliptic to elliptic-ovate, 20–40 cm long and 7–25 cm broad, petiole stout 1–2 cm long, stipules are present, conate or distinct. Flowers: tubular 80–90 born on ovoid or globose heads. Calyx: 5 truncated, corolla 5-lobed white in color, corolla tube greenish white, 7–9 mm long. Stamens: five in number, inserted on the mouth of the corolla, and alternate with the corolla lobes. Fruit: a syncarp, yellowish white, and fleshy 5–15 cm long and 3–5 cm in diameter, soft and fetid when ripe. Seeds: brown, 4–9 mm long, and have distinct air chamber.

Flowering and fruiting: Throughout the year

Chemical constituents: Fruits of noni contain americanin A, narcissoside, asperuloside, asperulosidic acid, borreriagenin, citrifolinin B epimer a, citrifolinin B epimer b, cytidine, deacetylasperuloside, dehydromethoxygaertneroside, epi-dihydrocornin, d-glucose, d-mannitol, methyl alpha-d-fructofuranoside, methyl beta-d-fructofuranoside, nicotifloroside, and beta-sitosterol 3-O-beta-d-glucopyranoside (Su et al., 2005). It also contain vitamins and minerals: magnesium, iron, potassium, selenium, zinc, copper, sulfur, and ascorbic acid (vitamin C).

Used in: Traditional

Parts used: Stem, leaves, bark, roots and fruits

Uses: *M. citrifolia* has wide range of therapeutic uses in ailments such as arthritis, burns, headache, wounds and skin infections. Various parts of the noni plant, such as stem, bark, root, leaves, and fruits, are used traditionally as medicines for the treatment of ailments like cough, cold, pain, liver diseases, hypertension, blood pressure, tuberculosis, malaria, intestinal worms, diabetes, loss of appetite, hernias, urinary tract infection, menstrual disorder, venereal diseases, cancer, cardiovascular diseases, arthritis, and the like. The seeds and their oil are applied topically on the scalp as insecticide and are used for the treatment of arthritis. The flowers are used to treat insect stings. It is reported for topical use in conditions such as swellings, sprains, bruises, and wounds. The literature prove that noni is pharmacologically active and is used in different forms of cancer, namely, colon, esophageal, breast, and colorectal cancers and cardiovascular diseases, diabetes, arthritis, and hypertension (Morton, 1992; Dixon et al., 1999; Blanco et al., 2006; Ali et al., 2016).

Methods of propagation: Seeds, stem, and root cuttings; air layering

DOI: 10.1201/9781003263500-173

Photo: Dr. Srinath Rao

O

Onosma bracteatum Wall.

Synonyms: *Borago officinalis*
Family: Boraginaceae
Common names: Bengali: Dadishaka, Gaozaban; **English:** Cow's tounge, Sedge; **Gujarati:** Bhopathsri; **Hindi:** Goaza, GanjabaGoziya; **Kannada:** Yettina naligae; **Marathi:** Pathari; **Sanskrit:** Kharapatra, Matsyagandhi, Darapatri, Godhumika, Gojika; **Telugu:** Yedunalaka chattu
Distribution: Found in the Mediterranean regions, Europe and Asia. In India, found abundantly in northwestern Himalayas from Kashmir up to 3500–4500 m
Flowering and fruiting: November–January
Description: An herb growing to a height of 30–60 cm. The plant is covered with small hairs. The leaves are covered with small hairlike structures with small nodules, giving a rough appearance like the tongue of a cow, hence the name Gojihwa. Flowers: blue or purple, trumpet-shaped, are symmetrical and grow in loose, forked cymes at interval on the stem. Calyx: consist of 5 separate, green linear lanceolate sepals. Corolla: star-shaped and composed of 5 blue petals, which are ovate lanceolate in shape, 10–15 cm in length and 5–8 cm wide with an obtuse apex and crenate margin. Nutlet: ovoid, roughly 4 mm, and acute.
Chemical constituents: The plant contains naphthoquinone, flavonoids, hydrocarbon, phenolic, ester), alkaloids, aromatics, carboxylic acid, fatty acids, terpenoids, while the most important ones are rosmarinic, ferulic, protocatechuic, chlorogenic, caffeic, and p-Coumaric acids and apigenin (Jabbar et al., 2022). The leaves contain lycopsamine and supindine viridiflorate as the predominant unsaturated pyrrolizidine alkaloids.
Used in: Ayurveda, Traditional, and Unani
Parts used: Leaf, flower, and root
Uses: It is used as tonic, alterative, demulcent, and diuretic and is considered cooling. It is useful as a spasmolytic. Sedge is beneficial in the treatment of kidney stones. As an expectorant, it alleviates the symptoms of the common cough and bronchitis. Ash prepared from the leaf is used to treat stomatitis and wounds. A decoction prepared from the leaves is used to treat jaundice and constipation. The cold infusion of the leaves is used as tonic in rheumatism, syphilis, leprosy, a hypochondriac's kidney disease, used as diuretic, demulcent and alternative. Cold infusion prepared from the leaves is used to treat psychiatric conditions and epilepsy. A cold infusion prepared from the flowers and fruit is given with *Glycyrrhiza glabra* in a dose of 50–60 ml to treat rhinitis and cough. The decoction of the plant relieves palpitations of the heart and irritation of the stomach and bladder. The root powder is used in cutaneous eruption as an external application. The plant is also useful in treating asthma and chronic respiratory disorders, blood disorders such as abscess, skin disorders, and bleeding disorders such as menorrhagia and nasal bleeding (Kirtikar and Basu, 1935).
Methods of propagation: Seeds

DOI: 10.1201/9781003263500-175

Operculina turpethum (Linn.) Silva Manso

Synonyms: *Ipomoea turpethum* R. Br.; *Merremia turpethum* (L.) Shah & Bhat; *Convolvulus turpethum* (L.)

Family: Convolvulaceae.

Common names: Bengali: Teudi, Tvuri, Dhdhakalami; **English:** Indian jalap; **Gujarati:** Kala Nasottara; **Hindi:** Nishothra, Pitohri, Trivrut; **Kannada:** Vili Tigade; **Malayalam:** Trikolpokanna; **Marathi:** Nisottar; **Sanskrit:** Ardhachandra, Sarvanubhuti, Trivrittika **Tamil:** Sivadai, Adimbu, Kumbam, Kumbanjan; **Telugu:** Tella, Tegada

Distribution: Throughout India up to 1000 m in damp shady places.

Description: Jalap is a stout perennial twining climber with long twisting pubescent stems. It exudes a milky juice on cutting. Leaves: simple, alternate without stipules, 7–11 cm long, 6.5–10 cm broad, broadly ovate or oblong-cordate, subacute, slightly repand, shallowly cordate at base, margin entire or slightly and undulately lobed, more or less pubescent on both sides, especially when young, minutely reticulate. Petioles 2–8 cm long, slightly pubescent. The flowers are regular, bisexual, rather large, white, campanulate, 4 cm long and as much in diameter, white in simple cymes, 2–5 flowers together, sepals 5, unequal, imbricate, the two outer 2.5 cm long and as broad, ovate, membranous, mucronate, hairy on both sides, sprinkled over with black dots, becoming much enlarged in the fruit, inner sepals 1.7–2 cm long and 1.4–1.7 cm broad, hairy; petals 5, fused into a more or less rotate corolla, 4 cm in diameter, segments obscure, convolute and contorted in bud; stamens 5, epipetalous, filaments dilated at base and hairy; ovary superior, 2 mm long, 2-locular with 2 ovules in each chamber, style simple, 1.6 cm long, stigma globose: fruit capsule over I.2 cm long, completely enclosed in enlarged, brittle, membranous, fleshy sepals, shortly stalked, depressed, strongly 4-lobed, slightly hairy. Seeds are large and glabrous. The roots are usually dull gray, long, slender, fleshy, cylindrical, and elongated and bear thin rootlets. The bark is three-layered, hence the name Trivrit.

Flowering: May–October; **Fruiting:** July–November

Chemical constituents: The roots of plant contain the glycosidic resin (about 10%), turpethin. Turpethin is made up of 1-turpethein containing jalapic, ipomoic, and tampicolic acids and f-turpenthein. It gives laxative action to the drug. The active principle glycosidic resin is mostly concentrated in the root bark. Turpethinic acids (A, B, C, D, and E) in a large quantity, which were isolated from the resin of the plant and contain different types of bioactive compounds such as albumin, lignin salts, volatile oil, starch, ferric oxide, lupeol, α- and β-turpethein, α- and β-rhamnose, fructose, β-sitosterol, and scopoletin (Mondal et al., 2006; Kohli et al., 2010).

Used in: Ayurveda

Parts used: Leaves and roots

Uses: The powdered root is given as a purgative; it also gives relief in habitual constipation, colic, and piles. Paste of the root is applied topically to treat vitiligo, other skin disorders, alopecia, cervical lymphadenitis, hemorrhoids, fistulas, and ulcers. Along with guggulu, ginger, and turmeric, it is effective in high cholesterol and obesity. Along with ginger, it is particularly beneficial in anemia, inflammation, jaundice, and rheumatic and paralytic affections.

The fresh juice of leaves is dropped into the eyes for inducing lachrymation in ophthalmia (Mondal et al., 2006; Kohli et al., 2010).

Methods of propagation: Seeds and vegetative propagation through stem cuttings

DOI: 10.1201/9781003263500-176

Origanum vulgare L.

Synonyms: *Origanum majorana*; *Origanum dictamnus*; *Origanum creticum; Origanum officinalis; Origanum orientale*

Family: Lamiaceae (Labiateae)

Common names: English: Marjoram, Oregano, Pot marjoram; **Gujarati:** Ajamo, Jungali marvo; **Hindi:** Ban tulsi; **Kannada:** Marug; **Malayalam:** Kattumaruva; **Marathi:** Jungali maruaa; **Sanskrit:** Maruvaka; **Telugu:** Mridumaruvamu

Distribution: It is endemic to Mediterranean region but is cultivated in Mexico, Italy, Turkey, Dominican Republic, and Greece. In India, it is found in temperate Himalayas from Kashmir to Sikkim.

Description: Its small evergreen subshrub in mild climates typically reaches 30–60 cm. The young stems are typically square and hairy and become woody with age. Its compact oval leaves are arranged oppositely and are covered with glandular trichomes, which are usually entire but sometimes have slightly toothed margins. The leaves are dotted with small glands containing the volatile oil, The flowers are small, 2-lipped, and borne in clusters about 4–8 mm long; they range in color from white to pink or pale purple. Calyx is 5-toothed, stamens 4. Fruit has 4 nutlets.

Flowering and fruiting: June–October

Chemical constituents: Martins et al. (2014) reported the major components to be luteolin 7-*O*-glucoside (20.88%), rosmarinic acid (14.62%), luteolin *O*-glucuronide (12.48%), and apigenin-7-*O*-glucuronide (5.78%). The other major phytochemicals identified include carvacrol, linalyl acetate, thymol, and cis-sabinene hydrate. Other compounds identified in an ethanolic extract of oregano are 4-hydroxy-4-methyl-2-pentanone, rosmarinic acid, thymol, luteolin 7-*O*-glucoside, and caffeic acid (Rodrigues et al., 2004; Martins et al., 2014; Veenstra and Johnson, 2019).

Used in: Traditional

Parts used: Leaves and oil

Uses: Possesses antioxidant, antimicrobial, anti-inflammatory, and analgesic properties. In traditional medicine, oregano has been used for respiratory conditions, that is, asthma, bronchitis, and cough; gastrointestinal, that is, diarrhea; indigestions; and stomachache (Veenstra and Johnson, 2019).

Marjoram is commonly used for runny nose, the common cold, cough, and other infections; dry and irritating coughs; swollen nose and throat; and ear pain. Marjoram tea is also used for various digestion problems including poor appetite, liver disease, gallstones, intestinal gas, and stomach cramps. Marjoram tea is used for relieving symptoms of menopause, treating mood swings related to menstrual periods, starting menstruation, and promoting the flow of breast milk. Other uses include treating diabetes, sleep problems, muscle spasms, headaches, sprains, bruises, and back pain. It is also used as a "nerve tonic" and a "heart tonic" and to promote better blood circulation. Marjoram oil is used for coughs, gallbladder complaints, stomach cramps and digestive disorders, depression, dizziness, migraines, nervous headaches, nerve pain, paralysis, coughs, and runny nose.

Methods of propagation: Stem cuttings

DOI: 10.1201/9781003263500-177

Orthosiphon aristatus (Blume) Miq.

Synonyms: *Ocimum aristatum* Blume.; *Ocimum grandiflorum* Blume; *Orthosiphon stamineus* Benth.; *Orthosiphon spiralis* (Lour) Merr.; *Orthosiphon* spicatus (Thunb); *Clerodendranthus stamineus* (Benth) Kudo.; *Clerodendrum spicatum* Thunb.

Family: Lamiaceae (Labiatae)

Common names: English: Cat's whiskers, Java tea; **Hindi:** Billi ke Muchae; **Kannada:** Mutrapindadagida, Kidney tea plant; **Malayalam:** Poochameesa

Distribution: Found mainly throughout southern China, the Indian subcontinent, Southeast Asia, and tropical Queensland, Australia

Description: Diffusely branched perennial herbs. Leaves: 3–6.5 cm long and 1.2– 3.5 cm broad, rhomboid to ovate-deltoid, base cuneate, margins serrate from about the middle upwards, apex acute, glandular-punctate beneath. Racemes: 8–16 cm long; verticils lax; peduncle strigulose with purplish hairs. Bracts: about 1 × 2 mm, broadly ovate. Calyx: tube to 3 mm long, 10-nerved; upper lip to 3 mm long, broadly ovate, purplish; lower lip longer. Corolla: white, tomentose without; tube 1–1.5 cm long; upper lip orbicular; lower lip as long as upper, boat-shaped. Stamens: exserted, filaments to 3 cm long. Nutlets: about 2 mm long, ovoid-oblong.

Flowering and fruiting: September–December

Chemical constituents: The major constituents were sesquiterpenes, including β-elemene, β-caryophyllene, orthochromene A, acetovanillochromene, sinensetin, tetramethyl scutellarein, eupatorin, neo-Orthosiphonis A and B, and orthosiphones A and B (Shibuya et al., 1999).

Used in: Folk and Traditional

Parts used: Leaves

Uses: Used in traditional medicine to prevent different diseases such as diabetes, kidney stone, edema, rheumatism, hepatitis, hypertensive, and jaundice. Antidiabetic, anti-inflammatory, antioxidant, hepatoprotective, analgesic, and nephroprotective activities have been reported in the plant extract and phytoconstituents (Matsubara et al., 1999). It is also used for increasing excretion of urine, lowering uric acid, protecting kidney, reducing oxidative stress, reducing inflammation, protecting the liver, protecting the stomach, lowering blood pressure, ameliorating diabetes, ameliorating hyperlipidemia, fighting microorganisms, and fighting anorexia (https://en.wikipedia.org/wiki/Orthosiphon_aristatus). The leaves exhibit excellent pharmacological activities such as antioxidant, antibacterial, hepatoprotective, anti-inflammatory, cytotoxic, antihypertensive, and vasodilatation (Ashraf et al., 2018).

Methods of propagation: Seeds and stem cuttings

DOI: 10.1201/9781003263500-178

Photo: C.T. Chandralekha

P

Pepromia pellucida (L.) Kunth

Synonyms: *Piper pellucidum* L.; *Micropiper exiguum* (Bl.) Miq.; *Micropiper pellucidum* (L.) Miq.; *Pepromia concinna* (Haworth) A. Dietrich; *Pepromia ephemera* Ekman.

Family: Piperaceae

Common names: English: Shiny bush, Pepper elder; **Hindi:** Chamakdar jhadi; **Kannada:** Neeru kaddi gida; **Malayalam:** Mashitandu chedi, Kannadippacha, Makapacha; **Marathi:** Godhadi **Sanskrit:** Varshabhoo, Toyakandha

Distribution: South America, Mexico, West Indies, and India. The species grows well in humid and loose soils in places with little sunlight in rainy periods.

Description: A small erect herb, up to 6 in. in height, stem pale green. Leaves cordate, glabrous, pale green, almost transparent when dry. Leaves are broadly ovate (egg-shaped) or ovate-triangular (1–3.5 cm long, 1–3.5 cm wide). Branched, hairless stems are erect or rising upward. Spikes terminal, flowers and fruits are minute. Flowers bisexual, minute in a cord like spike 3–6 cm long, arising from leaf axis.

Flowering and fruiting: Almost throughout the year

Chemical constituents: The essential oil from leaves of *P. pellucida* revealed the presence of the phenylpropanoids dillapiole and apiol as major constituents, in addition to mono- and sesquiterpenes (Silva et al., 1999). Samples of roots from *P. pellucida* from Nigeria were shown to contain β-farnesene as a major compound in the essential oil (Usman and Ismaeel, 2020).

Used in: Ayurveda and Folk

Parts used: Whole plant

Uses: In treating diseases, including antihypertensive, anti-inflammatory, antipyretic, analgesic, antibacterial, anti-amoeba, antioxidant, gastroprotective, (Ahmad et al., 2019) the herb is used by Indonesians to treat headaches accompanied by fever, and the leaves extract is used to treat stomach pain, gout, and hypertension. In Bolivia, all parts of this plant are used to stop bleeding. The herbaceous root of *P. pellucida* is used to treat fever and wounds. In northeastern Brazil, this herb is used for blood pressure and lowering cholesterol levels. In Guyana and the Amazon, it is used as a cough medicine, emollient, and diuretic and can be used to treat proteinuria (Nwokocha et al., 2012). In the Philippines, the decoction of this plant is used to reduce uric acid levels and kidney disorders. In various areas of Bangladesh, the leaves of this plant are used by local people as a treatment for mental illness. In South America, the fresh juice of the stems and leaves of this plant can be used as a treatment for inflammation of the eyes (Majumder, 2011).

Methods of propagation: Seeds and stem cuttings

DOI: 10.1201/9781003263500-180

Phyllanthus acidus (L.) Skeels

Synonyms: *Averrhoa acida* L.; *Cicca acida* (L.) Merr.; *Cicca acidissima* Blanco.; *Cicca disticha* L.; *Cicca nodiflora* Lam.; *Phyllanthus acidissimus* (Blanco) Mull.Arg. *Phyllanthus cicca* Müll.Arg., *Phyllanthus distichus* Mull. Arg.

Family: Euphorbiaceae

Common names: Bengali: Noyal, Harphal; **English:** Indian gooseberry, Malaya gooseberry, Star gooseberry; **Hindi:** Harfarauri; **Kannada:** Harparlevdi, Kirunelli; **Malayalam:** Arinelli, Bimbool, Arinellika, Kiru Nerle mara; **Marathi:** Harpharrori; **Sanskrit:** Lavali; **Tamil:** Arnelli; **Telugu:** Rachyusarikae

Distribution: Native of Malaya and Madagascar. Widely cultivated in India

Description: A small deciduous tree or shrub, reaching 2–9 m in height. The leaves are ovate or lanceolate with short petioles and pointed ends about 2–7.5 cm long. They are green and smooth on the upper side and blue green on the underside. The flowers are small and pinkish and appear in clusters in 5–12.5-cm-long panicles at leafless parts of the main branches, at the upper part of the tree. The fruits are numerous, oblate, with 6–8 ribs, and densely clustered. They are pale yellow or white, waxy, crisp, juicy, and very sour. About 4–6 seeds are contained in a stone at the center of each fruit.

Flowering and fruiting: Almost throughout the year

Chemical constituents: The air-dried of dichloromethane extract of *P. acidus* resulted in the isolation of the mixture of phyllanthol, α-amyrin, β-amyrin, and lupeol from the bark and mixture of β-sitosterol and stigmasterol from the leaves of the plant (Alimboyoguen et al., 2022).

Used in: Folk and Traditional

Parts used: Fruit and bark

Uses: Used to treat a wide spectrum of diseases such as inflammatory, rheumatism, bronchitis, asthma, respiratory disorder, hepatic diseases, and diabetes in India. In Indonesia, the bark of *P. acidus* is mixed and heated with coconut oil and then spread on skin rashes to relief from itching. In the Philippines, a decoction of bark is used to treat bronchial catarrh (Tan et al., 2020). In folk medicine, it is used for the treatment of several diseases such as cough, fever, bronchitis, asthma, respiratory disorders, hypertension, diabetes, rheumatism, pain, psoriasis, and other skin disorders. The unripe fruit is also used as sour flavoring and for making vinegar, jams, and jellies (Janick and Pauli, 2008; Leeya et al., 2010; Alimboyoguen et al., 2022).

Methods of propagation: Seeds

DOI: 10.1201/9781003263500-181

Phyllanthus reticulatus Poir

Synonyms: *Anisonema reticulatum* (Poir.) A. Juss.; *Cicca reticulata* (Poir.) Kurz.; *Diasperus reticulatus* (Poir.) Kuntze.; *Kirganelia reticulata (*Poir) Baill

Family: Euphorbiacea

Common names: English: Black honey shrub; **Gujarati:** Datwan; **Hindi:** Kale madhu ka paed, Panjuli; **Kannada:** Ari huli, Pulaveri, Anamsule; **Malayalam:** Niruri, Nireli; **Marathi:** Pavana; **Sanskrit:** Pulika, Krishna-kamboji; **Tamil:** Abaranji, Karunelli, Kattukilanelli; **Telugu:** Nallapuli Nallapurugudu, Pulaguwa, Phulsar

Distribution: The plant grows throughout tropical areas of India, Bangladesh, China, and the Malay Islands.

Description: It is a large, often scandent, shrub, much branched, 15–30 ft tall monoecious. Leaves on the main stems are reduced to scale, bark pale reddish brown in color, longitudinally fissured, branches slender, drooping, leafy shoots up to 25 cm long. Leaves: simple, 1.3–3.2 cm long 1–2cm broad, oblong or elliptical, thin glabrous, petiole 1.5–3mm long, stipules 1.5 mm. Flowers: in bisexual cymules, each with 1 female flower and 2–3 male flowers, which were axillary; the males in fascicles of 2–6; the females solitary, pedicels 1.5–6 mm, long slender. Calyx: glabrous, reaching 2.5mm. Long; segments oblong, very obtuse, alternating with the glands of the disk. Male flower: Stamens 5, unequal, the 3 inner connate into a column the 2 outer free, 0.5–1.2mm long, anthers erect vertically and shorter. Female flowers: ovary 5–10—celled; ovules 2 in each cell, styles 3, very short, bilobed. Fruits: small, purple fleshy berry 4–6mm diameter, baccate, subglobose, blackish or purplish color of fruits, smooth and shining. Seeds: 8–10 (usually 10), irregularly 3-gonous seeds, 1.5–2 mm long, nearly smooth, finely granulate (Khare, 2007).

Flowering and fruiting: March–July

Chemical constituents: The plant contains alkaloids, flavonoids, glycosides, steroids, phenols, tannins, and saponins (Manimegalai and Umavathi, 2010). Eight compounds, β-sitosterol-3-O-β-glucoside, stigmasterol-3-O-β-glucoside, methyl gallate, ellagic acid, corilagin, methyl brevifolincarboxylate, kaempferol, and astragalin, including two flavonoid glycosides rutin (quercetin 3-rutinoside), and quercetin 3-O-β–D-glucopyranoside (isoquercitrin), were isolated from leaves (Hong et al., 2007; Kharat et al., 2013). Three compounds (lupeol, lupeol acetate, stigmasterol) were isolated and identified by phytochemical study conducted on the leaves of *P. reticulatus* (Jamal et al., 2008). Other isolated compounds are tannic acid, friedelin, epifriedelinol, betulin, taraxerone, beta-sitosterol, glochidonol, octacosanol, taraxeryl acetate, and 21alpha-hydroxyfriedelan-3-one, and betulinic acid. Coumarin was isolated from stem bark (Begum et al., 2007).

Used in: Folk and Traditional

Parts used: Leaves, bark of the stem, and fruit

Uses: The leaves and bark are used as astringent and diuretic. The juice of leaves is used for the treatment of diarrhea in children. The bark showed significant antiviral and antiplasmodial activity. The plant possess antioxidant, analgesic and anti-inflammatory, antiarthritic, antinociceptive, hepatoprotective, hypoglycemic, and hypolipidemic activities (Kharat et al., 2013). The fruit of the plant is used in inflammation. *P. reticulatus* has been proved to show antidiabetic, antiviral, anticancer, anti-plasmodial, hepatoprotective, antibacterial, and anti-inflammatory activities. Root infusion is used for gonorrhea, and a root decoction is used as a purgative and for hookworm (Sharma and Kumar, 2013). The leaves

DOI: 10.1201/9781003263500-182

are used as diabetic and cooling medicine. It is reported to be used as diuretic, alternative, attenuant, astringent, and antidiarrheal (Kirtikar and Basu, 1991; Anonymus, 2005; Nadkarni and Nadkarni, 2009).

Methods of cultivation: Seeds

Photo: Dr. Srinath Rao

Pimenta dioica (L) Merr.

Synonyms: *Myrtus aromatica*; *Myrtus pimenta* L.; *Myrtus dioica* L.; *Pimenta officinalis* Lindl.

Family: Myrtaceae

Common names: Bengali: Kabab chini; **English:** Allspice, Jamaican pepper, New spice; **Hindi:** Kabab chini; **Kannada:** Mara menasu, Ganda menasu, Malayala munchi

Distribution: *P. dioica* is native to the Caribbean region, especially Jamaica and Cuba. Now cultivated widely in India and other countries

Description: Allspice is a small evergreen tree up to 15 m tall with a pale brown bark. Leaves are simple, opposite, entire, oblong—elliptical, 6–20 cm long, punctuate with pellucid glands. The flowers are small and whitish with a peculiar aroma. They are present in groups in cymes. They are structurally hermaphrodite but functionally dioecious. Those trees that bear no fruit are male trees, wherein the flowers will have above 100 stamens, and the flowers in bearing female trees have around 50 stamens. The receptacle has 4 cream-colored calyx lobes, spreading at anthesis and persistent in the fruit. Petals are 4, whitish and quickly deciduous. The style is white with a yellow stigma. In females, the style is slightly shorter and the stigma longer than in the barren trees. The ovary is inferior and 2-celled, usually with 1 ovule in each cell. Plants flower during March–June and the fruit, which is a berry, matures 3–4 months later. The fruits have two kidney-shaped seeds (Parthasarathy and Kandiannan, 2007; Rao et al., 2012).

Flowering and fruiting: March–June

Chemical constituents: Ferulic acid, rutin, gallic acid, and chlorogenic acid were isolated from the leaves. Essential oil form berries contain limonene 1,8 cineole terpinolene β-caryophyllene β-selinene methyl eugenol. The leaf oil contain eugenol, methyl eugenol, β-caryophyllene, and myrcene (Jirovetz et al., 2007; Rao et al., 2012).

Used in: Ayurveda, Folk, and Traditional

Parts used: Leaves and fruits

Uses: All parts of *P. dioica* are used in folk medicine. Jamaicans drink hot tea with allspice for colds, dysmenorrhea (menstrual cramps), and dyspepsia (upset stomach). Costa Ricans are known to use allspice to treat dyspepsia and diabetes. Guatemalans are known to apply crushed allspice berries to bruises and sore joints and for myalgia (muscle ache). In Cuban medicine, allspice along with other herbal mixtures is used to relieve indigestion. In Ayurveda, its use includes to relieve respiratory congestion and assorted odontalgia (toothache). In Europe, anecdotal uses of allspice extract for dyspepsia and as purgative exist. In modern herbal medicine, allspice extract has been used for neuralgic pain. Allspice essential oil, when added to massage oils and baths, is known to promote circulation so as to relieve pain from muscle cramps and strains. Also, it is used for headache and to combat stress and depression and overcome fatigue because of its comforting scent (Zhang and Lokeshwar, 2012). In India, it is used as an aromatic stimulant in digestive troubles, as an adjuvant to tonics and purgatives, as an anodyne against rheumatism and neuralgia (Anonymous, 1969). In modern herbal medicine, allspice extract has been used for neuralgic pain. Allspice essential oil, when added to massage oils and baths, is known to promote circulation so as to relieve pain from muscle cramps and strains. Also, it is used for headache, to combat stress and depression, and to overcome fatigue because of its comforting scent (Rao et al., 2012).

Methods of propagation: Cuttings, grafting, and budding

DOI: 10.1201/9781003263500-183

Photo: Dr. Srinath Rao

Pimpinella anisum L.

Synonymys: *Anisum odoratum* Raf.; *Anisum officinale* DC.; *Anisum officinarum* Moench *Anisum vulgare* Gaertn.; *Apium anisum* (L.) Crantz; *Carum anisum* (L.) Baill.

Family: Apiaceae (Umbelliferae)

Common names: Bengali: Mauri; **English:** Anie, Aniseed; **Gujarati:** Variyaja; **Hindi:** Saumph; **Kannada:** Sompu; **Malayalam:** Soppu, **Marathi:** Soump; **Sanskrit:** Shatapuspa; **Tamil:** Compu; **Telugu:** Soumpu

Distribution: Grows in the Eastern Mediterranean region, West Asia, the Middle East, Mexico, Egypt, and Spain

Description: Anise is an herbaceous annual plant growing to 60–90 cm or more. The leaves at the base of the plant are simple, 1–5 cm long and shallowly lobed, while leaves higher on the stems are feathery or lacy, pinnate, divided into numerous small leaflets. Both leaves and flowers are produced in large, loose clusters. The flowers are either white or yellow, approximately 3 mm in diameter, produced in dense umbels. The fruit is a dry oblong and curved schizocarp, 4–6 mm long.

Flowering: May–June **Fruiting:** July–September

Chemical constituents: The most important compounds of aniseeds essential oil were trans-anethole, estragole, γ-himachalene, para-anisaldehyde, and methyl chavicol (Shojaii and Fard, 2012). Other studies have demonstrated the presence of eugenol trans-anethole, methyl chavicol, anisaldehyde, estragole, coumarins, scopoletin, umbelliferone, estrols, terpene hydrocarbons, polyenes, and polyacetylenes as the major compounds of the essential oil of anise seed (Gulcin et al., 2003).

Used in: Ayurveda and Traditional

Parts used: Seeds

Uses: Anise has been used as a spice and medicine for thousands of years. It possesses various properties such as antimicrobial, antifungal, antiviral, antioxidant, muscle relaxant, analgesic, and anticonvulsant activity, as well as different effects on gastrointestinal system have been reported of aniseeds. It can also reduce morphine dependence and has beneficial effects on dysmenorrhea and menopausal hot flashes in women. In diabetic patients, aniseeds showed hypoglycemic and hypolipidemic effect and reduce lipid peroxidation Anise seeds are used as analgesic in migraine and as a carminative, aromatic, disinfectant, and diuretic in traditional medicine. Aniseed has warm and dry nature and can increase milk production, menstruation, urine, and sweat secretion and make the complexion good. It is also effective in polishing teeth. In some traditional texts, anise is mentioned for melancholy and nightmares and in the treatment of epilepsy and seizure (Mirheydar, 2001; Amin, 2005; Shojaii and Fard, 2012).

Methods of propagation: Seeds

DOI: 10.1201/9781003263500-184

Photo: Dr. Srinath Rao

Piper betle L.

Synonyms: *Artanthe hexagyna* Miq.; *Betela mastica* Raf.; *Chavica betle* Miq.; Chavica canaliculata C. Presl.

Family: Piperaceae

Common names: Bengali: Pan; **English:** Betel; **Gujrati:** Nagarbel; **Hindi:** Pan; **Kannada:** Viladele, eli; **Malayalam:** Vetulakkoli, **Marathi:** Nagaveli, pan; **Sanskrit:** Thambulvalli; **Tamil:** Vathalaikkodi; **Telugu:** Tamalapaku

Distribution: It is originated from Malaysia but is distributed extensively in South and Southwest China. This crop is usually cultivated in India, Sri Lanka, Malaysia, Thailand, Taiwan, and other Southeast Asian countries.

Description: It is a perennial climber with woody stem. The leaves are heart-shaped and bright green in color. Leaves are 4–7 in. long and 2–4 in. broad. It bears both male and female flowers. Fruits are rarely formed immersed in fleshy spike forming noodle-like structures.

Flowering and fruiting: Flowering and fruiting are rare, mainly because the plants are replanted every 4 or 5 years under cultivation. Plants flower when they are 8–10 years old.

Chemical constituents: Betel leaves consist of the most abundant phytochemicals, that is, hydroxychavicol, which contributes to the antiproliferative efficacy of betel leaf extract. This phenolic compound of betel leaf inhibits prostate cancer through reactive oxygen species–driven DNA damage and apoptosis. Some of the important constituents of the leaf oil are α-pinene, safrole, eugenol, selinene, hydroxychavicol cadinene, and caryophyllene (Gundala et al., 2014).

Used in: Ayurveda, Traditional, and Folk

Parts used: Leaves

Uses: It is an anthelmintic, astringent, aphrodisiac, aromatic, carminative, expectorant, laxative, febrifuge, and a tonic. Betelvine leaves are important because of their medicinal, religious, and ceremonial uses. It prevents halitosis, improves vocalization, strengthens gum, and treats indigestion, constipation, congestion, coughs, and asthma.

It also possesses antioxidant, anti-inflammatory, anti-apoptotic, anticancer and antimicrobial properties (Das et al., 2016). Betel leaves are mainly used as mouth freshener and are well known for curing many communicable and noncommunicable diseases like cold, cough, bronchial asthma, rheumatism, stomachalgia and used to treat other diseases like bad breath, boils and abscesses, conjunctivitis, constipation, swelling of gums, cuts, and injuries. The essential oil of betel leaves possesses antibacterial, anti-protozoan, and antifungal properties. From studies, it is known that the aqueous extract of betelvine reduces the adherence of early dental plaque bacteria. It is also found that betel has significant antiproliferative activity in vitro and in vivo prostate cancer models (Punuri et al., 2012; Paranjpe et al., 2013; Gundala et al., 2014).

Methods of propagation: Stem cuttings and suckers

DOI: 10.1201/9781003263500-185

Plantago asiatica L.

Synonyms: *Plantago depressa* Wild.; *Plantago major*; *Plantago lanceolata*
Family: Plantaginaceae
Common names: English: Asian plantain, Broad leaf plantain; **Kannada:** Sirapotla gida
Distribution: The plant is native to East Asia (China, Japan, Korea, etc.). It grows really well in disturbed areas like roadsides or even dirt roads.
Description: This plant is a perennial herb that grows to 20–60 cm high and has big roots. Its leaves, which gather on the ground, have oval blades and are 11–12 cm long and 4–9 cm wide, with obtuse apex and base, and 5 smooth main veins from the leaf base. The petioles are almost the same length as the blades. The plant has erect spikes of 20–45 cm high, with many small, white, hairless flowers, and oval sepals that are 1.8–2 cm long, tube corolla with 5 oval lobes, and 4 stamens. The fruits are oval-shaped pyxis of 3.5 mm high and 2 mm wide, which have 4 black seeds inside that are up to 1.8 mm long.
Flowering and fruiting: July–August
Chemical constituents: The chemical analysis of *Plantago major* reveals the remarkable glycoside aucubin. Aucubin has been reported as a powerful antitoxin. There are many more highly effective constituents in this plant, including ascorbic acid, apigenin, baicalein, benzoic acid, chlorogenic acid, citric acid, ferulic acid, oleanolic acid, salicylic acid, and ursolic acid.
Plantaginis semen has long been used in traditional Chinese medicine to treat diseases. *Plantaginis* semen is a dry mature seed of *Plantago asiatica* L. or *P. depressa* Willd (Lan et al., 2021). It is primarily composed of iridoids and phenylethanoid glycosides, flavonoids, and polysaccharides (Nie et al., 2017; Wang et al., 2016; Turgumbayeva et al., 2022).
Used in: Ayurveda, Folk, and Traditional
Parts used: Seeds
Uses: The seeds promote urination and are cold in nature. The seeds clear phlegm from the lungs. The are given in case of wasting disease and to improve fertility in men by improving secretion of semen. The seeds have diuretic action and help in dribbling of urine, strangury (blockage or irritation at the base of the bladder, resulting in severe pain and a strong desire to urinate), edema, and urinary tract infections. The seeds are indicated in case of diarrhea. The seeds remove dampness and heat from liver. The seeds and leaves are anti-inflammatory antiseptic, antitussive, cardiac, and hemostatic and are a diuretic expectorant. They assist in the repair of joints. A decoction of roots is used in the treatment of coughs. The seeds contain 30% mucilage, which swells up in the gut, acting like a bulk laxative and soothing the membrane.
Methods of propagation: Seeds
Further reading: https://www.bimbima.com/herbs/plantain-seeds/4111/#

DOI: 10.1201/9781003263500-186

Pogostemon cablin Benth.

Synonyms: *Mentha auricularia* Blanco.; *Pogostemon battakianus* Ridl.; *Pogostemon javanicus* Backer ex Adelb.; *Pogostemon mollis* Hassk

Family: Lamiaceae (Labiatae)

Common names: English: Patchouli, Stinkweed, Pucha pot; **Hindi:** Pachouli; **Kannada:** Patche tene; **Malayalam:** patchilla; **Marathi:** patchapan or patcha; **Godhadi Sanskrit:** Tamala patra; **Tamil:** pacchilai

Distribution: Native of the Philippines and grows wild in many South Asian countries and is presently cultivated on a commercial scale in India, Indonesia, Malaysia, China, Singapore, West Africa, and Vietnam (Swamy et al., 2010).

Description: Patchouli is a shrubby perennial plant that grows up to 3 ft in height. The large fragrant leaves are simple, oppositely arranged, roughly oval in shape, and irregularly toothed. Both the stems and the leaves are densely hairy. Flowers are small, pale purple to white, arranged in dense woolly spikes, and have long stamens. The leaves are typically simple and oppositely arranged; most are fragrant and contain volatile oils. The flowers are usually arranged in clusters and feature 2-lipped, open-mouthed, tubular corollas (united petals) with 5-lobed bell-like calyxes (united sepals). The fruit is commonly a dry nutlet.

Flowering and fruiting: During fall

Chemical constituents: More than 140 compounds, including terpenoids, phytosterols, flavonoids, organic acids, lignins, alkaloids, glycosides, alcohols, and aldehydes have been isolated and identified from patchouli. The main phytochemical compounds are patchouli alcohol, α-patchoulene, β-patchoulene, α-bulnesene, seychellene, norpatchoulenol, pogostone, eugenol, and pogostol (Swamy and Sinniah, 2015).

Used in: Ayurveda and Traditional

Parts used: Leaves, stem, and oil

Uses: It is an important herb which possesses many therapeutic properties and is widely used in the fragrance industries. In traditional medicinal practices, it is used to treat colds, headaches, fever, nausea, vomiting, diarrhea, abdominal pain, insect and snakebites. In aromatherapy, patchouli oil is used to relieve depression and stress, calm nerves, control appetite, and improve sexual interest (Swamy and Sinniah, 2015). In China, Japan, and Malaysia, it is used to treat colds, headaches, nausea, vomiting, diarrhea, abdominal pain, and insect and snakebites (China Pharmacopoeia Commission, 2010).

Methods of propagation: The plants are commonly propagated from stem cuttings.

DOI: 10.1201/9781003263500-187

Premna serratifolia L.

Synonyms: *Premna intigrifolia* L.; *Premna obtusifolia* R. Br.

Family: Lamiaceae (Verbenaceae)

Common names: English: Headacha tree; **Kannada:** Agnimanthha, Bachanige mara; **Malayalam:** Appel, Ben-moenja; **Marathi:** Arani; **Sanskrit:** Agnimantha; **Tamil:** Pasumunnai; **Telugu:** Gadanelli, Kanika

Distribution: A large shrub or a small tree widely distributed throughout the dry parts of India, Bangladesh, Sri Lanka, Myanmar, and southeast Asia.

Description: Evergreen straggling shrub or small trees, up to 7 m high. Leaves: simple, opposite, estipulate; petiole 4–14 mm, slender, pubescent, grooved above; lamina 2.5–8.5 × 2–7.2 cm, elliptic, elliptic-oblong, base acute, obtuse, subcordate or rounded, apex acuminate, mucronate, obtuse, margin entire, glabrous above except along the appressed midrib, chartaceous; lateral nerves 3–5 pair, pinnate, prominent, puberulous beneath; intercostal reticulate, obscure. Flowers: bisexual, greenish white, in terminal corymbose panicled cymes; bracts small; calyx small campanulate, 2-lipped, 5-lobed; corolla tube short, villous inside, lobes 5; stamens 4, didynamous, inserted below the throat of the corolla tube; anther ovate; ovary superior, 2–4-celled, ovules 4; style linear; stigma shortly bifid. Fruit: a drupe, seated on the calyx, globose, purple; seeds oblong or tree

Flowering and fruiting: May–November

Chemical constituents: Phytochemical investigations of *P. integrifolia* revealed the presence of flavonoid luteolin, sterols, triterpenes, alkaloids premnine, ganikarine, and premnazole (Azad et al., 2018).

Used in: Ayurveda, Siddha, Traditional, and Unani

Parts used: Leaves, stem, bark, and roots

Uses: Plant root of arni is anti-inflammatory, stomachic, and antipyretic and is prescribed in liver complaints, cold, obstinate fevers, flatulence, urticaria, rheumatism, and neuralgia. Per Ayurvedic system of medicine, the roots of this plant are an important constituent of the 10-herb formulation "Dasamula". The roots are said to have many medicinal properties, such as anti-inflammatory, antiparasitic, and antioxidative activities. The leaves of this tree were used to treat headache, hence, the tree's common name, "Headache tree". Leaves have potential in treating cancer, Parkinson's disease, rheumatoid arthritis, skin melanogenesis, diabetes, and obesity. The fruits are used for preparing an herbal drink.

The roots are widely used in the treatment of diabetes, inflammation, swellings, bronchitis, dyspepsia, liver diseases, fever, constipation, piles, and the like. Also, this plant is widely used by the traditional practitioners as cardiotonic, antibiotic, carminative, hepatoprotective, and antitumor. The pharmacological properties reported in the literature include its antidiabetic, anti-atherosclerotic, hypolipidemic, analgesic, antimicrobial, and antioxidant activities (Azad et al., 2018).

Methods of propagation: Seeds

DOI: 10.1201/9781003263500-188

R

Rauvolfia tetraphylla L.

Synonyms: *Rauvolfia canescence*; *Rauvolfia heterophylla*; *Rauvolfia hirsuta*

Family: Apocynaceae

Common names: Bengali: Bara chandrika, Gandhanakuli; **English:** Wild snakeroot, Devel pepper, Milkbush; **Hindi:** Badachandrika, Chandrabhaga; **Kannada:** Dodda Chandrike; **Malayalam:** Pampumkolli, Kattamalpori; **Sanskrit:** Vanasarpagandha; **Tamil:** Pampukaalaachchedi; **Telugu:** Papataku

Distribution: The plant is native to Mexico, Central America, West Indies, and northern South America. Found throughout India and cultivated for its medicinal uses.

Description: A small, much-branched woody shrub, 0.6–1.2 m high. Leaves are whorled, ovate-elliptic, and dark green in color and occur in groups of 4 unequal-sized leaves at each node. Flowers greenish white or creamy white in umbellate cymes. About 5 mm long, tube 3–4 mm long. Fruits (drupes): ovoid, deep red or purple/black when ripe; look like large pepper corns. Seeds oblong, rugose.

Flowering and fruiting: October–April

Chemical constituents: A vast number of secondary metabolites such as reserpine, reserpiline, yohimbine, ajmaline, lankanescine, rauvotetraphyllines, alstonine, deserpidine, aricine, isoreserpine, sarpagine, and others have been detected in the plant among which reserpine is the most important one (Mahalakshmi et al., 2019).

Used in: Traditional and Folk

Parts used: Leaves, fruits, root, and latex

Uses: The plant is used as a remedy for snake and other poisonous bites, blood pressure, diabetes, piles, malaria, wound, helminthiasis, hypertension, vomiting, insomnia, skin diseases, mental disorders, cough, fever, and others. The latex from the plant is reported to be cathartic, emetic, and expectorant besides its use in the treatment of dropsy. The leaf juice is used in eye troubles and decoction prepared from leaves is used for toothache. The root is reported to have antihypertensive, sedative, and hypnotic effects. A paste prepared from roots is useful in stomach pain and snakebite. The roots of the plant are used in nervous disorders and insomnia in Gopalganj district of Bihar, India. Roots are used in high blood pressure and mental disorders in Coimbatore district, Tamil Naud, India. The leaves and roots are used against nervous disorders, anxiety, excitement, cough and fever in Barpeta district, Assam, India. Roots are used for treating nervous disorders in Villupuram district, Tamil Nadu, India. The Nyishi tribe of Arunachal Pradesh, India, uses the roots of the plants as anthelmintic, antidote to snakebite, and to cure vomiting, insomnia, insanity, blood pressure, and malaria (Khare, 2007; Arulappan et al., 2015; Das, 2016; Mahalakshmi et al., 2019).

Methods of propagation: Seeds

DOI: 10.1201/9781003263500-190

Photo: C.T. Chandralekha

Rehum emodi Wall ex Meissn

Synonyms: *Rehum australe* D. Don

Family: Polygonaceae

Common names: English: Himalayan rhubarb, Indian rhubarb; **Hindi:** Revand chini; **Sanskrit:** Amalavetasa

Distribution: The species occurs in the temperate Himalayas, from Kashmir to Sikkim, at an elevation of 2800–4000 m. It is found in alpine zone on rocky soil, moraines, and crevices, between boulders and near streams in specific pockets.

Description: Rewand chini is a perennial stout herb, 1.5–3.0 m tall. Leaves are radical, orbicular, or broadly ovate, very large, and 30–45 cm in diameter with long petioles. Flowers are small, dark purple or pale red in axillary panicles. The plant has 3–5 years of juvenile phase, followed by reproductive phase. Seed production potential varies from 300 seeds to 950 seeds per plant (https://vikaspedia.in/agriculture/crop-production/package-of-practices/medicinal-and-aromatic-plants/rheum-emodi).

Flowering: June–July **Fruiting:** August–September

Chemical constituents: A number of anthraquinone derivatives including emodin, aloe-emodin, physcion, chrysophanol, rhein, emodin glycoside, and chrysophanol glycoside occur as the main chemical constituents. In the past few years, new components such as sulfemodin 8-*O*-β-D-glucoside, revandchinone-1, revandchinone-2, revandchinone-3, revandchinone-4, 6-methyl-rhein, and 6-methyl aloe-emodin have been reported (Zargar et al., 2011).

Used in: Traditional

Parts used: Roots and rhizome

Uses: *R. emodi* is a mild purgative, astringent, tonic, laxative, stomachic, and aperient. Powdered rhizomes are sprinkled over ulcers for quick healing. It is also reported to be a potent anti-inflammatory drug (https://vikaspedia.in/agriculture/crop-production/package-of-practices/medicinal-and-aromatic-plants/rheum-emodi).

It has been traditionally used as a diuretic, liver stimulant, purgative/cathartic, stomachic, anticholesterolemic, antitumor, antiseptic, and tonic (Zargar et al., 2011).

Methods of propagation: Seeds

DOI: 10.1201/9781003263500-191

Rhinacanthus nasutus (L.) Kurz

Synonyms: *Rhinacanthus communis* Nees; *Rhincanthus nasuta* L.; *Justicia nasuta* L.; *Pseuderanthemum connatum* Lindau.; *Dianthera paniculata* Lour.

Family: Acanthaceae

Common names: English: Snake jasmine; **Hindi:** Kabutar ka phul, Palakjuhi; **Kannada:** Nagamallige; **Marathi:** Gajkarni; **Malayalum:** Nagamulla; **Sanskrit:** Yuthikaparni; **Tamil:** Uragamalli; **Telugu:** Nagamalli

Distribution: Native to Thailand, much of South and Southeast Asia, and India

Description: A small slender erect, branched somewhat hairy shrub. 1–2 m in height; Leaves, oblong, 4–8 mm in length, narrowed and pointed at both the ends, panicle inflorescent, flowers born in clusters, white, bi-lipped, upper lip white, erect, oblong or lancelike, 2-toothed at the apex, about 3mm both in length and width, lower lip broadly ovate, 3-lobed, 1–1.5 cm in both length and breadth with few minute brownish dots near the base. The fruit is a capsule club shaped, with 4 seeds.; Calyx: green, hairy, about 5 mm long. Corolla: tube is greenish, slender, cylindrical, about 2 cm long.

Flowering and fruiting: Throughout the year

Chemical constituents: Rhinacanthone, a main bioactive naphthoquinone, can be isolated from roots. Several phytochemicals including β-amyrin, glutinol, lupeol, stigmasterol, sitosterol, umbelliferone, 2-methylanthraquinone, wogonin, oroxylin A, (+)-praeruptorin, p-hydroxy benzaldehyde, methyl vanillate, syringaldehyde, Dehydro-α-lapachone and novel compounds rhinacanthin A-D, G-Q, (RnA-D and RnG-Q) rhinacanthone, 4-acetonyl-3,5-dimethoxy-p-quinol have been isolated from the roots, stem, and leaves of this plant (Su et al., 1991; Wu et al., 1998a, 1998b).

Used in: Folk and Traditional

Parts used: Leaves and root

Uses: Roots and leaves brushed with lime juice is useful in ringworm infestation (Dhobie's itch) of skin and snakebites, aphrodisiac, useful in diabetes, hepatitis, hypertension. Tea prepared from its leaves recommended for persons who smoke, people working in polluted environment, and persons infected with lung infection, excessive phlegm, tuberculosis, pneumonia, and whooping cough; it contains anticancer and anti-inflammatory activity.

The ethanol extract of the root of this herb has been shown to be protective in vitro against the effects of hypoxia and reoxygenation in the mouse neuron cell line (HT-22 cells). Furthermore, various extracts from both the leaf and the root have been shown to protect the HT-22 cell line against the detrimental effects of both glutamate and amyloid beta protein fragments, suggesting a possible use as a treatment for Alzheimer's disease. Traditionally, the seeds, roots, and leaves of the plant have been used against scabies, eczema, and various skin conditions. The leaves are used for prickly heat as well as scurf, and the roots are boiled, along with milk, and used as an aphrodisiac. A decoction of roots is used as an antidote for snakebites. The stem and leaves of the plant are used to treat pulmonary tuberculosis and high blood pressure (Quattrocchi, 2012). The roots are used in Thai traditional medicine to treat snakebites, and the roots and/or leaves can be made into a balm and applied to the skin for the treatment of skin infections such as ringworm, or they may be brewed to form an infusion for the treatment of inflammatory disorders.

DOI: 10.1201/9781003263500-192

Its leaves are available to the public for purchase in the form of "tea bags" as a natural herbal remedy for many disorders, including diabetes, skin diseases (antifungal, ringworm, eczema, scurf, herpes), gastritis, raised blood pressure, improved blood circulation, early-stage tuberculosis, antitumor activity, and as an antipyretic. There have been many studies investigating the roles of this plant or compounds isolated from the herb regarding diseases such as Alzheimer's and other neurodegenerative diseases, cancer, diabetes, and infection with bacteria, fungi, or viruses (Brimson et al., 2020).

Methods of propagation: Stem cuttings

Rosmarinus officinalis L.

Synonyms: *Rosmarnus angustifolius* Mill.; *Rosmarinus communis* Noronha; *Rosmarinus flexuosus* Jord. & Fourr.; *Rosmarinus latifolius* Mill.

Family: Lamiaceae

Common names: English: Rosemary

Distribution: It grows wildly along the north and south coasts of the Mediterranean Sea and in the sub-Himalayan areas.

Description: A woody perennial herb with fragrant, evergreen, needle-like leaves and white, pink, purple, or blue flowers, Forms range from upright to trailing; the upright forms can reach 1.5 m (5 ft) tall, rarely 2 m (6 ft 7 in.). The leaves are evergreen, 2–4 cm (0.8–1.6 in.) long and 2–5 mm broad, green above, and white below, with dense, short, woolly hair. The flowers are purplish white and strongly 2-lipped and have two long-exserted (protruding) stamens. The fruit consists of 4 dry nutlets (1-seeded sections).

Flowering and fruiting: April–June

Chemical constituents: Phytochemical studies revealed that rosemary contains terpenoids, essential oils, alkaloids and flavonoids. Chemical analysis of different kinds of rosemary extracts composition reveals that the most potent active components are triterpenes, phenolic diterpenes and phenolic acids, including rosmarinic acid, carnosic acid, rosmanol, carnosol, ursolic acid, and betulinic acid (Okamura et al., 1994a, 1994b; Borras-Linares et al., 2014).

Parts used: Leaves and flowers

Uses: Rosemary is used as a decorative plant in gardens and has many culinary and medical uses. The plant is said to improve memory. The leaves are used to flavor various foods and as a stimulant and memory enhancer in folk medicine; it is used as an antispasmodic in renal colic and dysmenorrhea, in relieving respiratory disorders, and to stimulate growth of hair. Extract of rosemary relaxes smooth muscles of trachea and intestine, and has choleretic, hepatoprotective and antitumorigenic activity. Rosmarinic acid has a therapeutic potential in treatment or prevention of bronchial asthma, spasmogenic disorders, peptic ulcer, inflammatory diseases, hepatotoxicity, atherosclerosis, ischemic heart disease, cataract, cancer, and poor sperm motility (al-Sereiti et al., 1999). It has been used in folk medicine to alleviate several diseases including headache, dysmenorrhea, stomachache, epilepsy, rheumatic pain, spasms, nervous agitation, improvement of memory, hysteria, and depression, as well as physical and mental fatigue (Batanouny et al., 1999).

Methods of cultivation: It can be propagated from an existing plant by clipping a shoot (from a soft new growth) 10–15 cm (4–6 in.) long, stripping a few leaves from the bottom, and planting it directly into soil. Propagation is simple from cuttings in early to midsummer. Cut pieces of new growth, around 10 cm long, and insert them into a pot of compost. Keep them moist, and they should root by the autumn, when they can be potted up and grown before planting out in spring.

DOI: 10.1201/9781003263500-193

Rubia cordifolia (L.)

Synonyms: *Galium cordifolium* (L.) Kuntze; *Rubia cordata* Thunb.; *Rubia clematifolia* Reinw. ex Miq.; *Dioscorea verticillata* Lam.

Family: Rubiaceae

Common names: Bengali: Manjistha**; English:** Indian madder; **Gujrati:** Majith; **Hindi:** Manjith; **Kannada:** Manjistha; **Malayalam:** Mancatti, Punta; **Marathi:** Manjistha; **Sanskrit:** Manjistha; **Tamil:** Sevealli; **Telugu:** Tamrsvalli

Distribution: Throughout India, ascending to an altitude of 3700 m

Description: A branched climber with bristles. The stem is slender and 4-angled. The leaves are rough and arranged in a whorl of four per node and oval to heart-shaped with long leaf stalk. The leaves have 5–7 main nerves. The stipules are short living (leafy structures found at the nodal region). The flowers are very small, greenish white, and arranged in branched cluster called dichasial cyme (forked flowering stalk, which bears specific number of flowers in each branch). The fruit is round fleshy drupe. When ripe, it is dark purple. The roots have brownish red bark and give red dye.

Flowering: July–September **Fruiting:** September–November

Chemical constituents: Munjistin is a specific anthraquinone found in Rubia cordifolia. Other anthraquinones such as purpuroxanthin, rubiatriol, rubicoumaric acid, rubifolic acid, pseudoparpurin, alizarin, rubiadin, rubimallin, purpurin, xanthopurpurin, and ruberythric acid contribute to its diverse medicinal properties. The roots are rich in anthraquinones and their glycosides (around 20), the important ones include purpurin and munjistin in addition to xanthopurpurin, peudopurpurin-free alizarin, as well as its glucoside. The whole plant yields pentacyclic triterpenic acids—rubicoumaric and rubifolic acids. Antitumor cyclic hexapeptides have been isolated from the root. The plant's roots contain an organic compound called alizarin.

Used in: Ayurveda, Siddha, and Folk

Parts used: Root and stem

Uses: Although manjistha is, at the same time, bitter, astringent; sweet in taste; and heavy, dry, and hot in effect, it is a pacifier of kapha and pitta. Besides the glucosides known as manjisthin and purpurine, its chemical composition consists of various other components, which include resins, lime salts, and coloring agents. Manjistha is used in a number of diseases. It is a drug of choice for treating various systemic problems like raised uric acid and gouty arthritis, glandular swellings, recurrent skin infections, and other diseases of the skin like pigmentation anomalies and leukoderma. It is also included in various formulations to treat uterine and urinary infections, diarrhea, dysentery, and chronic fevers. Manjistha holds the reputation of a very good skincare herb. Used externally and internally, it helps one to gain luster and glow (of the skin) and aids to remove pimples, freckles and discoloration. Manjistha promotes the healing of skin tissues damaged by injury or infection. Its finely crushed powder can be simply applied on the face after mixing it with little honey. A combination of dried and crushed orange peels, 100 gm, and sandal, turmeric, and manjistha powders, each 50 gm, makes an excellent face pack. Because manjistha has a marked effect on the female reproductive organs, it is useful treating various gynecological problems, like white discharge and irregular menstruation. Action: roots and dried stem—blood purifier, astringent, diuretic, emmenagogue, deobstruent, antidysenteric, antiseptic, and alterative. According to Ayurveda, manjistha is probably the best alternative or blood-purifying herb in Ayurveda. It cools and detoxifies the blood, dissolves obstructions in

DOI: 10.1201/9781003263500-194

blood flow, and removes stagnant blood. It is having healing action; thus, it is helpful in erosions of gums and bleeding gums. In Ayurvedic medicine, it is used as an immune regulator. Its antioxidant properties are also being investigated. Its role in supporting heart health is evidenced by studies that show that it regulates blood pressure, blood vessel constriction, and the tendency of blood to form clots. Many research teams have suggested that *R. cordifolia* has anticancer activity, hepatoprotective effects, antimicrobial activity, and antihyperglycemic, antistress, and nootropic activity. It also suppresses the activation of mast cells through the inhibition of Syk for anti-allergic activity. Its paste is applied in various skin disorders like itching, black spots on the face, pimples, leukoderma, and others. Internal use of manjistha in the form of powder or decoction is also beneficial in various skin and blood disorders.

Methods of propagation: Seeds

Further reading: https://en.wikipedia.org/wiki/Rubia_cordifolia

Ruellia tuberosa L.

Synonyms: *Cryphiacanthus barbadensis* Nees; *Dipteracanthus clandestinus* C. Presl
Family: Acanthaceae
Common names: English: Fever root, Minnie root, Popping pod; **Hindi:** Chatpati; **Tamil:** Pattaskai; **Telugu:** Chetapatakaayala mokka
Distribution: *R. tuberosa* originates from tropical America but is naturalized in Southeast Asia (Thailand, Peninsular Malaysia, Java) and elsewhere in the tropics (India, Sri Lanka, Africa) as an escape from cultivation and as an ornament
Description: Description: An erect, up to 60–70-cm tall herb with stout, 4-angled stems, swollen and purplish at the nodes, tender parts sparsely hairy. Roots: slender, elongated, tuberous. Leaves: with up to 2 cm long petiole; lamina oblong-obovate to oblanceolate, 5–9(–10) × 2–4 cm, shining, basally cuneate to attenuate, obtuse to somewhat acute, often mucronulate. Flowers shortly pedicellate, blue-violet, showy, 5–5.5 cm long, about 3.5 cm across, solitary or in 1–3-flowered, terminal or axillary cymes; bracts and bracteoles linear-lanceolate, about 5 mm long. Calyx: linear, 1.3–2 cm long, ciliate, acute. Corolla: pubescent outside, tube about 3.5 cm long, abruptly narrowed at base, limb 5-lobed, ovate, about 1.5 cm long, patent. Stamens: with oblong-sagittate, about 4 mm long, puberulous anthers. Ovary: oblong, about 4 mm long, glabrous; style 2.3–2.5 cm long. Capsule: oblong, 2.2–3 cm long, glabrous or scabrescent, 24–28-seeded, minutely beaked at tip. Seeds: flattened, orbicular, about 2 mm across, hairy.
Flowering and fruiting: Throughout the year
Chemical constituents: *R. tuberosa* was found to contain apigenin, 7-β-D glucoronoid, and malvidine 3,5-diglucoside. Aerial part of *R. tuberosa* was found to contain phenylethanoid glycoside (e.g. isocassifolioside), flavone glycosides (e.g. hispidulin 7-O-α-L-rhamnopyranosyl-(1'″→2″)-O-β-D-glucuronopyranoside and pectolinaringenin 7-O-α-L-rhamnopyranosyl-(1'″→2″)-O-β-D-glucuronopyranoside), and verbascoside (e.g. isoverbascoside, nuomioside, isonuomioside, forsythoside B, paucifloside, cassifolioside, hispidulin 7-O-β-D-glucuronopyranoside, and comanthoside B (Panda, 2004; Phakeovilay et al., 2013).
Used in: Traditional
Parts used: Leaves, roots, and whole plant
Uses: Traditionally used as diuretic, anti-pyretic, analgesic, anti-hypertensive, anthelmintic, abortifacient, emetic, in bladder disease, kidney disorder, bronchitis, gonorrhea, and syphilis. In the Siddha system of medicine, the leaves are given with liquid copal as a remedy for gonorrhea and ear diseases and used for stomach cancer (Reddy et al. 1991). Dried and ground roots in a dose of 2 oz cause abortion and are used for sore eyes (Kirtikar and Basu, 1935).

A decoction is given in chronic bronchitis; also used as a diuretic for the treatment of stones in the bladder. Whole plant: to treat bladder diseases and frequent micturition; decoction with *Petiveria alliacea* is drunk to "clean out" uterine tract (dilation and curettage) or as an abortifacient. Root: infusion for kidney diseases; in a syrup for whooping cough; infusion or decoction for a diabetes remedy; tubers in a tea used for cleansing the blood. Root and leaf: in a tea used for alleviating retention of urine and to remedy weakness. Diuretic, antipyretic, analagesic, antihypertensive. Peruvians suggest the infusion for bronchitis, flu, gonorrhea, headache, inflammation, leprosy, oliguria, pain, and wounds. Venezuelans use a root decoction for asthma, blennorrhagia, catarrh, and pulmonosis (Chothani et al., 2010; Arirudran et al., 2015).
Methods of propagation: Seeds

DOI: 10.1201/9781003263500-195

Rumex vesicarius L.

Family: Polygonaceae

Synonyms: *Acetosa vericaria* (L.) A. Love; *Lapathum vesicarium* (L.) Moench.; *Rumex roseus* Auch. ex Meisn.

Common names: Bengali: Bun palung, Takpalong, Chukapalong; **English:** Bladder Dock, Blister Sorrel, Country Sorrel, Rosy Dock, and Ruby Dock; **Hindi:** Chukra, Chooka; **Kannada**: Chukki soppu; **Marathi:** Chukki, **Sanskrit:** Amlavetasa, Chukram, Shatavedhi; **Tamil:** Chukkan-karai; **Telugu:** Chukkakura, Pulla prabba

Distribution: Africa, Arab, Persia, and Afghanistan. In India cultivated for its vegetable

Description: An annual, pale green, glabrous herb, branched from the root, rather fleshy, 15–30 cm high. Leaves: 2.5–7.5 cm, elliptic, ovate or oblong, obtuse or acute, base cuneate. Racemes: 2.5–3.8 cm, terminal and leaf-opposed. Flowers: sometimes 2-nate and connate, valves large, orbicular, 2-lobed at each end, very membranous and reticulate without a marginal nerve. Fruit: 1.3 cm diameter, white or pink, valves hyaline.

Flowering and fruiting: December–February

Chemical constituents: Leaves are rich in ascorbic, citric, and tartaric acids. Roots contain rumicin and lapathin, which is identical to chrysophanic acid. The aerial parts of this and other species of *Rumex* also contain anthraquinone derivatives and flavonoids like emodin, aloe-emodin, chrysophanol, chrysophanic acid, physcion; isovitexin, iso-orientin, quercetin, kaempferol and luteolin glucosides; chromone and flavone derivatives, tannins, mucilage, calcium oxalate, and starch (Ghani, 2003). Cystine, glutamic acid, proline, phenylalanine, and histidine have been detected in the protein extract (Rastogi and Mehrotra, 1993).

Used in: Traditional

Parts used: Leaves, fruits, and seeds

Uses: The plant is cooling, tonic, analgesic, laxative and stomachic; useful in heart troubles, tumors, constipation, hiccup, flatulence, asthma, bronchitis, and piles. Leaf juice is astringent, cooling, aperient, and a strong diuretic; it relieves toothache, checks nausea, and promotes appetite. Fruits are aperient and diuretic; eaten fresh against jaundice, hepatic conditions, constipation, and indigestion. Roasted seeds are prescribed in dysentery (Yusuf et al., 2009). Traditionally used as aperient, astringent, diuretic, and cooling agent. The plant juice is useful in curing stomach heat, toothache, and checks nausea.

Methods of cultivation: Seeds

DOI: 10.1201/9781003263500-196

***Rumex vesicarius* L.**

Photo: Dr. Srinath Rao

S

Sarcostemma acidum Wight. & Arn.

Synonyms: *Sarcostemma brevistigma*
Family: Asclepiadaceae
Common names: English: Moon plant; **Gujarati: Hindi:** Somalata; **Kannada:** Hambu balli; **Sanskrit:** Soma, Somavalli; **Tamil:** Kodikkalli, Somamum; **Telugu:** Aaku jemudu
Distribution: Grown in India, Europe, and the United States. It is found in dry, rocky places in Bihar, Bengal, Konkan, Deccan, Tamil Nadu, Maharashtra, Madhya Pradesh, and Kerala.
Description: A leafless, jointed, trailing perennial shrub with green, cylindrical, fleshy glabrous and twining branches that have milk white latex; leaves reduced to scales, opposite, flowers white or pale greenish white, fragrant, in umbels on branch extremities, fruits follicles, tapering at both ends, seeds flat, ovate, and comose.
Chemical composition: Phytochemical studies have revealed the presence various pregnane glycosides, which are diglycosides and triglycosides. Some important pregnane glycosides are sarcogenin (triglycoside), brevobiose (disaccharide), sarcobiose (nonreducing disaccharide), tigmobiose (nonreducing disaccharide), brevine (triglycoside), and brevinine (pregnane ester diglycoside; Khare et al., 1980a, 1980b; Lalitha and Sethuraman, 2009).
Uses: The herb is highly used by rural and tribal people in curing various disorders like asthma, swelling, fever and cold, dyspepsia, inflammatory infection, and gastric problems, among others. The plant is bitter, acrid, cooling, alternate, narcotic, emetic, antiviral, and rejuvenating. Whole part of the plant is used in vitiated condition of pith, dipsia, viral infection, hydrophobia, psychopathy, and general debility (Lalitha and Sethuraman, 2009).
Used in: Folk
Parts used: Whole plant
Methods of propagation: Seeds

DOI: 10.1201/9781003263500-198

Saussurea lappa C. B. Clarke

Synonyms: *Aucklandia lappa* DC.; *Saussurea costus*

Family: Asteraceae (Compositae)

Common names: English: Costus; **Gujrati:** Upaleta; **Hindi:** Kutha; **Kannada:** Koshta; **Malayalam:** Sepuddy; **Marathi:** Kushta; **Sanskrit:** Kushtha; **Tamil:** Kostum; **Telugu:** Changal, Kostum

Distribution: The plant is found at an altitude of 2600–3600 m in the Himalayan region and in Kashmir

Description: An erect, robust, pubescent, perennial herb, with a stout simple stem 1–2 m high. Leaves are lobate, membranous, scaberulous above, glabrate beneath, auricled at base, irregularly toothed with characteristic odor. The upper leaves are smaller, subsessile or shortly petioled; two small lobes at the base of these leaves almost clasping the stem. The inflorescence is arranged in terminal and axillary clusters, with dark bluish-purple to black flowers. Flower: bluish-purple, stalkless flowers are present in clusters of 2–5. Flowers are borne either terminally or in axils of leaves. Fruits are curved and cupped and about 3 mm long. Fruits contain hairs on them (Kamalpreet et al., 2019).

Chemical constituents: Its chemical constituents such as costunolide, dehydrocostus lactones, cyanropicrin, and others have been isolated and their potential as bioactive molecules substantiated. The root oil contains mainly high boiling sesquiterpenes and sesquiterpene alcohols. The following constituents have been reported: costunolide ($C_{15}H_{20}O_2$), primary sesquiterpene lactone, dehydrocostus lactone (DL), dihydro-dehydrocostus lactone, camphene, costol, β-sitosterol, stigmasterol, betulin, aplotaxene, β-selinine, β-elemene, α- and β-ionones, a bicyclic sesquiterpenic acid, phellandrene, C13-ketone, and certain unidentified constituents, including hydrocarbons, ketones, and alcohols (Ansari, 2019).

Used in: Ayurveda, Siddha, and Unani

Parts used: Root and root oil

Uses: The herb is known to boost metabolic rates, kill helminths, treat bacterial infections, relieve spasms, increase sexual libido, treat diarrhea, relieve flatulence, induce urination, promote expectoration, repel insects, maintain body healthiness, and stimulate its activity. The essential oil of the roots has strong antiseptic and disinfectant properties, which help heal wounds, cuts and bruises quickly. Costus is useful in the treatment of bronchial asthma.

It is a potent anti-inflammatory, which relieves joint aches and pains, as a result of gout or rheumatism. Costus is applied in the following range of conditions: skin conditions, rheumatism, phlegm accumulation, leprosy, jaundice, hiccupping, flatulence, edema, dyspepsia, cough, cholera, and bronchial asthma. As a topical remedy, the herb is applied to heal tumors, skin conditions, severe sores, and wounds. The roots are bitter, acrid sweet, thermogenic, aphrodisiac, anodyne, carminative, digestive, stomachic, diaphoretic, stimulant, depurative, galactagogue, diuretic, disinfectant, antibacterial, antispasmodic, expectorant, febrifuge, skin stimulant, and tonic. Roots are useful in chronic and foul ulcers, cough, asthma, flatulence, cholera, leprosy, leukoderma, jaundice, ring worm, diarrhea, hiccough, hysteria, cardiac disorder, and gout. The root oil is used in perfumes, incense, and in hair tonic (Sun et al., 2003).

Further reading: https://www.easyayurveda.com/2013/09/19/kushta-saussurea-lappa-benefits-usage-dose-side-effects/; https://www.planetayurveda.com/library/kushta-saussurea-lappa/).

Methods of cultivation: Seeds

DOI: 10.1201/9781003263500-199

Sida cordifolia Linn.

Synonyms: *Sida herbacea; Sida rotundifolia; Sida althaeitolia.*

Family: Malvaceae

Common names: Bengali: Brela; **English:** Country mallow; **Guajarati:** Junglimethi; **Hindi:** Bariar, Kungyi, Kharenti; **Kannada:** Hethutti, Bili kurunthotti, Bili kadeeru; **Malayalam:** Anakurunthotti, Kurunthotti, Velluram; **Marathi:** Bala, Chikana; **Sanskrit:** Bala, Batyalaka, Beeiband, Suvarna; **Tamil:** Arivalmanaippundu, Mayir-manikham; **Telugu:** Chirubenda, Tellagorra

Distribution: All over India and Sri Lanka, commonly found growing as a weed along the roadsides

Description: Much branched under shrubs, branchlets densely stellate-tomentose. Leaves: heart-shaped, serrate and truncate, 2.5–7 cm long and 2.5–5 cm broad, with 7–9 veins. Flowers: small, yellow or white in color, solitary and axillaries. Calyx: campanulate; lobes 5, triangular, densely pubescent outside. Petals 5, pale yellow. Stamens: monadelphous. Fruits: a loculicidal capsule, with 8–10 strongly reticulated mericarps, ciliate on the upper margins. Roots: 5–15 cm long with few lateral roots of smaller size; tap roots are generally branched at the tip; outer surface of the root is off to grayish yellow; odor less with slightly bitter taste.

Flowering: August–December **Fruiting:** October–January

Chemical constituents: Alkaloid asparagine, ephedrine (mainly in seeds and aerial parts), phytosterols, and mucin

Used in: Ayurveda

Parts used: Seeds, leaves, and roots

Uses: Oil prepared using its root is applied to cure arthritis and rheumatism. Poultice or paste of the herb is applied to promote suppuration of boils, abscesses, and ulcers. Seeds, known as "Beejaband", are used as aphrodisiac. Leaves are cooked and eaten to cure bleeding piles. The juice of leaves is given to cure diarrhea in pregnant women. The strong decoction of the root is given with ginger to cure fever. The juice of the whole plant pounded with a little water is given to treat gonorrhea. The juice of the leaves is applied to wounds to hasten healing, or the juice of the roots is dropped on the wound and a compress soaked with the juice is bandaged on the wound. The paste made of the powder of the roots with honey is kept in the mouth and slowly swallowed in case of hoarseness of the throat and to improving voice. The roots are boiled in milk until all the water is evaporated; this is taken to reduce inflammation of testicles. A poultice of leaves is applied to reduce inflammations. The paste of flowers and unripe fruits is taken orally with water, to cure painful urination. Root juice is taken until paralysis is cured. Ghee boiled with the paste of the herb is repeatedly taken in large quantities after meals to cure phthisis, headache, loss of voice, cough, The herb is used to treat bronchial asthma, cold, flu, chills, lack of perspiration, headache, nasal congestion, aching joints, and bones, cough and wheezing, infertility, nervous debility, emaciation, and fluid retention in the body. The roots are prescribed in the treatment of fever, urinary, and nervous diseases. The roots are considered cooling, tonic, and astringent.

Methods of propagation: Seeds

DOI: 10.1201/9781003263500-200

Smilax zeylanica L.

Synonyms: *Smilax macrophylla* Roxb.; *Smilax ovalifolia*

Family: Liliaceae

Common names: Bengali: Kumarika; **English:** Rough bindweed; **Hindi:** Kumarika; **Kannada:** Kaadu hambu thavare; **Malayalam:** Kaltmara; **Marathi:** Ghotvel; **Sanskrit:** Ushava, Vanamadhusnahi; **Tamil:** Avadi, Kal thamarai; **Telugu:** Kondadatamara

Distribution: Throughout India, Burma, Malaysia, Myanmar, and Sri Lanka

Description: Armed or unarmed climber. Leaves: coriaceous, lustrous, 7–15 × 4–11 cm, broadly ovate to elliptic, cuspidate, base rounded or shortly cuneate; 3–5-nerved. Petiole: about 15 mm long, base sheathing, with tendrils at the end. Flowers: unisexual, white, in dense umbels; umbels axillary, 1–3 on a common peduncle. Bracts: ovate. Perianth recurved in mature flowers, outer 3 segments, 4 mm long, oblong, inner narrower. Stamens: about as long as the perianth. Female flower: staminodes 3. Fruit: a berry small round and contain 2–4 small seeds.

Flowering and fruiting: January–April

Chemical constitution: The preliminary phytochemical analysis of the plant revealed the presence of glycosides, diosgenin, saponins, β-sitosterol, sarsapogenin/smilagenin, flavonoids, phytosterols, alkaloids, carbohydrates, fixed oils, fats, gums, mucilage, polyphenolic compounds, and tannins (Hooda et al., 2011).

Used in: Ayurveda

Parts used: Leaves

Uses: The roots are used for venereal diseases, also applied in rheumatic swellings and given in urinary complaints and dysentery. It is used for the treatment of syphilis, gonorrhea, and skin diseases and acts as a blood purifier.

Methods of propagation: Seeds

DOI: 10.1201/9781003263500-201

Soymida febrifuga (Roxb.) A. Juss.

Synonyms: *Swietenia febrifuga* Roxb.; *Swietenia obtusifolia* Stokes; *Swietenia soymodia*A. Duncan

Family: Meliaceae

Common names: English: Indian redwood, Bastard cedar; **Hindi:** Rohini; **Sanskrit:** Mamsa Rohini; **Tamil:** Somi, Sittathi; **Telugu:** Somi, Somidha

Distribution: Dry deciduous forest India: Uttar Pradesh, Bihar. Orissa, Madhya Pradesh, Rajasthan, Gujarat, Maharashtra, Andhra Pradesh, Karnataka, Tamil Nadu, and Kerala; and it is found up to an elevation of 200–800 m.

Description: A large, semideciduous tree growing up to 10–20 m tall, 2.5–3 m in girth, with dark brown bark exfoliating into large stripes of scales with persistent leaf scars. Leaves: alternate, distichous, oblong, margin entire crowded at the ends of branches, leaflets 7–13, sessile or sub-sessile, elliptic or oblong, coriaceous, glabrous. Flowers: greenish white in axillary and terminal panicles. Fruits: capsule, ovoid, woody, reddish purple, at length blackish brown.

Flowering: February–April **Fruiting:** May–June

Chemical constituents: The various compounds were isolated from the different parts of plants of *S. febrifuga*, like phenols, flavonoids, alkaloids, steroids, tannins, fatty acids, and others (Singh et al., 2022).

Used in: Ayurveda and Traditional

Parts used: Bark

Uses: It is used in the treatment of diarrhea and dysentery and is also often used as a febrifuge instead of quinine. It is said to be an efficient remedy for the dangerous jungle fever of India, when quinine produces no effect. It has also been employed successfully against typhus fever and has proven to be effective in treating bad cases of gangrene. The deep red bark is very astringent and bitter. The bark is used to treat chronic jaundice. The bark of this plant possess a resinous bitter principle used in Ayurveda (Yoganarasimhan, 1996), adapted for gargles, vaginal infections, enemata, rheumatic swellings, diarrhea, dysentery, and fevers (Chopra et al., 1956). The bark is antimalarial (Kirtikar and Basu, 2003) and anti-inflammatory. A strong, red fiber is obtained from the bark. It is made into ropes; clear gum is obtained from the bark in large pieces. It affords a good mucilage. The wood is used in furniture, buildings, wells, and so on. The bark is used for tanning and dyeing and for intoxicating fish.

Methods of propagation: Seeds. Seeds lose viability very quickly.

DOI: 10.1201/9781003263500-202

Spondias pinnata (L. f.) Kurz.

Synonym: *Mangifera pinnata* L.f.; *Spondias mangifera* Willd; *Spondias amara* Lamk.; *Spondias paniculata* Roxb.

Family: Anacardiaceae

Common name: Bengali: Aamada; **English:** Wild mango, Indian Hog plum; **Gujarati:** Abeda, Ambaada; **Hindi:** Ambara; **Kannada:** Amateykai, Ambatimara; **Marathi:** Ambada; **Malayalam:** Ampalam, Ampazham; **Sanskrit:** Aamaraata; **Tamil:** Mampulicci; **Telugu:** Adavimamidi; **Tamil:** Adavimamadi, Pulima

Distribution: A deciduous tree, distributed in India, Siri Lanka, and Southeast Asian countries

Description: It is a deciduous trees, reaching up to 25 m high, bark 15–25 mm thick, surface gray to pale brown; exudation colorless, gummy. Leaves: alternate, imparipinnate, clustered at the end of branches, estipulate; rachis 25–45 cm, slender, glabrous, swollen at base; leaflets 5–21, opposite or sub-opposite; petiole 2–8 mm long, slender, glabrous; lamina 4–23 × 2.5–10 cm, oblong, elliptic or elliptic-oblong, base obtuse, subacute, oblique or round, apex acuminate or obtusely acuminate, margin entire, glabrous, chartaceous or membranous; lateral nerves many, parallel, close, slender, prominent, connected by an intramarginal nerve, intercostal reticulate. Flowers: polygamous, yellowish white, subsessile, in terminal spreading panicles; calyx small; lobes 5, imbricate aestivation, deciduous; petals 5, oblong, glabrous, spreading, valvate aestivation; disc thick, annular, 10-crenate; stamens 10, inserted below the disc; filaments slender; anthers versatile; ovary superior, ovoid, immersed in the disc, 5-celled, 1 ovule in each cell, pendulous; styles 5, connivant; stigma spreading. Fruit: a fleshy drupe, 3.8–5 cm long, yellow, endocarp woody surround by longitudinal interwoven fibers; seeds 1–3, of which only one is perfect, pendulous, oblong.

Flowering: March–April

Chemical constituents: Triterpenoid compounds, including β-amyrin and oleanolic acid, were isolated from the methanolic extract of *S. pinnata* fruit. Sterols such as stigmast-4-en-3-one, 24-methylenecycloartanone, lignoceric acid, β-sitosterol, and β-sitosterol β-D-glucoside were isolated from the ethanolic extract of *S. pinnata* aerial parts. Ergosterol triterpenes 1 and 2 were isolated from the chloroform/methanol extract of *S. pinnata* bark (Singh and Saxena, 1976; Tandon and Rastogi, 1976; Lai et al., 2014).

Used in: Ayurveda and Traditional

Parts used: Fruits, stem, bark, and roots

Uses: Fruits are astringent and ant scorbutic and used in bilious dyspepsia; unripe fruit is used as aphrodisiac. Bark astringent and refrigerant used in diabetes, diarrhea, and dysentery; a paste of the bark is applied to cure rheumatism, and the roots are used in regulating menstruation. In India, the bark of *S. pinnata* is used as a rubefacient for the treatment of painful joints. It is also used to treat diarrhea and dysentery and to prevent vomiting. A decoction prepared from the root bark is used to regulate menstruation and treat gonorrhea. The gum produced from *S. pinnata* is used as a demulcent (Kritikar and Basu, 1975; Benthall, 1993).

Methods of propagation: Shoot cuttings

DOI: 10.1201/9781003263500-203

Swertia chirayita (Roxb. ex Fleming) H. Karst

Family: Gentianaceae

Common names: Bengali: Chireta **English: Hindi: Kannada:** Nelabevu; **Malayalam:** Nilaveppa; **Marathi:** Chiraita; **Sanskrit:** Chiratitka; **Tamil:** Nilavembu; **Telugu:** Charayatah

Distribution: Indigenous t temperate Himalaya. Found at an altitude of 1200–3000 m (4000–10,000 ft), from Kashmir to Bhutan, and in the Khasi hills at 1200–1500 m (4000–5000 ft). It can be grown in subtemperate regions between 1500- and 2100-m altitude.

Description: Chiraita is an erect, biennial herb, 50–125 cm tall. Stem is robust, branching, cylindrical below, 4-angled upward, containing large pith. Leaves are opposite, broadly lanceolate, 5-nerved, and subsessile. Flowers are lurid greenish yellow, tinged with purple in large panicles. Capsules are egg-shaped; seeds are minute, smooth, and many-angled.

Flowering and fruiting: August–October

Chemical constituents: Xanthones, flavonoids, terpenoids, iridoids, and secoiridoid glycosides (Pant et al., 2000). It also contain amarogentin, amaroswerin, gentiopicirin, mangiferin, sweroside, swerchirin, and swertiamarin (Kumar and VanStaden, 2016).

Parts used: Entire plant dried

Uses: The plant has wide range of medicinal uses includes treatment of anemia, bronchial asthma, chronic fever, constipation dyspepsia, diabetes, epilepsy, gastritis, hepatitis, hypertension, liver disorders, malaria, melancholia, mental disorders, skin disorders, ulcers, and worms (Banerjee et al., 2000; Rai et al., 2000). Its extract is reported to show anti hepatitis B virus activities (Zhou et al., 2015).

The plant is used as a bitter tonic in treatment of fever and for curing various skin diseases. The bitterness, anthelmintic, hypoglycemic, and antipyretic properties are attributed to amarogentin (most bitter compound isolated to date), swerchirin, swertiamarin, and other active principles of the herb. Herbal medicines such as AYUSH-64, diabecon, mensturyl syrup, and Melicon V ointment 2–4 contain extract in different amounts for its antipyretic, hypoglycemic, antifungal, anticancer, and antibacterial properties. Hirayata removes the toxins (Ama) and detoxifies the body completely, and it makes it a good herb for rejuvenation. Chirayata is one of the herbs that reduces the toxins and gives best nutrition to the body. It reduces the extra body fat and weight and gives proper energy to the body. Being strong bitter herb, chirayata is a good analgesic, antipyretic, and anti-inflammatory, and anti-cancerous. A decoction is used to clean discharging wounds. Persons suffering from generalized itching are asked to take bath with its decoction. It improves the digestive system to stimulate liver for proper secretion of bile, increases appetite, relieves thirst, helps in amapachan, and as a laxative; therefore, it is beneficial in all digestive problems. It is useful for loss of appetite and dyspepsia and purifies the blood. It is beneficial to reduce inflammation and is used to reduce blood sugar levels to control diabetes. It is hemostatic and therefore useful in bleeding piles. is helpful in jaundice and anemia. It is an expectorant and anti-asthmatic. It is useful in asthma produced by obstruction of air passage. Breast milk is purified by digesting rasa dhatu; it produces pure and easily digestible milk. It has worm-destroying properties and is used in killing intestinal worms. An infusion of the herb is taken for this purpose. It is useful in malarial fever (Kumar and VanStaden, 2016).

Methods of propagation: Seeds

DOI: 10.1201/9781003263500-204

T

Thunbergia grandiflora (Roxb. ex Rottl.) Roxb.

Synonyms: *Flemingia grandiflora* Roxb. ex Rottl.; *Pleuremidis grandiflora* (Roxb. ex Rottl.) Raf.; *Thunbergia adenophora* W. W. Sm.; *Thunbergia chinensis* Merr.; *Thunbergia cordifolia* Nees; *Thunbergia lacei* Gamble

Family: Acanthaceae

Common names: Bengali: Neel lota; **English:** Bengal clock vine, Bengal trumpet vine, Blue sky flower; **Hindi:** Neel lata; **Gujarati:** Tumakhlung

Distribution: Native to China and India, now found in Australia, Myanmar, Nepal, Thailand, and Vietnam

Description: Woody vine, twining, 10–15 m in length. Stems: quadrangular, pubescent, up to 2.5 cm in diameter. Leaves: simple, opposite; blades 15–26 × 13–30 cm, ovate or broadly ovate, chartaceous, the apex acute or acuminate, the base cordiform or subcordiform, the margins lobate-dentate, ciliate; upper surface dark green, shiny, puberulous, with slightly prominent venation, petioles 6–12 cm long, sulcate. Inflorescence: axillary or terminal drooping racemes. Flowers: whitish blue or violet, pedicels robust, cylindrical, 4–6 cm long; bracts light green, membranaceous, ovate, 4–6 cm long, covering the calyx and the corolla tube. Calyx green, in the form of a ring, 4–5 mm long; corolla lilac or white, infundibuliform, with 5 lobes, the tube 6–7 cm long, light yellow inside, narrow at the base, the limb 6–7 cm in diameter. Stamens: 4 didynamous, ovary bilocular, fleshy, style long stout, stigma bifid. Fruit: a capsule about 3 cm long, subglobose at the base, the upper half in the form of a beak, dehiscent in two halves.

Flowering and fruiting: Throughout the year

Chemical constituents: The plant contains apigenin, carboxylic acid, flavonoids, iridoid glycosides, luteolin, luteolin glucosides, and malvidin (Ibrahim et al., 2017).

Used in: Traditional

Parts used: Leaves, stem, and roots

Uses: All parts of this plant are used medicinally; different parts are used for different diseases. Leaves are antipyretic; used as a poultice in stomach complaints, cuts, and boils; and as an antidote for scorpion bite and snakebite. Dried leaves are used in an herbal tea to treat alcohol and drug abuse; the leaf sap is used to treat ear infections, deafness, and menorrhagia. The leaves also possess antibacterial, antifungal, anti-inflammatory, antioxidant, and antitoxic properties. The roots are used to treat rheumatic arthralgia and swellings caused by falls and fractures (Chatterjee et al., 2015).

Methods of propagation: Seeds and stem cuttings

DOI: 10.1201/9781003263500-206

Thunbergia fragrans Roxb.

Synonyms: *Thunbergia volubilis*

Family: Acanthaceae

Common names: English: White clock vine, White lady, White thunbergia; **Hindi:** Chimin; **Kannada:** Hegala balli, indrapushapa; **Malayalam:** Noorvan-valli; **Marathi:** Chimine; **Tamil:** Indrapushpam, Kaasanangaai, Thannikkodi; **Telugu:** Palatheega

Description: A slender, herbaceous twiner or climber with quadrangular strigose young shoots. Leaves: 2–4.5 cm long, slender petiole; lamina broadly elliptic to ovate or lanceolate-ovate, 4–10 × 2–5 cm, entire or few toothed, sparsely strigose on both sides near the truncate to cordate base, palmately 3–5-nerved, acute-acuminate. Flowers: white, 4–5 cm long, about 5 cm across, solitary axillary, on 2–7-cm-long stout pedicels; bracts connate and spathaceous, oblong-lanceolate, 1.5–2 cm × about 6–8 mm, acute, hairy. Calyx: cupular with 12–16, subulate, 1–3 mm long, glandular hairy teeth. Corolla tube cylindrical, 2–2.5 cm long; lobes 5, subequal, nearly as long as tube, puberulous outside. Staminal filaments are glandular, hairy below; anthers oblong, slightly beaked above, without spur at base. Style: approximately 2 cm long. Capsule: depressed globose, about 8 mm across, scabrid, puberulent, with 1–1.5 cm long beak. Seeds: reticulate.

Flowering and fruiting: Throughout the year

Chemical constituents: The entire plant contains trigonelline; the flowers contain apigenin 7-O-β-glucuronide and daucosterol-4 (Sheshadri Shekhar et al., 2007).

Used in: Traditional and Folk

Parts used: Twigs, leaves, and roots

Uses: Tender twigs are made into a paste and applied all over the body against the fever. Leaves used as poultice in skin diseases; tender leaf juice is poured in the nose and the paste applied on face for giddiness and on forehead for headache; a crushed-leaf decoction is drunk for paralysis; a leaf paste is applied on cuts and wounds; leaves ground with cumin mixed in water and taken orally as abortifacient. Fresh root juice is used as eye drops. In magico-religious beliefs, it is used to keep off evil spirits and ghosts; a flower paste is applied all over the body, and fruits are tied as a necklace.

Methods of propagation: Seeds and stem cuttings

DOI: 10.1201/9781003263500-207

Photo: C.T. Chandralekha

Thymus vulgaris L.

Synonyms: *Thymus aestivus* Reuter ex Willk. & Lange; *Thymus chinensis* K. Koch.; *Thymus ilerdensis* F. Gonzalez ex Costa; *Thymus webbianus* Rouy

Family: Lamiaceae (Labiatae)

Common names: English: Garden thyme, German thyme; **Hindi:** Jangali ajwain. Banajwain; **Malayalam:** Thottathulasi

Distribution: Thyme is a native to Southern Europe from Spain to Italy. Apart from Europe, it is grown in Australia, North Asia, North Africa, Canada, and the United States. In India, it is cultivated in the western temperate Himalayas and Nilgiris.

Description: *T. vulgaris* L. is an aromatic, perennial, straight growing plant that measures up 10–30 cm in height with woody base. The leaves are small, opposite, grayish-green-colored, oblong–lanceolate to linear, and are gland-dotted. They are measured up to 5–10 mm long and 0.8–2.5 mm wide with recurved margins. The flowers are light violet in color, two-lipped, and possess a hairy glandular calyx. They measure up to 5 mm long with leaf-like bracts in loose whorls arranged in axillary clusters on the branchlets or in terminal oval or rounded heads.

Flowering and fruiting: April–October

Chemical constituents: The major constituents of commercial *T. vulgaris* essential oil are thymol (23%–60%), γ-terpinene (18%–50%), *p*-cymene (8%–44%), carvacrol (2%–8%), and linalool (3%–4%; Duke, 1992).

Used in: Traditional

Parts used: Aerial parts and oil extract

Uses: The different extracts and essential oil obtained from the plant have been assessed and reported to treat ailments like microbial infections, inflammation, noncommunicable diseases like cancer, and sexually transmitted diseases like HIV-1 and herpes. The plant was chiefly used for the treatment of wounds, as it possesses healing and antiseptic properties. Its aerial parts are used for fumigation and for, treating skin and respiratory diseases. The plant possesses the ability to cure gastrointestinal aberrations, as the extract was given to treat worms in children. The oil reduces rheumatic aches and sciatica. Flower shoots are used to treat mouth and skin infections, high blood pressure, heart problems, fluid retention, cystitis, digestive system disorders, rheumatism, and arthritis (Basch et al., 2004; Kuete, 2017; Li et al., 2019; Patil et al., 2021).

Thyme has been thought of to be antiseptic, antimicrobial, medication, astringent, anthelmintic, medicinal drug, carminative, disinfectant, medicinal drug, and tonic. Thyme is incredibly useful in cases of assorted intestinal infections and infestations, like hookworms, ascarids, Gram-positive and -negative bacteria, fungi, and yeasts, as well as *Candida albicans*. Its active constituent, thymol, is active against enterobacteria and cocci bacteria. Thyme may also improve liver functioning, and act as an appetite stimulant. It will be used in treatment of cartilaginous tube, bronchial and urinary infections. It is used as a gargle, thyme is helpful in treatment of laryngitis and inflammation. The main component of the volatile oil of thyme, thymol, is active against enterobacteria and cocci bacteria.

Methods of propagation: Propagated from cuttings.

DOI: 10.1201/9781003263500-208

Trichopus zeylanicus Gaertn.

Synonyms: *Podianthus arifolius* Schnizl.; *Steireya angustifolia* (Lindl.); *Trichopodium angustifolium* Lindl.; *Trichopodium travancoricum* Bedd.

Family: Dioscoreaceae (Trichopodaceae)

Common names: Malayalam: Arogya pacha

Distribution: Native to India. The herb grows on sandy soil near rivers and streams in shady places in lowland and intermediate altitude forests. grows in Malaysia, Singapore, Sri Lanka, Thailand, and the Southwestern Ghats mountains of South India.

Description: A small herbaceous plant. The leaves are about 20 cm (7.9 in.) long and grow from a rhizome. The shape of the leaves can be highly variable even within one location, but the most common shape is cordate. The unusual flowers are purplish black.

Flowering: It flowers yearlong.

Chemical constituents: The leaf of the plant contains flavonoid glycosides, glycolipids, and some other nonsteroidal compounds.

Parts used: Leaves and fruits

Uses: The medical properties of *T. zeylanicus* were discovered by modern Indian scientists only recently on a scientific expedition to the Agastya Hills in the Western Ghats in December 1987. They noticed that their guides, belonging to the Kaani tribe, were very energetic in sharp contrast to themselves. They had walked for several hours with the scientists, but the difference was that they ate the fruits of a wild plant *T. zeylanicus* as they walked. It was found from the Kaani men that it was indeed the fruits they were eating that made them energetic, a fact about the plant well known to the tribe for ages. The Malayalam name of the plant is "Arogya pacha", literally meaning "the green that gives strength". The plant is also used to treat liver disease, stomach ulcers, fatigue, and sexual performance problems (Pushpangadan et al., 1988).

Methods of cultivation: Shoot tip culture

Further reading: https://en.wikipedia.org/wiki/Trichopus zeylanicus

DOI: 10.1201/9781003263500-209

Trichosanthes tricuspidata Lour.

Synonyms: *Trichosanthes bracteata* (Lam.) Voigt; *palmata* Roxb.; *Trichosanthes pubera* Blume.; *Modecca bracteata* Lam.

Family: Cucurbitaceae

Common names: English: Redball snakegourd; **Hindi:** Lal indrayan; **Kannada:** Kagemari balli; **Malayalam:** Kalayar; **Marathi:** Kaundal; **Sanskrit:** Vishala; **Telugu:** Avuduta

Distribution: It ranges from the Eastern Himalayas in India and southern China through southern Japan, Malaysia, and tropical Australia, which is found at an elevation of 1200–2300 m.

Description: It is a large climber, often attaining a height of 9–10 m. It has a robust stem that is woody below and has 3-cleft tendrils. The leaves are variable, palmately 3–5-lobed with a cordate base, and the lobes are ovate to oblong with serrate or dentate margins. Male flowers are in axillary 5–10 flowered racemes with large bracts, while the female flowers are solitary. The corolla petals are wedge-shaped, fringed and white in color. The fruits are globose, and when ripe, they are red with 10 orange streaks.

Flowering and fruiting: May–November

Chemical constituents: Mohamed (1974) isolated a tetrahydroxy pentacyclic triterpene "trichotetrol" from the root extract of this vine. From the fruits of *T. tricuspidata*, 14 cucurbitane glycosides were isolated (Kanchanapoom et al., 2002). An extract of the fruits of this plant was found to be cytotoxic in KB cells, and 2 new cucurbitacins were reported: tricuspidatin and 2-O-glucocucurbitacin J. (Maile et al., 1994). The root contains methyl palmitate, palmitic acid, suberic acid, α-spinasterol, stigmast-7-en-3-beta-ol, α-spinasterol 3-o-beta-D-glucopyranoside, stigmast-7-en-3-beta-ol-3-O-beta-D-glucopyranoside, glyceryl 1-palmitate, glyceryl 1-stearate, bryonolic acid, cucurbitacin B, isocucurbitacin B, 3-epi-isocucurbitacin B, 23,24-dihydrocucurbitacin D, isocucurbitacin D and D-glucose (Bhandari et al., 2008).

Used in: Ayurveda, Unani, and Traditional

Parts used: Fruits, seeds, and root

Uses: In Ayurvedic medicines, the fruits are used in the treatment of asthma, earache, and ozena (intranasal crusting, atrophy and fetid odor). In the Unani system of medicine, the fruits are used as a carminative (an agent that relieves flatulence), a purgative, and an abortifacient, to lessen inflammation, cure migraines, and reduce heat of the brain, as a treatment for ophthalmia (inflammation of the eye), leprosy, epilepsy, and rheumatism (painful local inflammation of joints and muscles), as well as other uses. The seeds are emetic and a good purgative. In the Thai traditional system of medicine, the plant is used as an antifever remedy, a laxative, and an anthelmintic, as well as in migraine treatments (Kanchanapoom et al., 2002). The roots of the plant are used to treat lung diseases in cattle and for the treatment of diabetic carbuncles and headaches (Chopra et al., 1956) has reported the use of this plant in curing bronchitis, and the application of seed paste for hoof and mouth disease in cattle. The practitioners of Ayurveda also use the fruits in treating stomatitis. The oil extracted from the root is used as a pain killer. In Bastar district, Chhattisgarh, India, the plant is used for curing snakebite poisoning and the juice of the plant is applied externally for skin eruptions, An extract of the fruits of this plant was found to be cytotoxic in KB cells (Maile et al., 1994).

Methods of Propagation: Seeds

DOI: 10.1201/9781003263500-210

U

Uraria crinata (L.) Desv. ex DC.

Synonyms: *Abrus crinita*; *Hedysarum crinitum* (Lin); *Doodia crinita* (L.) Roxburgh; *Hedysarum comosum* Vahl; *Uraria comosa* (Vahl) Candolle; *Uria pecta* Wight; *Uraria macrostachya* (Wal-lich) Prain.

Family: Fabaceae (Leguminosae)

Common names: Bengali: Sankarjata; **English:** Asian foxtail, Cat's tail plant; **Hindi:** Dabra, Pitvan, Shankraj; **Malayalam:** Muvila; **Marathi:** Pitvan, Prisniparni; **Sanskrit:** Prushthiparni

Distribution: China, Taiwan, India, Southeast Asia. Growing in grassland/savanna/scrubland and disturbed areas, at elevations up to 1500 m. In India found in Assam and Meghalaya

Description: A woody shrub 1–1.5 m tall, with gray pubescent erect stems and long bushy inflorescence. Leaves are pinnately compound with about 2–3 pairs of lateral leaflets and 1 slightly larger terminal leaflet. Leaflets are oblong to narrowly ovate (6–15 cm long, 3–8 cm wide). The petioles are often long (5.5–15 cm) gray-white pubescent on midvein, base rounded to slightly cordate, apex slightly acute, obtuse, or rounded. The inflorescence is light purple and shaped like a cone (15–30 cm long). Calyx: shallowly cup-shaped, long white hirsute; lower lobes ca. 3.5 mm, upper lobes about 3 mm. Corolla: purple, 6–9 mm, have one large ovate petal that hangs over 3 smaller petals (0.6–0.9 cm wide). Fruit is an elliptic pod. Pod: 2–4(–7)-jointed, constricted between joints, indehiscent, shiny black, hairy. Seed: reniform, globose, compressed, 1.5–2 mm in diameter, brown.

Flowering and fruiting: April–September

Chemical constituents: A new isoflavanone, (3S)-5,7-dihydroxy-2′,3′,4′-trimethoxy-6,5′ diprenylisoflavanone, and 8 known compounds, including five flavones (2–6), two triterpenes and a steriod, were isolated from the whole plant.

Used in: Traditional

Parts used: Whole plant

Uses: This species is often used in traditional medicine, with different plant parts being used to cure a range of conditions including dysentery and diarrhea, enlarged spleen and liver, and tumors. It is also applied against pustules and fistulae. The whole plant is used medicinally for stopping bleeding, reducing fever, and relieving coughs. It is effective in driving out intestinal worms and other parasites and is sometimes used as a carminative for children (http://proseanet.org/). A decoction of the roots is used in the treatment of diarrhea (Uphof, 1959). The crushed leaves are applied externally to rid the hair of lice.

Methods of cultivation: Seeds

DOI: 10.1201/9781003263500-212

V

Vernonia anthelmintica (L.) Willd.

Synonyms: *Baccharoides anthelmintica* (L.) Moench.;*Conyza anthelmintica* L.; *Dolosanthus sylvaticus* Klatt.; *Vernonia stenolepis* Oliv.

Family: Asteraceae (Compositae)

Common names: English: Purple fleabane; **Hindi:** Kaleejeera; **Kannada:** Kadu jiriga; **Malayalam:** Kalajirakam, Kattujeeragum; **Marathi:** Kalijiri; **Sanskrit:** Aranyajeeraka, Brihatpali, Vaakuchee; **Tamil:** Kattu shiragam; **Telugu:** Adavijilakara

Distribution: Throughout India in hilly areas

Description: Stout herbs, growing about 1.5 m high; stems scabrid. Leaves: alternate about 13 × 5 cm, elliptic, acute at apex and base, scabrous; sharply serrate; nerves 10–11 pairs; petiole 1–2 cm long. Heads: 15 × 12 mm, in lax corymbose panicles; bracts 10 mm long, leafy, oblong, obtuse; inner bracts smaller, 8 × 2 mm. Flowers: 30–40; corolla 10 mm long, with a narrow basal part, widen above, lobes acute, glabrous; anthers 2 mm long. Fruit: an achene 4 × 1 mm, glandular between ribs, pubescent; outer pappus paleaceous, fimbriate, inner pappus 6 mm long, reddish.

Flowering and fruiting: September–March

Chemical constituents: phytochemical investigations of the herb have shown that it contains flavones, caffeoylquinic acid derivatives, sesquiterpene, triterpenoids, elemanolide sesquiterpenes, and guaianolide sesquiterpenoids (Wang et al., 2012).

Used in: Folk and Traditional

Parts used: Seeds

Uses: This plant has a long history of traditional use for the management of several disorders related to skin, central nervous system, kidney, gynecology, gastrointestinal, metabolism, and general health. It has been used in local medicine to cure a wide spectrum of disorders including asthma, sores, inflammatory swellings, skin ailments, kidney troubles, itching of the eyes, and hiccough (Kirtikar and Basu, 1993; Manvar and Desai, 2012). The medicinal plant has been reported to possess a variety of pharmacological activities, such as antimicrobial, anticancer, antidiabetic, anti-inflammatory, analgesic, and antipyretic, as well as a diuretic and larvicidal activities (Amir and Chin, 2011; Dogra et al., 2020). In the Uyghur traditional system, which is one of the conventional systems followed in Central Asia, the plant is used to improve vitiligo, pain, stomachache, and liver ailments and as an anthelmintic and detumescent.

Methods of propagation: Seeds

DOI: 10.1201/9781003263500-214

W

Wikstroemia indica (L.) C. Meyer

Synonyms: *Daphne indica* L. *Daphne cannabina* Lour. *Daphne viridiflora* Wall. *Wikstroemia viridiflora* Meisn. *Wikstroemia valbrayi* H. Lév. *Capura purpurata* L.

Family: Thymelaeaceae

Common names: English: Tie bush, Indian stringbush

Distribution: Native in China, Taiwan, India, Indonesia, Malaysia, Papua New Guinea, the Philippines, Thailand, Vietnam, and Australia

Description: A shrub growing 1–2 m tall or more. Branches: reddish brown, glabrous. Leaves: opposite; petiole about 1 mm; leaf blade reddish brown on both surfaces when dried, obovate, elliptic-oblong, or lanceolate, 2–5 × 1–2 cm, papery to thinly leathery, both surfaces glabrous, base broadly or narrowly cuneate, apex obtuse or acute; lateral veins dense, slender, at narrow angle to midrib. Inflorescences: terminal, capitate, several flowered; peduncle 1 cm, glabrous. Pedicel 1–2 mm. Calyx: yellowish green, 7–12 mm, exterior glabrescent; lobes 4, broadly ovate to oblong, about 3 mm, apex acute or obtuse. Stamens, 8. Disk: scales often 2 or 4. Ovary: obovoid or ellipsoid, glabrous or apex sparsely pubescent; style very short; stigma capitate.

Flowering and fruiting: Rainy season

Chemical constituents: An alcoholic extract of the plant was found to contain daphnoretin, chrysophanol, myricitrime, and rutin.

Used in: Folk

Parts used: Dried foliage and whole herb

Uses: As a traditional Chinese herb, this plant has long been employed as an antipyretic, detoxicant, expectorant, vermifuge, and abortifacient. Traditionally, it is widely employed against anemia, arthritis, cancer, syphilis, and whooping cough. Also, it is used to cure fever, pains, snakebites, liver disorders, skin eruptions, blotches, pimples, diarrhea, and sore throats.

Methods of propagation: Seeds

DOI: 10.1201/9781003263500-216

X

Xanthium strumarium L.

Synonyms: *Xanthium canadense* Mill.; *Xanthium occidentale* Bertol.; *Xanthium pennsylvanicum* Grand.; *Xanthium sibricum* Patrin ex Widder.; *Xanthium indicum Klatt*

Family: Asteraceae (Compositae)

Common names: English: Common cocklebur, Broad bur; **Gujarati:** Godrian; **Hindi:** Bada gohura, Chota dathura Sankhauuli; **Kannada:** Maruluumatti; **Malayalam:** Marul-umattai; **Marathi:** Ghagara, Shankeshrvar; **Sanskrit:** Arishta, Medhya, Sarpakshi; **Tamil:** Marulumattai; **Telugu:** Marulutige

Distribution: Mostly found in temperate zones but also found in subtropical and Mediterranean climates, along seashores and rivers. In India, it is commonly found on roadsides, in rice fields, hedges, and waterways throughout tropical parts.

Description: An annual herb, up to 1 m in height, with a short, stout, hairy stem, Stems are round or slightly ribbed, often speckled with purple and have short white hairs scattered across the surface; leaves, broadly alternate, are triangular-ovate or suborbicular, light and bright green in color in an alternate pattern with irregular lobes and relatively inconspicuous teeth, 5–15 cm long, often three-lobed, with prominent veins, long petiole, scabrous on both sides. Inflorescence is a raceme in leaf axils or at the end of branches. The flowers are monoecious male heads are situated above the female heads and are white or green; numerous male uppermost, female ovoid, covered with hooked bristles. Fruits are obovoid, enclosed in a hard involucre, with two hooked beaks and hooked bristles.

Flowering: July–September **Fruiting:** August–October

Chemical constituents: The aerial parts of the plant contain a mixture of unidentified alkaloids. Besides alkaloids, the aerial parts of the plant contain sesquiterpene lactones, namely, xanthinin; its stereoisomer, xanthumin, xanthatin; carboxyatractyloside; phytosterols, xanthanol, isoxanthanol, xanthinosin, 4-oxo-bedfordia acid, hydroquinone; xanthanolides; sulfated glycoside, atractyloside and xanthostrumarin. The fruits contain a number of medically active compounds, including glycosides and phytosterols (Parveen et al., 2017).

Used in: Ayurveda

Parts used: Leaves and root

Uses: The plant has alexiteric, anthelmintic, antipyretic, antirheumatic, anti-syphilitic, cooling, digestive, fattening, laxative, tonic, activities and improves appetite, complexion, memory and voice. It cures biliousness, epilepsy, fever, leukoderma, poisonous bites of insects, and salivation. The leaves and roots are used for their anodyne, appetizer, diaphoretic, diuretic, emollient, laxative, and sedative activities. An infusion of the plant has been used in the treatment of rheumatism, diseased kidneys, and tuberculosis. It has also been used as a liniment on the armpits to reduce perspiration.

The plant is anodyne, antibacterial, and antifungal. The root is a bitter tonic and febrifuge. It has historically been used in the treatment of scrofulous tumors and used locally on ulcers, boils, and abscesses. The paste of green spiny fruits is used against migraine, the juice of leaves and fruits is believed to be useful for smallpox, and the roots are used for cancer. The burs are used in China as a tonic, diuretic, and sedative. A decoction of the root has been used in the treatment of high fevers and leucorrhea and to help a woman expel the afterbirth. A decoction of the seeds has been used in the treatment of bladder complaints. A poultice of the powdered seed has been applied as a salve on open sores. Seed oil is used for treating bladder infections, herpes, and erysipelas (Chopra et al., 1986; Kamboj and Saluja, 2010; Parveen et al., 2017).

DOI: 10.1201/9781003263500-218

Methods of propagation: Seeds

Photo: Dr. Srinath Rao

References

Adewole S, Adewole J. 2008. Protective effects of Annona muricata linn. (Annonaceae) leaf aqueous extract on serum lipid profiles and oxidative stress in hepatocytes of streptozotocin-treated diabetic rats. *African Journal of Traditional, Complementary and Alternative Medicines* 6:30–41.

Adewole SO, Caxton-Martins EA. 2006. Morphological changes and hypoglycemic effects of Annona muricata Linn. (Annonaceae) leaf aqueous extract on pancreatic B-cells of streptozotocin-treated diabetic rats. *African Journal of Biomedical Research* 9:173–187.

Adhikari PK, Paul SB. 2018. Medicinally important plant cleome gynandra: A phytochemical and pharmacological explanation. *Asian Journal of Pharmacy and Clinical Research* 11(1):21–29.

Afroz S, Oliur Rahman M, Abul Hassan MD. 2018. Taxonomic revision of the genus crinum L. (LILIACEAE) of Bangladesh. *Bangladesh Journal of Plant Taxonomy* 25(2):257–271.

Agnihotri VK, Thappa RK, Meena B, Kapahi BK, Saxena RK, Qazi GN, Agarwal SG. 2004. Essential oil composition of aerial parts of Angelica glauca growing wild in North-West Himalaya (India). *Phytochemistry* 65(16):2411–2413. https://doi.org/10.1016/j.phytochem.2004.07.004. PMID: 15381015.

Ahirwar B, Ahirwar D. 2021. Lepadigathiscristata wild. Traditional plant of Chhattisgarh tribes: An alternative cure. *Research Journal of Pharmacy and Technology* 14(6):3430–3432.

Ahmed HM, Nour BY, Mohammed YG, Khalid HS. 2010. Antiplasmodial activity of some medicinal plants used in Sudanese folk-medicine. *Environmental Health Insights* 4:1–6.

Ahmad I, Ambarwati N, Elya B, Omar H, Mulia K, Yanuar A, Negishi O, Mun A. 2019. A new angiotensin-converting enzyme inhibitor from Peperomia pellucida (L.) Kunth. *Asian Pacific Journal of Tropical Biomedicine* 9:257–262.

Ahmad M, Butt MA, Zhang G, Sultana S, Tariq A, Zafar M. 2018. Bergenia ciliata: A comprehensive review of its traditional uses, phytochemistry, pharmacology and safety. *Biomedicine & Pharmacotherapy* 97:708–721.

Al-Dosari MS. 2010. The effectiveness of ethanolic extracts of Amaranthus tricolor Linn: A natural hepatoprotective agent. *American Journal of Chinese Medicine* 38:1051–1064.

Ali M, Kenganora M, Manjula SN. 2016. Health benefits of Morinda citrifolia (Noni): A review. *Pharmacognosy Journal* 8(4):321–334.

Ali S, Zameer S, Yaqoob M. 2017. Ethnobotanical, phytochemical and pharmacological properties of Galinsoga parviflora (Asteraceae): A review. *Tropical Journal of Pharmaceutical Research* 16(12):3023–3033.

Alimboyoguen AB, De Castro-Cruz KA, Fidel EM, Shen CC, Tsai PW. 2022. Chemical constituents of Phyllanthus acidus (L.) skeels. *Indian Journal of Pharmaceutical Education and Research* 56(2):455–460.

Almeida I, Fernandes E, Lima J, Costa P, Bahia M. 2008. Walnut (Juglans regia) leaf extracts are strong scavengers of pro-oxidant reactive species. *Food Chemistry* 106:1014–1020.

Al-Quran S. 2008. Taxonomical and pharmacological survey of therapeutic plants in Jordan. *Journal of Natural Products* 1:10–26.

al-Sereiti MR, Abu-Amer KM, Sen P. 1999. Pharmacology of rosemary (Rosmarinus officinalis Linn.) and its therapeutic potentials. *Indian Journal of Experimental Biology* 37(2):124–130.

Al-Shamma A, Miyscher LA. 1979. Comprehensive survey of indigenous Iraqi plants for potential economic value. 1. Screening results of 327 species for alkaloids and antimicrobial agents. *Journal of Natural Products* 42(6):634–642.

Al-Snafi AE. 2014. The pharmacology of Anchusa italica and Anchusa Strigosa: A review. *International Journal of Pharmacy and Pharmaceutical Sciences* 6(4):7–10.

Al-Snafi AE. 2015a. The chemical constituents and pharmacological effects of Ammania baccifera: A review. *International Journal of Pharmacy* 5(1):28–32.

Al-Snafi AE. 2015b. The pharmacological and therapeutic importance of agrimonia eupatoria: A review. *Asian Journal of Pharmaceutical Science & Technology* 5(2):112–117.

Al-Snafi AE. 2016c. The pharmacological activities of Cuminum cyminum: A review. *IOSR Journal of Pharmacy* 6(6):46–65.

Al-Snafi AE. 2019a. Constituents and pharmacology of Luffa cylindrica: A review. *Journal of Pharmacy* 9(9):68–79.

Al-Snafi AE. 2019b. A review on Lycopus europaeus: A potential medicinal plant. *IOSR Journal of Pharmacy* 9(7):80–88.

Al-Snafi AL. 2015. The chemical constituents and pharmacological effects of Ammania baccifera-A review. *International Journal of Pharmacy* 5(1):28–32.

Amin GR. 2005. *Popular Medicinal Plants of Iran*. Tehran, Iran: Vice-Chancellorship of Research, Tehran University of Medical Science Press, pp. 38–162.

Amir F, Chin KY. 2011. The chemical constituents and pharmacology of Centratherum anthelminticum. *International Journal of PharmTech Research* 3(3):1772–1779.

Amir F, Yam WS, Chin KY. 2011. Chemical constituents and biological applications of Lippia nodiflora. *Archives of Pharmacy Practice* 2(3):101–105.

Amritha CK, Suma C, Ananya VM, Thushara PV, Deepak VS. 2020. Phytochemical and pharmacological activities of justicia procumbens L.: A review. *International Journal of Research Ayurveda and Pharmacy* 11(5):190–193.

Aneja KR, Joshi R, Sharma C. 2010. Potency of Barleria prionitis L. bark extracts against oral diseases causing strains of bacteria and fungi of clinical origin. *New York Science Journal* 3:5–12.

Aneja S, Vats M, Sardana S, Aggarwal S. 2011. Pharmacognostic evaluation and phytochemical studies on the root of Amaranthus tricolor (Linn.). *International Journal of Pharmaceutical Sciences and Research* 2:2332–2336.

Anjaneyulu AS, Rao SS. 1998. A new ketosteroid from the bark of Couroupita guianensis Aubl. *Indian Journal of Chemistry* 37(4):382–386.

Anjaria J, Parabia M, Dwivedi S. 2002. *Ethnovet Heritage Indian Ethnoveterinary Medicine – An Overview*. Ahmedabad, India: Pathik Enterprise.

Anonymous. 1948–1976. *The Wealth of India, Raw Materials*. Vols. I–XI. New Delhi: Publication and Information Directorate, CSIR.

Anonymous. 1969. *The Wealth of India: A Dictionary of Indian Raw Materials and Industrial Products*. Vol. 8. New Delhi: Publication and Information Division, CSIR, pp. 58–59, 142–143.

Anonymous. 1985. *The Wealth of India: Raw Materials*. New Delhi, India: National Institute of Science Communications and Information Resources, Council of Scientific and Industrial Research, pp. 114, 121.

Anonymous. 1987. *Medicinal Plants of India*. Vol. II. New Delhi: Indian Council of Medical Research, p. 122.

Anonymous. 1997. *Standardization of Single Drugs of Unani Medicine*. Part III. New Delhi: CCRUM, Ministry of H & FW, Government of India, pp. 295–300.

Anonymous. 1998. *The Wealth of India*. Hillside, New Delhi, India: Publication and Information Directorate, CSIR.

Anonymous. 2003. The wealth of India: Raw material. In *Actinodaphne*. New Delhi: Publication and Information Directorate CSIR, pp. 70–71.

Anonymous. 2005. *The Wealth of India*. Vol. 7. New Delhi: Council of Scientific and Industrial Research.

Anonymous. 2009. *The Wealth of India*. Vol. 6. New Delhi: Council of Scientific and Industrial Research, pp. 12–15.

Ansari S. 2019. Ethnobotany and pharmacognosy of Qust/ Kut (Saussurea lappa, C. B. Clarke) with special reference of Unani medicine. *Pharmacognosy Reviews* 13(26):71–76.

Anwar HM, Amirul IM, Suman S, Mushfiqur R, Afjalus SM. 2013. Assessment of phytochemical and pharmacological properties of ethanolic extract of Cerbera manghas L. leaves. *International Research Journal of Pharmacy* 4:120–123.

Arirudran B, Vijayalakshmi K, Saraswathy A. 2015. Study on the effect of Ruellia tuberosa L. (whole plant) treatment on total body and organ weight in den induced hepato carcinoma. *World Journal of Pharmaceutical Research* 4(1):1262–1276.

Arulappan TM, Britto JS, Ruckmani K, Kumar MR. 2015. An ethnobotanical study of medicinal plants used by ethnic people in Gingee hills, Villupuram district, Tamil Nadu, India. *American Journal of Ethnomedicine* 2(2):84–102.

Ashraf K, Sultan S, Adam A. 2018. Orthosiphon stamineus Benth. Is an outstanding food medicine: Review of phytochemical and pharmacological activities. *Journal of Pharmacy and Bioallied Science* 10(3):109–118.

Asolkar LV, Kakkar KK, Chakre OJ. 1956. *Glossary of Indian Medicinal Plants with Active Principal*. Part I. New Delhi: Publications and Information Directorate, Council of Scientific and Industrial Research Publication, p. 165.

Ata A, Kalhari KS, Samarsekera R. 2009. Chemical constituents of Barleria prionitis and their enzyme inhibitory and free radical scavenging activities. *Phytochemistrty Letters* 2:37–40.

Awasthi YC, Mitra CR. 1968. Madhuca latifolia: Triterpenoid constituents of the trunk bark. *Phytochemistry* 7(8):1433–1434.

Azad R, Babu NK, Gupta AD, Reddanna P. 2018. Evaluation of anti-inflammatory and immunomodulatory effects of Premna integrifolia extracts and assay-guided isolation of a COX-2/5-LOX dual inhibitor. *Fitoterapia* 131:189–199.

Azeez MA, Bello OS, Adedeji OA. 2013. Traditional and medicinal uses of Luffa cylindrica: A review. *Journal of Medicinal Plants Studies* 1(5):102–111.

Badukale NA, Panchale WA, Manwar JV, Gudalwar BR, Bakal RL. 2021. Phytochemistry, pharmacology and botanical aspects of Madhuca indica: A review. *Journal of Pharmacognosy and Phytochemistry* 10(2):1280–1286.

Bal LM, Meda V, Naik SN, Satya S. 2011. Sea buckthorn berries: A potential source of valuable nutrients for nutraceuticals and cosmoceuticals. *Food Research International* 44(7):1718–1727.

Balakrishna K, Kundu AB, Patra A. 1990. Roxburghiadiol A and roxburghiadiol B, two 14 alpha-methylsterols from Aglaia roxburghiana. *Journal of Natural Products* 53(2):523–526. https://doi.org/10.1021/np50068a044

Baliga MS, Kurian PJ. 2012. Ixora coccinea Linn.: Traditional uses, phytochemistry and pharmacology. *Chinese Journal of Integrative Medicine* 18(1):72–79.

Balogun AM, Fetuga BL. 2004. Fatty acid composition of seed oils of some members of the leguminosae family. *Food Chemistry* 17:175–182.

Balraj P, Nagarajan S. 1982. Apigenin-7-O-glucuronide from the flowers of Asteracantha longifolia Nees. *Indian Drugs* 19:150–152.

Baltina LA, Flekhter OB, Nigmatullina LR, Boreko EI, Nikolaeva NI, Tolstikov OV. 2003. Lupane triterpenes and derivatives with antiviral activity. *Bioorganic & Medicinal Chemistry Letters* 13(20):3549–3552.

Banerjee D, Maji AK, Mahapatra S, Banerji P. 2012. Barleria prionitis Linn.: A review of its traditional uses, phytochemistry, pharmacology and toxicity. *Research Journal of Phytochemistry* 6(2):31–41.

Banerjee S, Sur TP, Das PC, Sikdar S. 2000. Assessment of the antiinflammatory effects of Swertia chirata in acute and chronic experimental models in male albino rats. *Indian Journal of Pharmacology* 32:21–24.

Baruah D, Neog B. 2016. Botanical, phytochemistry and pharmacological review of Flacourtia jangomas (Lour) Raeusch. *International Journal of Current Medical and Pharmaceutical Research* 2:244–247.

Basch E, Ulbricht C, Hammerness P, Bevins A, Sollars D. 2004. From natural standard thyme (Thymus vulgaris L.). *Journal of Herbal Pharmacotherapy* 4(1):49–67. https://doi.org/10.1300/J157v04n01_07

Batanouny KH, Aboutabl E, Shabana MC, Soliman F. 1999. *Wild Medicinal Plants in Egypt: An Inventory to Support Conservation andSustainable Use*. Cairo, Egypt: Academy of Scientific Research & Technology.

Baytop T. 1999. *Therapy with Medicinal Plants in Turkey (Past and Present)*. 2nd ed. Turkey: Nobel Medicine Publisher.

Begum T, Rahman SM, Rashid AM. 2007. Phytochemical and biological investigations of Phyllanthus reticulatus. *Dhaka University Journal of Pharmaceutical Sciences* 5(1–2):21–23.

Benthall AP. 1993. *The Tree of Calcutta and Its Neighbourhood*. Thaker Spink and Co.

Bentley R, Trimen H. 2006. *Medical Plants*. Vol. 1. New Delhi: Ajay Book Services, pp. 5–7.

Bhagat RB, Kulkarni DK. 2003. Phytochemical evaluation and in vitro antimicrobial activity of Jatropha glandulifera Roxb. (Family Euphorbiaceae). *International Journal of Pharmacy & Life Sciences* 4(12):3178–3184.

Bhagyasri Y, Siva Subramanian N, Karthikeyan R, Jagadesh K, Vaseem Akram M. 2017. A pharmacological review on Breynia retusa. *Journal of Pharmacognosy and Phytochemistry* 6(3):511–513.

Bhakuni DS, Dhar ML, Dhar MM, Dhawan BN, Gupta H, Srimal RC. 1971. Screening of Indian plants for biological activity. Part III. *Indian Journal of Experimental Biology* 2:91.

Bhamarapravati S, Pendland SL, Mahady GB. 2003. Extracts of spice and food plants from Thai traditional medicine inhibit the growth of the human carcinogen Helicobacter pylori. *In Vivo* 17(6):541–544.

Bhandari S, Dobhal U, Sajwan M, Bisht NS. 2008. Trichosanthes Tricuspidata: A medicinally important plant. *Trees for Life Journal* 3:5.

Bhaskara Rama Murti P, Seshadri TR. 1941, August. A study of the chemical components of Decalepis Hamiltonii (Makali Veru). *Proceedings Mathematical Sciences* 14(2).

Bhattacharjee SK, De LC. 2005. *Medicinal Herbs and Flowers*. 1st ed. Jaipur, India: Aavishkar Publishers, p. 36.

Bhavamisra A. 2010. Bhavaprakash Nigantu, commentary by Chunekar. Varanasi: Choukhambha Bharati Academy. Vol. 1.

Bhavana GP, Chandrika R, Saraswathi KJT. 2013. Quantitative determination of secondary compounds in populations of Eryngium foetidum L. from India. *International Journal of Current Science* 9:24–28.

BHMA. 1996. *British Herbal Pharmacopoeia*. BHMA Publications. ISBN 0 903032 10 4.

Bibi Y, Arshad M, Ahmad N, Riaz I, Chaudhari SK. 2015. An insight into medicinal and ethnopharmacological potential of Crotalaria burhia. *Asian Pacific Journal of Tropical Disease* 5:511–514.

Bilia AR, Palme E, Catalano S, Pistellil L, Morelli I. 1993. Constituents and biological assay of Agrimonia eupatoria. *Fitoterapia* 64:549–550.

Biller A, Boppre M, Witte L, Hartmann T. 1994. Pyrrolizidine alkaloids in Chromolaena odorata. chemical and chemoecological aspects. *Phytochemistry* 35:615–619.

Bina F, Rahimi R. 2017. Sweet marjoram: A review of ethnopharmacology, phytochemistry, and biological activities. *Evidence Based Complementary and Alternative Medicine* 22(1):175–185.

Bisset NG, Wichtl M. 1994. *Herbal Drugs and Phytopharmaceuticals*. Stuttgart: Medpharm GmbH Scientific Publishers; Boca Raton: CRC Press, pp. 91–95.

Biswas D, Kumar D, Kumar S. 2014. Acute and sub-acute toxicity studies of Actaea acuminata H. Hara roots. *Indian Journal of Research in Pharmacy and Biotechnology* 2(4):1263–1270.

Biswas D, Kumar D, Kumar S. 2017. Neuropharmacological investigations on Actaea acuminata Wall. ex Royle roots. *Indian Journal of Experimental Biology* 55(1):27–35.

Biswas K, Ashima, Biswas R. 2017. In-vitro clonal propagation of an Indian medicinal plant Justicia procumbens. *International Journal of Applied Agricultural Research* 12(2):239–245.

Blanco YC, Vailant F, Perez, AM, Reynes M, Brilliote JM, Brat P. 2006. The noni fruit (Morinda citrifolia L.). A review of agricultural research. Nutritional and therapeutic properties. *Journal of Food Composition and Analysis* 19(6–7):645–654.

Blumenthal M. 1998. *The Complete German Commission E Monographs*. Austin: American Botanical Council, pp. 233–234.

Board N. 2005. *Compendium of Medicinal Plants*. New Delhi: National Institute of Industrial Research, pp. 205–207.

Boit HG, Dopke W, Stender W. 1957. Alkaloids from species of Crinum, Zephyranthes, Leucojum and Clivia. *Chemische Berichte* 90:2203–2206.

Borras-Linares I, Stojanovic Z, Quirantes-Pine R, Arraez-Roman D, Svarc-Gajic J, Fernandez-Gutierrez A, Segura-Carretero A. 2014. Rosmarinus officinalis leaves as a natural source of bioactive compounds. *International Journal of Molecular Sciences* 15:20585–20606.

Brahmanand T. 2009. *Astanghrdayam*. 2nd ed. New Delhi, India: Chaukhambha Sanskrit Pratishthan.

Brimson JM, Prasanth MI, Malar DS, Brimson S, Tencomnao T. 2020. Rhinacanthus nasutus "tea" infusions and the medicinal benefits of the constituent phytochemicals. *Nutrients* 12(12):3776.

Brown D. 1996. *Encyclopedia of Herbs and Their Uses*. London: Dorling Kindersley. ISNB: 9-780751-303148.

Brown D. 2000. *Plants of the Arum Family*. Portland, OR: Timber Press. ISBN: 0-88192-485-7.

Brown SH, Frank MS. 2020. Railroad vine (Ipomoea pes-caprae): Identification and uses. *Edis* 1–7.

Bunney S. 1984. *The Illustrated Encyclopedia of Herbs: Their Medicinal and Culinary Uses*. New York: Dorset Press.

Burdock GA. 2016. *Fenaroli's Handbook of Flavor Ingredients*. 6th ed. CRC Press. ISBN: 9781420090864.

Burkill IH. 1966. *Ministry of Agriculture (Malaysia). A Dictionary of the Economic Products of the Malay Peninsula*. London: Crown Agents for the Colonies.

Bussmann RW, Paniagua Zambrana NY, Romero C, Hart RE. 2018. Astonishing diversity-the medicinal plant markets of Bogotá, Colombia. *Journal of Ethnobiology and Ethnomedicine* 14(1):43.

Butola JS, Vashistha RK. 2003. An overview on conservation and utilization of Angelica glauca Edgew. in three Himalayan States of India. *Medicinal Plants—International Journal of Phytomedicines and Related Industries* 5(3):171–178. https://doi.org/10.5958/j.0975-6892.5.3.028

Cano L. 1998. *Inventario Etnobotánico*. Vol. 1. Xalapa, Mexico: Universidad Veracruzana, Flora Medicinal de Veracruz, pp. 48–49.

Carnat A, Lamaison JL, Petttjean-Freyret C. 1991. L'aigremoine: Etude compare d'Agrimonia eupatoria L. et Agrimonia procera Wallr. *Plantes Medicinales et Phytotherapie* 25:202–211.

Carvalho M, Ferreira PJ, Mendes VS, Silva R, Pereira, JA, Jerónimo C, Silva BM. 2010. Human cancer cell antiproliferative and antioxidant activities of Juglans regia L. *Food and Chemical Toxicology* 48:441–447.

Castro A, Coll J, Arfan M. 2011. Neo-Clerodane diterpenoids from Ajuga bracteosa. *Journal of Natural Products* 74(5):1036–1041.

Chaevallier A. 1996. *The Encyclopedia of Medicinal Plants*. London: Dorling Kindersley. ISBN: 9-780751-303148.

Chan EWC, Wong SK, Chan HT, Baba S, Kezuka M. 2016. Cerbera are coastal trees with promising anticancer properties but lethal toxicity: A short review. *Journal of Chinese Pharmaceutical Sciences* 25:161–169.

Chandel S, Bagai U. 2010. Antiplasmodial activity of Ajuga bracteosa against Plasmodium berghei infected BALB/c mice. *Indian Journal of Medical Research* 131:440–444.

Chandel S, Bagai U. 2011. Screening of antiplasmodial efficacy of Ajuga bracteosa Wall ex. Benth. *Parasitology Research* 108(4):801–805.

Chang ZC, Hu P, Xiumin. 2012. Screening for antitumor activity of six extract and preliminary chemical composition analysis. *Shizhen National Medicine* 23(03):580–581.

Chatterjee S, Sultana KW, Roy A, Chandra I. 2015. An overview on ethnopharmacological and phytochemical properties of Thunbergia sp. med aromat plants. *Journal of Medicinal and Aromatic Plants* 4(5):1–6.

Chaudhary A, Anil Bhandari A, Pandurangan A, Koul S. 2015. Madhuca Indica J. F. Gmel. (Sapotaceae): An overview. *International Journal of Pharmaceutical Sciences Letters* 5(2):539–545.

Chaudhary K, Aggarwal B, Singla RK. 2012. Ichnocarpus Frutescens: A medicinal plant with broad spectrum. *Indo Global Journal of Pharmaceutical Sciences* 2(1):63–69.

Chawla P, Chawla A, Vasudeva N, Sharma SK. 2012. A review of chemistry and biological activities of the genus Aerva-desert plant. *Journal of Acta Poloniae Pharmaceutica* 2:171–177.

Chen JL, Blanc P, Stoddart CV, Bogan M, Rozhon EJ, Parkinson N, Ye Z, Cooper R, Balick M, Nanakorn W, Kernan MR. 1998. New iridoids from the medicinal plant Barleria prionitis with potent activity against respiratory syncytial virus. *Journal of Natural Products* 61:1295–1297.

Cheng F, Zhou Y, Wang M, Guo C, Cao Z, Zhang R, Peng C. 2020. A review of pharmacological and pharmacokinetic properties of stachydrine. *Pharmacological Research* 155:104755. ISSN 1043-6618, doi: 10.1016/j.phrs.2020.104755

Chewonarin T, Kinouchi T, Kataoka K, Arimochi H, Kuwahara T, Vinitketkumnuen U, Ohnishi Y. 1999. Effect of rosell (Hibiscus sabdariffa), a Thai medicinal plant, on the mutagenicity of various mutagens in Salmonella typhnumuriumand on formation of aberrant Crypt Foci induced by the colon carcinogens azoxymethane and 2-amino-1-methyl-6-phenylimidazo [4,5-b] pyridine in F344 rats. *Food and Chemical Toxicology* 37(6):591–601.

Chiej R. 1984. *Encyclopaedia of Medicinal Plants*. MacDonald. ISBN: 0-356-10541-5.

China Pharmacopoeia Commission. 2010. *Pharmacopoeia of the People's Republic of China*. Vol. 1. Beijing, China: Chemical Industry Press, pp. 42–46.

Choi JS, Lee HJ, Park KY. 1998. In vitro antimutagenic effect of alateranin and rubrofusarin gentiobioside from roasted Cassia tora. *Natural Product Sciences* 4(2):100–104.

Chopra RN, Nayar SL, Chopra IC. 1956. *Glossary of Indian Medicinal Plants*. Vol. 1, 3rd ed. New Delhi: Council of Scientific and Industrial Research, pp. 4, 7–8.

Chopra RN, Nayar SL, Chopra IC. 1956. Glossary of Indian medicinal plants. In *Actinodaphne Hookeri*. New Delhi: Publication & Information Directorate CSIR Press, p. 6.

Chopra RN, Nayar SL, Chopra IC. 1986. *Glossary of Indian Medicinal Plants (Including the Supplement)*. New Delhi: Council of Scientific and Industrial Research, p. 259.

Chopra RN, Nayer SL, Chopra IC. 1992. *Glossary of Indian Medicinal Plants*. New Delhi: CSIR, pp. 50–52.

Chothani DL, Patel MB Mishra SH, Vaghasiya HU. 2010. Review on Ruellia tuberosa (cracker plant). *Pharmacognosy Journal* 2(12):506–512.

Choudhary GP, Jain AP. 2016. A review on Mitragyna parvifolia (Roxb.) Korth.—an Indian medicinal Plant. *International Journal of Pharmacy and Pharmaceutical Research Human* 7(1):175–184.

Chowdhary S, Verma D, Kumar H. 2009. Biodiversity and traditional knowledge of Bergenia spp. in Kumaun Himalaya. *Science Journal* 2:105–108.

Chua M, Baldwin TC, Hocking TJ, Chan K. 2010. Traditional uses and potential health benefits of Amorphophallus konjac K. Koch ex N.E.Br. *Journal of Ethnopharmacology* 128(2):268–278. https://doi.org/10.1016/j.jep.2010.01.021

Cissé M, Dornier M, Sakho M, Ndiaye A, Reynes M, Sock O. 2009. The bissap (Hibiscus sabdariffa): Composition and principal uses. *Fruits* 64:179–193.

Conn, BJ. 1992. *PlantNET—FloraOnline: Callicarpa Pedunculata*. Plantnet.rbgsyd.nsw.gov.au. Archived from the original on 2005 [April 2020].

da Silva Almeida JRG, de Menezes Barbosa J, Cavalcante NB, Delange DM. 2018. A review of the chemical composition and biological activity of Leonotis nepetifolia (Linn.) R. Br.(lion's ear). *Revista Cubana de Plantas Medicinales* 23(4).

Dandge S, Rothe P, Pethe A. 2010. Antimicrobial activity and pharmacognostic study of Luffa acutangula Roxb var amara on some deuteromycetes fungi. *International Journal of Innovation Science* 2:191–196.

Dang NH, Van Anh LT, Dat NT. 2020. Anti-inflammatory effects of essential oils of amomum aromaticum fruits in lipopolysaccharide-stimulated RAW264.7 cells. *Hindawi Journal of Food Quality* 1–5, Article ID 8831187. https://doi.org/10.1155/2020/8831187

Daniel M, Sabnis SD. 1978. Chemotaxonomical studies on Apocynaceae. *Indian Journal of Experimental Biology* 16:512.

Das R. 2016. Biodiversity of ethnomedicinal plants used by the ethnic tribal people of Barpeta district of Assam, North East India. *Asian Journal of Pharmaceutical Science and Technology* 6(1):27–32.

Das S, Parida R, Sandeep IS, Nayak S, Mohanty S. 2016. Biotechnological intervention in betelvine (Piper betle L.): A review on recent advances and future prospects. *Asian Pacific Journal of Tropical Medicine* 9(10):938–946.

Das S, Suresh Kumar P. 2014. GC-MS based phytochemical screening in ethyl acetatroot extract of Blepharispermum subsessile DC. *Journal of Chemical and Pharmaceutical Sciences* 2:113–116.

Das SS, Gauri B, Mishra B, Dey S. 2013, July. Purification and characterization of a betanidinglucosyltransferase from Amaranthus tricolor l. Catalyzing non-specific biotransformation of flavonoids. *Plant Science* 211:61–69.

Dash SK, Padhy S. 2006. Review on ethnomedicines for diarrhoea diseases from Orissa: Prevalence versus culture. *Journal of Human Ecology* 20:59–64.

Dawid-Pać R. 2013. Medicinal plants used in treatment of inflammatory skin diseases. *Postepy Dermatologii I Alergologii* 30(3):170–177. https://doi.org/10.5114/pdia.2013.35620

De Feo V. 2004. The ritual use of Brugmansia species in traditional Andean medicine in Northern Peru. *Economic Botany* 58:S221–S229.

DeAlbuquerque UP, Monteiro JM, Ramos MA, DeAmorim EL. 2007. Medicinal and magic plants from a public market in northeastern. *Brazilian Journal of Ethnopharmacology* 110(1):76–91.

Deane J. 2014, 26 August. *Quackgrass*. http://www.eattheweeds.com/quack-grass/

Deliorman D, Çalis I, Ergun F. 2001. Iridoids from Galium aparine. *Pharmaceutical Biology* 39(3):234–235.

Department of AYUSH, Ministry of Health and Family Welfare, Government of India. 2004. *Ayurvedic Pharmacopeia of India*. Part I, Vol. 4. New Delhi: Department of AYUSH, Ministry of Health and Family Welfare, Government of India, pp. 4–5.

Deraniyagala SA, Ratnasooriya WD, Goonasekara CL. 2003. Antinociceptive effect and toxicological study of the aqueous bark extract of Barringtonia racemosa on rats. *Journal of Ethnopharmacology* 86(1):21–26.

Dey AA, Dey JN. 2011. Traditional use of medicinal plants in pediatric and maternal care practiced by the ethnic group of Purulia district, West Bengal. *International Journal of Medicinal and Aromatic Plants* 1(3):189–194.

Dey YN, Wanjari MM, Kumar D, Lomash V, Jadhav AD. 2016. Curative effect of Amorphophallus paeoniifolius tuber on experimental hemorrhoids in rats. *Journal of Ethnopharmacology* 192:183–191.

Dhalwal K, Shinde VM, Biradar YS, Mahadik KR. 2008. Simultaneous quantification of bergenin, catechin, and gallic acid from Bergenia ciliata and Bergenia ligulata by using thin-layer chromatography. *Journal of Food Composition and Analysis* 21(6):496–500.

Dhanapal R, Kavimani S, Bhyrapur Matha VR, Gupta M, Basu SK. 2005. Antisteroidgenic activity of ethanol extract of ammannia baccifera (L.) whole plant in female albino mice ovaries. *Iranian Journal of Pharmacology and Therapeutics* 4(1):43–45.

Dhar ML, Dhar MM, Dhawan BN, Mehrotra BN, Srimal RC, Tandon JS. 1973. Screening of Indian plants for biological activity: Part IV. *Indian Journal of Experimental Biology* 2:43–54.

Dixon AR, Heather M, Nina LK. 1999. Ferment this: The transformation of Noni, a traditional Polynesian medicine (Morinda citrifolia, Rubiaceae). *Ecological Botony* 53(1):51–68.

Dogra NK, Kumar S, Kumar D. 2020. Vernonia anthelmintica (L.) willd.: An ethnomedicinal, phytochemical, pharmacological and toxicological review. *Journal of Ethnopharmacology* 28(256):112777. https://doi.org/10.1016/j.jep.2020.112777

Duke JA. 1992. *Handbook of Phytochemical Constituents of GRAS Herbs and Other Economic Plants*. Boca Raton, FL: CRC Press.

Duke JA, Ayensu ES. 1985. *Medicinal Plants of China*. Reference Publications. ISBN: 0-917256-20-4

Duke JA, Duke PA, Ducellier JL. 2008. *Handbook of Medicinal Plants of the Bible*. New York: CRC Press.

El-Hossary GA, Selim MA, Sayed AE, Khaleel AE. 2000. Study of the flavonoid content of Bassia muricata and Bauhinia racemosa. *Bulletin of Faculty of Pharmacy, Cairo University* 38:93–97.

Estekhdami P, Nasiri Dehsorkhi A. 2019. Chemical composition of volatile oil of Ferula Assafoetida L. *International Journal of Research Studies in Agricultural Sciences (IJRSAS)* 5(8):9–14. https://doi.org/10.20431/2454-6224.0508002

Famina Jasmine S, Vinayaga Moorthi P. 2017. Couroupita guianensis: The reservoir of medicinal compounds of human welfare. *Asian Journal of Pharmceutical and Clinical Research* 10(3):50–52.

Fathiazad F, Mazandarani M, Hamedeyazdan S. 2011. Phytochemical analysis and antioxidant activity of Hyssopus officinalis L. from Iran. *Advanced Pharmaceutical Bulletin* 1(2):63–7. https://doi.org/10.5681/apb.2011.009

Faucounnier ML, Jaziri M, Homes J, Shimomura K, Marlier M. 1996. II Anthemis nobilis L. (Roman chamomile), in vitro culture, micropropagation, and the production of essential oils. *Biotechnology in Agriculture and Forestry* 37:16–20.

Ferdous F, Hossain MDK, Rahman MS, Hossain MDA, Kabir S, Rashid MA. 2010. Triterpenes and Sterols from Saman. *Dhaka University Journal of Pharmaceutical Sciences* 9(2):69–73.

Fernando T, Bean G. 1984. Fatty acids and sterols of Amaranthus tricolor L. *Science Direct Food Chemistry* 15(3):233–237.

Gairola Y, Biswas S. 2008. Bioprospecting in Garhwal Himalayas Uttarakhand. *Current Science* 94:1139–1143.

Garg S, Bhushan R, Kapoor R. 1979. Chrysin-7-O-galactoside: A new flavanoids from Aerva persica burm. f. *Indian Journal of Chemistry Section-B* 17:416–417.

Garg SN, Naqvi AA, Singh A, Ram G, Kumar S. 1999. Composition of essential oil from an annual crop of Hyssopus officinalis grown in Indian plains. *Flavour and Fragence Journal* 14:170–172.

Gayathri Y, Babu AN, Lakshmi JN, Kiranmai K, Kavya G. 2021. Review on phyto-pharmacological and medicinal uses of Hyptis suaveolens (L) Poit. *Lipids* 4(3):2.

George J, Ravishankar GA, Keshava N, Udhayashankar K. 1999a. Antibacterial activity of super critical extract from Decalepis hamiltonii roots. *Fitoterapia* 70:172–174.

Ghaima KK. 2013. Antibacterial and wound healing activity of some Agrimonia eupatoria extracts. *Baghdad Science Journal* 10(1):152–160.

Ghosh S, Parihar VS, Dhavale DD, Chopade BA. 2015. Commentary on therapeutic potential of Gnidia glauca: A novel medicinal plant. *Medicinal Chemistry* 5(8):351–353.

Ghosh T, Ahire M, Patil S, Jabgunde A, Bhat Dusane M, Joshi BN, Pardesi K, Jachak S, Dhavale DD, Chopade BA. 2011. Antidiabetic activity of Gnidia glauca and Dioscorea bulbifera: Potent amylase and glucosidase inhibitors. *Evidence Based Complementary and Alternative Medicine* 2012:1–10.

Ghule BV, Murugananthanb G, Yeolea PG. 2007. Analgesic and antipyretic effects of Capparis zeylanica leaves. *Fitoterapia* 78(5):365–369.

Godbole NN, Gunde BG, Shrivastav PD. 1941. An investigation of oil from seed of Hygrophila spinosa. *Journal of the American Oil Chemists' Society* 18:206–207.

Godghate AG, Patil RS, Rajaram S, Sawant RS. 2015. Gnidia glauca (FRESEN) Gilg.: Phytochemical and antibacterial view. *International Journal of Recent Scientific Research* 6(6):4854–4857.

Gogoi B. 2017. Baccaurea ramiflora Lour.: Biochemical and ethnobotanical value with scope for bio-prospection. *Annals of Plant Science* 7:1649–1652.

Goodla L, Manobolu M, Ravilla S, Poondamalli PR. 2010. Safety evaluation of the ethanol extract of Ammannia baccifera (Lythraceae): Assessment of acute and subacute toxicity. *Journal of Pharmacy Research* 3(11):2634–2637.

Govindarajan M, Vaseeharan B, Alharbi NS, Kadaikunnan S, Khaled JM, Al-Anbr MN, Alyahya SA, Maggi F, Benelli G. 2018. High efficacy of (Z)-γ-bisabolene from the essential oil of Galinsoga parviflora (Asteraceae) as larvicide and oviposition deterrent against six mosquito vectors. *Environmental Science and Pollution Research International* 25(11):10555–10566. https://doi.org/10.1007/s11356-018-1203-3

Grieve M. 1971. Clivers. In *A Modern Herbal: The Medicinal, Culinary, Cosmetic and Economic Properties, Cultivation and Folk-Lore of Herbs, Grasses, Fungi, Shrubs, & Trees with All Their Modern Scientific Uses*. Vol. 1. Dover Publications, p. 207.

Grubben GJH, Denton OA. (eds.). 2004. *Plant Resources of Tropical Africa 2. Vegetables. PROTA Foundation*, Wageningen, Backhuys Publishers, Netherlands, p. 668.

Gulcin I, Oktay M, Kirecci E, Kufrevioglu OI. 2003. Screening of antioxidant and antimicrobial activities of anise (Pimpinella anisum L.) seed extracts. *Food Chemistry* 83(3):371–382.

Guliyev VB, Gul M, Yildirim A. 2004. Hippophae rhamnoides L.: Chromatographic methods to determine chemical composition, use in traditional medicine and pharmacological effects. *Journal of Chromatography B* 812(1–2):291–307.

Gummadi VP, Battu GR, Keerthana Diyya MS, Manda K. 2016. A review on palmyra palm (Borassus flabellifer). *International Journal of Current Pharmaceutical Review and Research* 8:17–20.

Gundala SR, Yang C, Mukkavilli R, Paranjpe R, Brahmbhatt M, Pannu V, Cheng A, Reid MD, Aneja R. 2014. Hydroxychavicol, a betel leaf component, inhibits prostate cancer through ROS-driven DNA damage and apoptosis. *Toxicology and Applied Pharmacology* 280(1):86–96. https://doi.org/10.1016/j.taap.2014.07.012

Gupta A, Behl T, Panichayupakaranan P. 2019. A review of phytochemistry and pharmacology profile of Juglans regia. *Obesity Medicine* 16(22):100142.

Gupta RK, Gaur YD, Malhotra SP, Dutta BK. 1966. Medicinal plants of the Indian arid zone. *Journal of Agriculture Tropical & Botanical Application* 13:247–288.

Gyatso T, Hakim C. 2010. *Essentials of Tibetan Traditional Medicine*. Berkeley: North Atlantic Books, p. 223.

Hajra PK, Rao RR, Singh DK, Uniyal BP. 1995. *Flora of India*. Vol. 12. Calcutta: BSI, p. 187.

Hamill FA, Apio S, Mubiru, NK, Mosango M, Bukenya ZR, Maganyi, OW, Soejarto, DD. 2000. Traditional herbal drugs of Southern Uganda. *International Journal of Ethnopharmacology* 70(3):281–300.

Hardel DK, Das AK, Bhanja S. 2012. A comparison study of fluorescence characteristics of powder of Haritaki: Terminalia chebula (pericarp), Yavani: Trachyspermum ammi (fruit), Ajmoda: Apium leptophyllum (fruit), Sunthi: Zingiber officinale (rhizome). *International Research Journal of Pharmacy* 3:255–257.

Hariharan V, Rangaswami S, Sarangan S. 1972. Saponins of the seeds of Bassia latifolia. *Phytochemistry* 11(5):1791–1795.

Härmälä P, Vuorela H, Hiltunen R, Nyiredy SZ, Sticher O, Törnquist K, Kaltia S. 1992. Strategy for the isolation and identification of coumarins with calcium antagonistic properties from the roots of Angelica archangelica. *Phytochemical Analysis* 3(1):42–48. https://doi.org/10.1002/pca.2800030108

Hayashi K, Narutaki K, Nagaoka Y, Hayashi T, Uesato S. 2010. Therapeutic effect of arctiin and arctigenin in immunocompetent and immunocompromised mice infected with influenza. *Biological and Pharmaceutical Bulletin* 33(7):1199–1205. https://doi.org/10.1248/bpb.33.1199

Heiss EH, Tran TV, Zimmermann K, Schwaiger S, Vouk C, Mayerhofer B, Malainer C, Atanasov AG, Stuppner H, Dirsch VM. 2014. Identification of chromomoric acid C-I as an Nrf2 activator in Chromolaena odorata. *Journal of Natural Products* 77:503–508.

Helal IE, Galala AA, Saad HE, Halim AF. 2015. Chemical composition and α-amylase inhibitory activity of Apium leptophyllum essential oils. *Journal of American Science* 11:204–209.

Helal IE, Galala AA, Saad HE, Halim AF. 2016. Bioactive constituents from Apium leptophyllum fruits. *British Journal of Pharmaceutical Research* 14(3):1–8.

Hemadri K, Sharma PC, Narayaanaappa D, Rao SS, Murthy KS. 1996. *Medico-Botanical Exploration of Phulabani and Koraput Districts of Orissa. Folklore Claims*. New Delhi: CCRAS, Ministry of Health and Family Welfare, Government of India, p. 115.

Herrera-Arellano A, Flores-Romero S, Chávez-Soto MA, Tortoriello J. 2004. Effectiveness and tolerability of a standardized extract from Hibiscus sabdariffain patients with mild to moderate hypertension: A controlled and randomized clinical trial. *Phytomedicine* 11(5):375–382.

Hong LS, Yu WC, Kuang CC. 2007. Chemical Investigation of Phyllanthus reticulatus by HPLCSPE-NMR and conventional methods. *Phytochemical Analysis* 18, 251–255.

Hooda V, Sharma GN, Singla RK. 2011. Smilax Zeylanica Linn.—a natural therapeutic hub. *Pharmacology Online* 2:151–154.

Hossain MAASM, Sarkar S, Saha S, Hossain Hasan MLMM. 2013. Biological assessment on Cerbera manghas (linn.). *Pharmacology* 1:155–160.

Huang DF, Nie S, Jiang L, Xie M. 2024. A novel polysaccharide from the seeds of Plantago asiatica L. induces dendritic cells maturation through toll-like receptor 4. *International Immunopharmacology* 18:236–243.

Hussain A, Virmani OP, Popli SP, Misra LN, Gupta MM. 1992. *Dictionary of Indian Medicinal Plants*. Central Institute of Medicinal and Aromatic Plants.

Ibrahim H, Kaempferia Galanga L. 1999. Medicinal and poisonous plants. *Plant Resources of South-East Asia* 12(1):334–335.

Ibrahim M, Khan R, Malik A. 2010. Two new guaianolides from Amberboa ramosa. *Natural Product Communication* 5:1865–1868.

Ibrahim MT, El Wafa SAA, Sleem AA. 2017. Phytochemical and biological investigation of Thunbergia grandiflora. *Journal of Pharmacognosy and Phytochemistry* 6(2):43–51.

Ibrahim NH, Awaad AS, Alnafisah RA, Alqasoumi S, El-Meligy RM, Mahmoud AZ. 2018. In—Vitro activity of Desmostachya bipinnata (L.) Stapf successive extracts against Helicobacter pylori clinical isolates. *Saudi Pharmacology Journal* 26(4):535–540.

Imran I, Hussain L, Zia-Ul-Haq M, Janbaz KH, Anwar H, Gilani AH, De Feo V. 2011. Gastrointestial and respiratory activities of Acacia leucophloea. *Journal of Ethnopharmacology* 138(3):676–682.

Iqbal SM, Mushtaq A, Jabeen Q. 2014. Antimicrobial, antioxidant and calcium channel blocking activities of Amberboa divaricata. *Bangladesh Journal of Pharmacology* 9(1). ISBN: 0-917256-20-4

Isika OA, Tameka WM, William SN, Essien E. 2006. Volatile constituents from Samanae saman (Jacq.) Merr. Fabaceae. *African Journal of Biotechnology* 5(20):1890–1893.

Islam T, Rahman A, Islam AU. 2013. Effects of aqueous extract of fresh leaves of Abroma augusta L. on oral absorption of glucose and metformin hydrochloride in experimental rats. ISRN pharmaceutics. *Journal of Medicinal Plants Research* 7(41):3014–3029.

Jabbar AA, Abdullah FO, Hassan AO, Galali Y, Hassan RR, Rashid EQ, Salih MI, Aziz KF. 2022. Ethnobotanical, phytochemistry, and pharmacological activity of Onosma (Boraginaceae): An updated review. *Molecules* 27(24):8687.

Jadhav A, Acharya R, Harisha CR, Shukla VJ, Chandola H. 2015. Pharmacognostical and preliminary physico-chemical profiles of Blepharispermum subsessile DC. *Root Ayu* 36(1):73–76. https://doi.org/10.4103/0974-8520.169012

Jagtap SD, Deokule SS, Pawar PK, Harsulkar AM. 2009. Traditional ethnomedicinal knowledge confined to Pawara tribes of Satpura hills, Maharashtra, India. *Ethnobotanical Leaflet* 13:98–115.

Jain SK. 1965. Medicinal plant lore of the tribals of Bastar. *Economic Botany* 19:236–250.

Jain SK, DeFilipps RA. 1991. *Medicinal Plants of India*. 2 Vols. Algonac: Reference Publications, Inc.

Jain SK, Srivastava S. 2005. Traditional uses of some Indian plants among islanders of the Indian Ocean. *Indian Journal of Traditional Knowledge* 4:345–357.

Jamal AK, Yaacob WA, Din LB. 2008. A chemical study on Phyllanthus reticulates. *Journal of Physical Science* 19(2):45–50.

Jamkhande PG, Wattamwar AS. 2015. Annona reticulata Linn. (Bullock's heart): Plant profile, phytochemistry and pharmacological properties. *Journal of Traditional and Complementary Medicine* 5(3):144–152.

Janaki S, Vijayasekaran V, Viswanathan S, Balakrishna K. 1999. Anti-inflammatory activity of Aglaia roxburghiana var. beddomei extract and triterpenes roxburghiadiol A and B. *Journal of Ethnopharmacology* 67(1):45–51.

Jangam KK, Vikhe DN, Jadhav RS. 2022. Pharmacognostic, phytochemical and pharmacological study on Gardenia gummifera. *International Journal of Advanced Research in Science, Communication and Technology* 2(1):336–344.

Janick J, Pauli R. 2008. *The Encyclopedia of Fruits and Nuts*. Vol. 2. CABI Publishing, p. 373.

Jayaprakasam B, Damua AG, Gunasekara D, Blond A, Bodo B. 1999. Dihydroechioidinin, aflavone from Andrographis echioides. *Phytochemistry* 52:935–937.

Jayaweera DMA. 1981. *Medicinal Plants Used in Ceylon*. Part I. Sri Lanka, Colombo: National Science Council, pp. 1–61.

Jayaweera DMA. 1982. *Medicinal Plants (Indigenous and Exotic) Used in Ceylon*. Part 2, 1st ed. Sri Lanka, Colombo: National Science Council, pp. 108–109.

Jirovetz L, Buchbauer G, Stoilova I, Krastanov A, Stoyanova A, Schimdt E. 2007. Spice Plants: Chemical composition and antioxidant properties of Pimenta Lindl. Essential oils, part 1: Pimenta dioica (L.) Merr., leaf oil from Jamaica. *ErNährung/Nutrition* 2:55–62.

Johnson M, Wesely EG, Zahir Hussain MI, Selvan N. 2010. In vivo and in vitro phytochemical and antibacterial efficacy of Baliospermum montanum (Willd.) Muell. *Agriculture Asian Pacific Journal of Tropical Medicine* 3(11):894–897.

Jones P. 1991. *Just Weeds: History, Myths, and Uses*. New York: Prentice Hall Press.

Joshi AR, Joshi K. 2000. Indigenous knowledge and uses of medicinal plants by local communities of the Kali Gandaki Watershed Area, Nepal. *Journal of Ethnopharmacology* 73:175–183.

Joshi DY, Nariya MB, Barvaliya R. 2021. A Phyto-pharmacological review of Blepharis maderaspatensis (L.) B. Heyne ex Roth. *Journal of Ayurvedic and Herbal Medicine* 7(1):56–59.

Joshi SG. 2007. *Medicinal Plants*. New Delhi, Oxford: IBH Publishing Co. Pvt. Ltd, p. 243.

Jyothi A, Naga A, Rajalakshmi BS, Bharati AS. 2016. Vatsanabha: An agada perspective. *International Ayurvedic Medical Journal* 4(7).

Kachiguma NA, Mwase W, Maliro M, Damaliphetsa A. 2015. Chemical and mineral composition of Amaranth (AmaranthusL.). *Journal of Food Research* 4(4):92–102.

Kaikini Aakruti A, Dhande Swati R, Kadam Vilasrao J. 2013. Overview of Indian medicinal tree: Bambusa Bambos (DRUCE). *International Research Journal of Pharmacy* 4(8):52–55.

Kala CP. 2005. Ethnomedicinal botany of the apatani in the eastern Himalayan region of India. *Journal of Ethnobiology and Ethnomedicine* 1(1):11.

Kamaladevi P. 2017. Phytochemical and pharmacological analysis of Lepidagathis cristata, Willd. (Acanthaceae). *Journal of Scientific Transactions in Environment and Technovation* 11(2):73–79.

Kamalpreet KL, Singh A, Kaur J, Kaur N. 2019. A brief review of remedial uses of Saussurea lappa. *Journal of Pharmacognosy and Phytochemistry* 8(3):4423–4430.

Kamboj A, Saluja AK. 2010. Phytopharmacological review of Xanthiums trumarium L. (Cocklebur). *International Journal of Green Pharmacy* 129–139.

Kanaka R, Narasinga R, Venkateshwarlu M, Sammaiah D, Anitha U, Ugandhar T. 2013. Studies on the medicinal plant biodiversity in forest ecosystem of Mahadevpur forest of Karimnagar (A.P.) India. *Bioscience Discovery* 4:82–88.

Kanchanapoom T, Ryoji K, Yamasaki K. 2002. Cucurbitane, hexanorcucurbitane and octanorcucurbitane glycosides from fruits of Trichosanthes tricuspidata. *Phytochemistry* 59(2):215–228. https://doi.org/10.1016/s0031-9422(01)00430-7

Kanjanapothi D, Panthong A, Lertprasertsuke N, Taesotikul T, Rujjianawate C, Kaewpinit D, Sudthayakorn R, Choochote W, Chaithong U, Jitpakdi A, Pitasawat B. 2004. Toxicity of crude rhizome extract of Kaempferia galanga L. (ProhHom). *Journal of Ethanopharmacology* 90:359–365.

Kapoor LD. 2001. *Hand Book of Ayurvedic Medicinal Plants*. Boca Raton, FL: CRC Press. ISBN: 8493-2929-9.

Kapoor LD, Singh A, Kapoor SL, Srivastava SN. 1975. Survey of Indian plants for saponins, alkaloids and flavonoids. IV. *Lloydia* 38:221–224.

Kataria S, Shrivastava B, Khajuria R, Suri K, Sharma P. 2011. Pharmacognostic evaluation of Crotalaria burhia buch.-Ham. *Indian Journal of Traditional Knowledge* 10:629–635.

Katewa S, Chaudhary B, Jain A. 2004. Folk herbal medicines from tribal area of Rajasthan, Indian. *Journal of Ethnopharmacology* 92:41–46.

Katewa S, Galav P. 2006. Additions to the traditional folk herbal medicines from Shekhawati region of Rajasthan. *Indian Journal of Traditional Knowledge* 5:494–500.

Kaur AK, Jain SK, Gupta A, Gupta SK, Bansal M, Sharma PK. 2009. Phytopharmacognostic investigation of Bauhinia Tomentosa Linn. *Biosciences Biotechnology Research Asia* 6(2).

Kaur M, Singh G, Mohan C. 2013. Barringtonia acutangula: A traditional medicinal plant. *International Journal of Pharmaceutical Sciences Review and Research* 23(1):168–171.

Kavitha A, Deepthi N, Ganesan R, Gladwin Joseph SC. 2022. *Common Dryland Trees of Karnataka: Bilingual Field Guide*. Ashoka Trust for Research in Ecology and the Environment.

Kayani WK, Rani R, Ihsan-ul-Haq, Mirza B. 2014. Seasonal and geographical impact on the morphology and 20-hydroxyecdysone content in different tissue types of wild Ajuga bracteosa Wall. ex Benth. *Steriods* 87:12–20.

Kebriaee-Zadeh A. 2003. Overview of national drug policy of Iran. *Iranian Journal of Pharmaceutical Sciences* 2(1–2):6.

Khan B. 2008. *Encyclopedia of Flora and Fauna of Bangladesh*. 1st edn. Dhaka: Asiatic Society of Bangladesh, p. 7.

Khan FM. 2009. Ethno-veterinary medicinal usage of flora of greater Cholistan desert, Pakistan. *Pakistan Veterinary Journal* 29(2):75–80.

Khan MS, Javed K, Khan MH. 1995. Chemical constituents of the leaves of Ichnocarpus frutescens R. *British Journal of the Indian Chemical Society* 72:65–66.

Khanam UK, Oba S. 2013. Bioactive substances in leaves of two amaranth species, Amaranthus Tricilor and A. Hypochondriacus. *Canadian Journal of Plant Science* 93:47–58.

Khanra R, Dewanjee S, Tarun K Dua TK, Sahu R, Gangopadhyay M, De Feo V, Zia-Ul-Haq M. 2015. Abroma Augusta L. (Malvaceae) leaf extract attenuates diabetes induced nephropathy and cardiomyopathy via inhibition of oxidative stress and inflammatory response. *Journal of Translational Medicine* 13(1):6.

Khare CP. 2007. *Indian Medicinal Plants—An Illustrated Dictionary*. Springer Science and Business Media, LLC, p. 49.

Khare P, Kishore K, Sharma DK. 2018. Medicinal uses, phytochemistry and pharmacological profile of Madhuca longifolia. *Asian Journal of Pharmacy and Pharmacology* 4(5):570–581.

Khastir HN, Sengupta P. 1960. Investigation on Ichnocarpus frutescens R. *British Journal of Applied Chemistry* 23(2):111–112.

Khokar A, Menghani E. 2015. Screening of successive extracts of Amorphophallus konjac for antibacterial activity. *African Journal of Biotechnology* 14(34):2599–2603.

Kinghorn AD. 1979. Carcinogenic irritant Euphorbiaceae. In Kinghorn AD (ed) *Toxic Plants*. New York: Columbia University Press, pp. 137–160.

Kirtikar BD, Basu BD. 1935. *Indian Medicinal Plants*. Vol. 3. Dehradun, India: International Book Distributors.

Kirtikar KR, Basu BD. 1918. *Indian Medicinal Plants*. Allahabad: Sudhindra Nath Basu, M.B. Panini, pp. 986–987.

Kirtikar KR, Basu BD. 1975. *Indian Medicinal Plants*. Vol. 2. New Delhi: Jayad Press, p. 236.

Kirtikar KR, Basu BD. 1975a. *Indian Medicinal Plants*. Vol. 1, 2nd ed. Allahabad, India: International Book Distributors; Dehradune, India: M/S Bishen Mahendra Pal Singh, p. 51.

Kirtikar KR, Basu BD. 1975b. *Indian Medicinal Plants*. Vol. 2. New Delhi: Jayad Press; Dehradun: M/S Bishen Mahendra Pal Singh, pp. 236, 842–844.

Kirtikar KR, Basu BD. 1984. *Indian Medicinal Plants*. Vol. 4, 2nd ed. Dehra Dun: Bishen Singh Mahendra Pal Singh, pp. 2422–2423.

Kirtikar KR, Basu BD. 1987. Indian medicinal plants. In *Compositae*. Vol. 2. Dehradun: International Book Distributors, pp. 1420–1423.

Kirtikar KR, Basu BD. 1991. *Indian Medicinal Plants*. Vol. III, 2nd ed. Dehradun: Bishen Singh and Mahendra Pal Singh, Dehradun Publisher Ltd, India, pp. 1916–1917, 2219–2220.

Kirtikar KR, Basu BD. 1993. *The Indian Medicinal Plants*. Vol. I, 2nd ed. New Delhi: International Book Distributors.

Kirtikar KR, Basu BD. 1994. *Indian Medicinal Plants: Bishen Singh Mahendrapal Singh*. Vol. 3. Dehradun, India: International Book Distributors, pp. 2186–2188.

Kirtikar KR, Basu BD. 1995. Indian medicinal plants. In *Actinodaphne*. Vol. 3, 2nd ed. Dehradun, India: International Book Distributors, pp. 2156–2157.

Kirtikar KR, Basu BD. 1996. *Indian Medicinal Plants*. Vol. III. Dehradun, India: International Book Distributors, Booksellers and Publishers, pp. 2063–2068.

Kirtikar KR, Basu BD. 2003. *Indian Medicinal Plants*. Dehradun: Oriental Enterprises, pp. 778–780.

Kirtikar KR, Basu BD, An ICS. 1993. *Indian Medicinal Plants*. Vol. 1. Dehradun, India: Orient Enterprises, pp. 496–498.

Kiruthika N, Dhivya R, Kalaiselvi K, Kanimozhi P, Panneerselvam K. 2012. Phytochemical studies on E.acaulis. *International Journal of Pharma and Bio Sciences* 3(3):(B)1054–1062.

Kishor Kumar V, Lalitha KG. 2017. Pharmacognostical and phytochemical studies of helleborus niger L root. *Ancient Science of Life* 36(3):151–158.

Kohli KR, Nipanikar SU, Kadbhane KP. 2010. A comprehensive review on Trivrit [Operculina turpethum syn. Ipomoea turpethum]. *International Journal of Pharma Bio Science* 1:443–452.

Koul B, Kumar A, Yadav D, Jin J. 2020. Bergenia genus: Traditional uses. *Phytochemistry and Pharmacology Molecules* 25(23):1–19.

Kuchta K, Volk RB, Rauwald HW. 2013. Stachydrine in Leonurus cardiaca, Leonurus japonicus, Leonotis leonurus: Detection and quantification by instrumental HPTLC and 1H-qNMR analyses. *Die Pharmazie* 68(7):534–540.

Kuete Thymous vulgaris V Kuete (ed.). 2017. *Medicinal Spices and Vegetables from Africa*. 1st ed. Elsevier Inc., pp. 599–609.

Kuhn MA, Winston D. 2000. *Herbal Therapy Supplements: A Scientific Traditional Approach*. Philadelphia: Lippincott, p. 232. ISBN: 978-0-7817-2643-6.

Kumar GG, Gali V, Dwiwedi S. 2011. Phytochemical investigation of Crotalaria burhia Hamilt. *International Journal of Research in Pharmaceutical Biomedical Science* 2:1721–1724.

Kumar D, Bhat ZA, Singh P, Shah MY, Bhujbal SS. 2010. Ailanthus excelsa Roxb. is really a plant of heaven. *International Journal of Pharmacology* 6(5):535–550.

Kumar D, Sharma S, Kumar S. 2020. Botanical description, phytochemistry, traditional uses, and pharmacology of Crataeva nurvala Buch. Ham.: An updated review. *Future Journal of Pharmaceutical Sciences* 6:113.

Kumar MSB, Balakrishnan NP. 1996. Ethanobotanical studies of the genusEuphorbia L., (Euphorbiaceae). *Journal of Economic and Taxonomic Botany* 12:46–49.

Kumar V, VanStaden J. 2016. A review of Swertia chirayita (Gentianaceae) as a traditional medicinal plant. In McGaw LJ (ed.) *Frontiers in Pharmacology*. Pretoria, South Africa: University of Pretoria.

Kumari S, Anmol, Bhatt V, Patil SS, Sharma U. 2021. Cissampelos pareira L.: A review of its traditional uses, phytochemistry, and pharmacology. *Ethanopharmacology* 274.

Kumudhavalli MV, Jayakar B. 2011. Phytochemical and pharmacological evaluation of the dried leaves of E. acaulis. *Journal of Pharmacy Research* 4(9):3219–3221.

Kurian JC. 2004. *Plant that Heal*. 7th ed. Pune: Oriental Watchman Publishing House, p. 31.

Kurup PNV, Ramadas VNK, Joshi P. 1979. *Handbook of Medicinal Plants*. New Delhi: Central Council for Research in Ayurveda and Siddha.

Kylin M. 2010. *Swedish University of Agricultural Sciences The Faculty of landscape Planning, Horticulture and Agriculture Science Plant Breeding and Biotechnology*. http://stud.epsilon.slu.se

Lai TK, Acharya K, Chatterjee S, Sherpa NL. 2014. Antipseudomonal ergosteryl triterpenes from the paste of Spondias pinnata Kruz. bark, pre-treated with curd brew. *International Journal of Pharmaceutical Sciences Review and Research* 28(1):143–146.

Lakshmi M, Nandagopal S. 2017. Studies on the leaf essential oil of Coleus zeylanicus (Benth.) L.H. Cramer-a valuable medicinal plant. *Research Journal of Pharmaceutical, Biological and Chemical Sciences* 8(5):119, 120–125.

Lalitha KG, Sethuraman MG. 2009. Phytochemical and pharmacological evaluation of the flowers of Sarcostemma brevistigma wight. *Oriental Pharmacy and Experimental Medicine* 9(3):252–258.

Lan J-P, Tong R-C, Sun X-M, Zhang H-Y, Sun S, Xiong A-Z, Wang Z-T, Yang L. 2021. Comparison of main chemical composition of Plantago asiatica L. and P. depressa Willd. seed extracts and their anti-obesity effects in high-fat diet-induced obese mice. *Phytomedicine* 81.

Larsen T, Thilsted SH, Biswas SK, Tetenus I. 2007. The leafy vegetable amaranth (Amaranthusgaganticus) is a potent inhibitor of Calcium availability and retention in rice based diets. *British Journals of Nutrition* 90(3):521–527.

Launert E. 1981. *Edible and Medicinal Plants*. Hamlyn. ISBN: 0-600-37216-2.

Leboeuf M, Cavé A, Bhaumik P, Mukherjee B, Mukherjee R. 1980. The phytochemistry of the annonaceae. *Phytochemistry* 21:2783–2813.

Leeya Y, Mulvany MJ, Queiroz EF, Marston A, Hostettmann K, Jansakul C. 2010. Hypotensive activity of an n-butanol extract and their purified compounds from leaves of Phyllanthus acidus (L.) Skeels in rats. *European Journal Pharmacology* 649(1–3):301–313.

Li JX, Hareyama T, Tezuka Y, Zhang Y, Miyahara T, Kadota S. 2005. Five new oleanolic acid glycosides from Achyranthes bidentata with inhibitory activity on osteoclast formation. *Planta Medica* 71:673–679. https://doi.org/10.1055/s-2005-871275

Li TSC. 2006. *Taiwanese Native Medicinal Plants Phytopharmacology and Therapeutic Values*. CRC, Taylor and Francis.

Li X, He T, Wang X, Shen M, Yan X, Fan S, Wang L, Wang X, Xu X, Sui H, She G. 2019. Traditional uses, chemical constituents and biological activities of plants from the genus Thymus. *Chemistry and Biodiversity* 16(9):e1900254. https://doi.org/10.1002/cbdv.201900254

Lim SH, Lee, SH, Lee, CH, Choi C. 2021. Pharmacological activity of Garcinia indica (Kokum): An updated review. *Pharmaceuticals (Basel)* 14(12):1338. https://doi.org/10.3390/ph14121338

Lim TK. 2012. *Edible Medicinal and Non-Medicinal Plants*. Dordrecht: Springer. Vol. 3, p. 898. https://doi.org/10.1007/978-94-007-2534-8

Lim TK. 2013. Flacourtia jangomas. In *Edible Medicinal and Non-Medicinal Plants*. Dordrecht: Springer, pp. 771–775.

Liu L, Li W, Koike K, Zhang S, Nikaido T. 2004. Newalpha-tetralonylglucosides from the fruit of Juglans mandshurica. *Chemical and Pharmceutical Bulletin* 52:566–569.

Liu PY, Lin ZS, Guo ZX. 1998. Research and utilization of amorphophallus in China. *Acta Botanica Yunnanica* (Suppl. X):48–61.

Lohar DL, Chawan DD, Garg SP. 1975. Phytochemical studies on Cassia species of Indian arid zone. *Current Science* 44(2):67.

Lust J. 1983. *The Herb Book*. Bantam Books. ISBN: 0-553-23827-2.

Madaan, R, Bansal G, Kumar S, Sharma A. 2011. Estimation of total phenols and flavonoids in extracts of Actaea spicata roots and antioxidant activity studies. *Indian Journal of Pharmaceutical Science* 73(6):666–669.

Madhuri K, Elango K, Ponnusankar S, Madhuri K, Elango K, Ponnusankar S. 2012. Saussurea lappa (Kuth root): Review of its traditional uses, phytochemistry and pharmacology. *Oriental Pharmacy and Experimental Medicine* 12:1. https://doi.org/10.1007/s13596-011-0043-1

Maghdu NA, Palaniyappan KD. 2014. In vitro antifungal potentials of bioactive compound oleic acid, 3-(octadecyloxy) propyl ester isolated from Lepidagathis cristata Willd. (Acanthaceae) inflorescence. *Asian Pacific Journal of Tropical Biomedicine* 4(Suppl 2):S661–S664.

Mahalakshmi SN, Achala HG, Ramyashree KR, Prashith Kekuda TR. 2019. Rauvolfia tetraphylla L. (Apocynaceae)-A comprehensive review on its ethnobotanical uses, phytochemistry and pharmacological activities. *International Journal of Pharma and Bio Sciences* 9(2):664–682.

Maharana PK. 2021. Ethnobotanical, phytochemical, and pharmacological properties of Cerbera manghas L. *Journal of Bioscience* 46:25.

Mahbubar Rahman AHM. 2013. Ethno-medico-botanical studies on cucurbits of Rajshahi division, Bangladesh. *Journal of Medicinal Plants Studies* 1:118–125.

Mahbubur Rahman AHM, Iffat Ara Gulshana M. 2014. Taxonomy and medicinal uses on amaranthaceae family of Rajshahi, Bangladesh. *Applied Ecology and Environmental Sciences* 2(2):54–59, 878.

Maheshwari JK. 1996. *Ethnobotany in South Asia*. Jodhpur: Scientific Publishers, p. 346.

Maile P, Guenard D, Franck M, Van TM, Gaspard C, Sevenet T. 1994. New cytotoxic cucurbitacins from the pericarps of Trichosanthes tricuspidata fruits. *Natural Product Letters* 16(1):15–19.

Majumder P. 2011. Phytochemical, pharmacognostical and physicochemical standardization of Peperomia pellucida (L.) HBK. stem. *International Journal of Comprehensive Pharmacy* 2(8):1–4.

Manimegalai M, Umavathi S. 2010. Effect of Kirganelia reticulata (Euphorbiaceae) plant extract against the growth and development of the human filarial vector culex quinquefasciatus (Diptera: Culicidae). *EM Intern* 29(3):413–416.

Mann R, Agarwal K, Singh C, Aeri V, Nema RK. 2008. Adventitious shoot proliferation from aseptically germinated seedlings of Cuminum cyminum. *Pharmacognosy Magazine* 4(14):132–137.

Manvar MN, Desai TR. 2012. Pharmacognostical and phytochemical investigations on Vernonia anthelmintica Willd. *Planta Activa* (4):1–3.

Manzur-ul-Kadir Mia M, Fahim Kadir M, Shahadat Hossan M, Rahmatullah M. 2009. Medicinal plants of the Garo tribe inhabiting the Madhupur forest region of Bangladesh. *American-Eurasian Journal of Sustainable Agriculture* 3(2):165.

Mapura, A, Timberlake J. 2004. *A Checklist of Zimbabwean Vascular Plants*. Southern African Botanical Diversity Network Report No. 33, Sabonet, Pretoria and Harare, p. 40.

Martins N, Barros L, Santos-Buelga C, Henriques M, Silva S, Ferreira IC. 2014. Decoction, infusion and hydroalcoholic extract of Origanum vulgare L.: Different performances regarding bioactivity and phenolic compounds. *Food Chemistry* 158:73–80. 10.1016/j.foodchem.2014.02.099

Marya BH, Bothara SB. 2011. Ethnopharmacological properties of cocculus hirsutus (L.) diel-a review. *International Journal of Pharmaceutical Sciences Review and Research* 7(1):108–112.

Matsubara T, Bohgaki T, Watarai M, Suzuki H, Ohashi K, Shibuya H. 1999. Antihypertensive actions of methylripariochromene A from Orthosiphon aristatus, an Indonesian traditional medicinal plant. *Biological and Pharmaceutical Bulletin* 22:1083–1088.

Matsumoto T, Hosono-Nishiyama K, Yamada H. 2006. Antiproliferative and apoptotic effects of from Arctium lappa on leukemic cells. *Planta Medica* 72(3):276–278. https://doi.org/10.1055/s-2005-916174

Matsushige A, Matsunami K, Kotake Y, Otsuka H, Ohta S. 2012. Three new megastigmanes from the leaves of Annona muricata. *Journal of Natural Medicine* 66:284–291.

Maurya D, Adhikari C, Kumar T, Bishwas AM. 2021. A review: Aganosma dichotoma (Roth) K. Schum.: Traditional use, phytochemical constituents and pharmacological properties. *International Journal of Botany Studies* 6(6):1276–1279.

McIntyre M. 1993. *The British Herbal Compendium: A Handbook of Scientific Information on Widely Used Plant Drugs*. Vol. 1, p. 239. British Herbal Medicine Association.

Meena AK, Meena J, Jadhav A, Padhi MM. 2014. A review on Hiptage benghalensis (Madhavilata) used as an Ayurvedic drug. *Asian Journal of Pharmacy and Techology* 4(1):28–31.

Melinda Chua M, Baldwin TC, Trevor J, Hocking TJ, Kelvin Chan K. 2010. Traditional uses and potential health benefits ofAmorphophallus konjac K. Koch ex N.E.Br. *Journal of Ethnopharmacology* 128(2):268–278.

Merlin NJ, Parthasarathya V, Manavalana R, Kumaravelb S. 2009. Chemical investigation of aerial parts of Gmelina asiatica Linn by GC-MS. *Pharmacognosy Research* 1(1):152–156.

Miah M, Shimu AS, Mahmud S, Omar FB, Khatun R, Sumon SC, Faisal Hoque, KM, Abu Reza M. 2020. Methanolic bark extract of Abroma augusta (L.) induces apoptosis in EAC Cells through altered expression of apoptosis regulatory genes. *Evidence Based Complimentary and Alternative Medicine*. https://doi.org/10.1155/2020/9145626

Miguel MG. 2018. Betalains in some species of the Amaranthaceae family: A review. *Antioxidants* 7:53.

Mirheydar H. 2001. *Herbal Information: Usage of Plants in Prevention and Treatment of Diseases*. Tehran, Iran: Islamic Culture Press Center.

Mishra M, Shukla YN, Jain SP, Kumar S. 1999. Chemistry and pharmacology of some Hibiscus sp. A review. *Journal of Medicinal & Aromatic Plant Science* 21(4):1169–1186.

Mishra S, Ahmad S, Kumar N, Sharma BK. 2013. Annona muricata (the cancer killer): A review. *Global Journal of Pharmaceutical Research* 2:1613–1618.

Misra RC. 2004. Therapeutic uses of some seeds among the tribals of Gandhamardan hill range, Orissa. *Indian Journal of Traditional Knowledge* 3:105–115.

Mistry N, Silori CS, Gupta L, Dixit AM. 2003. Indigenous knowledge on animal healthcare practices in district Kachchh, Gujarat. *Indian Journal of Traditional Knowledge* 2(3):240–254.

Modi A, Vimal Kumar V. 2014. Luffa echinata Roxb.-A review on its ethanomedicinal, phytochemical and pharmacological perspective. *Asian Pacific Journal of Tropical Disease* 4(1):7–12.

Modi K, Shah MB. 2017. A complete pharmacognostical profile of Rungia repens. *Pharmacognosy Journal* 9(2):123–127.

Moghadamtousi SZ, Fadaeinasab M, Nikzad S, Mohan G, Mohd Ali H, Abdul Kadir H. 2015. Annona muricata (Annonaceae): A review of its traditional uses, isolated acetogenins and biological activities. *International Journal of Molecular Sciences* 16(7):15625–15658.

Mohamed PA. 1974. Isolation of "Trichotetrol"—A new tetrahydroxy pentacyclic triterpene from Trichosanthes bracteata (Cucurbitaceae). Linn., Voight. Syn. T. palmata (Roxb.). *Current Science* 43:116.

Mondal A, Kabir G, Ghosh GP, Yasmin N, Alam AM, Khatun HA. 2006. Morphological variation of ten Ipomoea species of Bangladesh. *Pakistan Journal of Biological Science* 9:1714–1719.

Mondal S, Moktan S. 2020. A paradoxically significant medicinal plant Carapichea ipecacuanha: A review. *Indian Journal of Pharmaceutical Education and Research* 54(2) (Supp 1):56–66.

Monroy-Ortiz C, Castillo-Espana P. 2007. *Plantas Medicinales Utilizadas en el estado de Morelos*. 1st ed., Vol. 1. Universidad Autónoma del Estado de Morelos. Cuernavaca, Mexico: Agave Angustifolia, pp. 265–266.

Morton JF. 1992. The ocean-going noni, or Indian mulberry (Morinda citrifolia, Rubiaceae) and some of its "colorful" relatives. *Economic Botany* 46(3):241–256.

Mostafa I, El-Aziz EA, Hafez S, El-Shazly A. 2013. Chemical constituents and biological activities of Galinsoga parviflora cav. (Asteraceae) from Egypt. *Zeitschrift fur Naturforschung – Section C Journal of Bioscience* 68(7–8):285–292.

Mothan RA, Gruenert R, Bednarski PJ, Lindequist U. 2009. Evaluation of the in vitro anticancer, antimicrobial and antioxidant activities of some Yemeni plants used in folk medicine. *Pharmazie* 64:260–268.

Movaliya V, Zaveri M. 2014. A review on the pashanbheda plant "Aerva javanica". *International Journal of Pharmaceutical Sciences Review and Research* 25(2):268–275.

Mujumdar AM, Naik DG, Dandge CN, Puntambekar HM. 2000. Anti-inflammatory activity of Curcuma amada Roxb. in albino rats. *Indian Journal of Pharmacology* 32:375–377.

Mundaragi A, Thangadurai D. 2015. Proximate composition, nutritive value and antioxidant activity of a wild fruit, Flacourtia montana J. Graham. (Salicaceae). *Vegetos* 28(4):181–187.

Muthu C, Ayyanar M, Nagappan R & Ignacimuthu S (2006) Medicinal plants used by traditional healers in Kancheepuram District of Tamil Nadu, India. *Journal of Ethnobiology and Ethnomedicine* 2(43):1–10.

Nadkarni A. 1996. Dr. K. M. Nadkarni's Indian Materia medica: With ayurvedic, Unani-tibbi, siddha, allopathic, homeopathic, naturopathic & home remedies, appendices & indexes. 1. *Popular Prakashan* 1:751–752.

Nadkarni KM. 2010. *Indian Plants and Drugs*. New Delhi: Ajay Book Services, pp. 158–159.

Nadkarni KM, Nadkarni AK. 2009. *Indian Materia Medica*. Vol. I. Mumbai: Popular Prakashan, pp. 719–721, 943–944.

Naf R, Velluz A, Brauchli R, Thommen W. 1995. Agarwood oil (Aquilaria agallocha Roxb.). Its composition and eight new valencane, eremophilane, vetispirane derivatives. *Flavour and Fragrance Journal* 10(3):147–152.

Naf R, Velluz A, Busset N, Gaudin JM. 1992. New norsesquiterpenoids with10 epi eudesmane skeleton from agarwood (Aquilaria agallocha Roxb.). *Flavour and Fragrance Journal* 7(6):295–298.

Nag A, Galav P, Kateva SS. 2007. Indigenous animal health practices from the Udaipur district, Rajasthan. *Indian Journal of Traditional Knowledge* 6(4):583–588.

Nagarajan SL, Rao JM, Gurudatta KN. 2001. Chemical composition of the volatiles of Decalepis hamiltonii (Wight & Arn). *Flavour and Fragrance Journal* 16:27–29.

Nagendra Prasad K, Shivamurthy GR, Aradhya SM. 2008. Ipomoea aquatica, an underutilized green leafy vegetable: A review. *International Journal of Botany* 1(4):12.

Neelambika HS, Leelavathi S. 2014. In vitro comparative study of membrane stabilization capacity of different extracts of Blepharis maderaspatensis (L.) Heyne Ex Roth. and Blepharis molluginifolia Pers. grown in the region of Mysore, Karnataka. *IJPSR* 5(7):2698–2702.

Neuwinger H. 2000. *African Traditional Medicine*. Stuttgart: Medpharm Scientific Publication, pp. 156–169.

Newall CA, Anderson LA, Phillipson JD. 1996. *Herbal Medicines: A Guide for Health-care Professionals*. London: The Pharmaceutical Press.

Nie Q, Hu N, Gao H, Fan L, Chen H, Nie S. 2017. Polysaccharide from Plantago asiatica L. attenuates hyperglycemia, hyperlipidemia and affects colon microbiota in type 2 diabetic rats. *Food Hydrocolloids*. https://doi.org/10.1016/j.foodhyd.2017.12.026

Niranjan P, Shrikanth P, Ravikrishna S. 2018. A review and therapeutic potentialities of Picchilataru—Persea Macrantha (Nees) kosterm. A medicinal plant. *International Journal of Ayurveda and Pharmaceutical Chemistry* 9(2).

Nirubama K, Kanchana G, Rubalakshmi G. 2014. Bioactive compounds in Andrographis echioides (L.) Nees. Leaves by GC-MS analysis. *International Journal of Current Research in Biosciences and Plant Biology* 1(3):92–97.

Niwa T, Etoh H, Shimizu A, Shimizu Y. 2000. Cis-N-(p-Coumaroyl) serotonin from Konnyaku, Amorphophallus konjac K. Koch. *Bioscience Biotechnology and Biochemistry* 64(10):2269–2271. https://doi.org/10.1271/bbb.64.2269

Nwokocha CR, Owu DU, Kinlocke K, Murray J, Delgoda R, Thaxter K, McCalla G, Young L. 2012. Possible mechanism of action of the hypotensive effect of Peperomia pellucida and interactions between human cytochrome P450 enzymes. *Medicinal Aromatic Plants* 1(4):1–6.

Ogura M, Kolke K, Cordell, GI, Farnsworth, NR. 1978. Potential anticancer agents. VIII. Constituents of Baliospermum montanum (Euphorbiaceae). *Planta Medica* 33:128.

Okamura N, Fujimoto Y, Kuwabara S, Yagi A. 1994. High-performance liquid chromatographic determination of carnosic acid and carnosol in Rosmarinus officinalis and Salvia officinalis. *Journal of Chromatography A* 679:381–386.

Okamura N, Haraguchi H, Hashimoto K, Yagi A. 1994. Flavonoidsin Rosmarinus officinalis leaves. *Phytochemistry* 37:1463–1466.

Oliveira DM, Melo FG, Balogun SO, Flach A, de Souza ECA, de Souza GP, Rocha IdNA, da Costa LAMA, Soares IM, da Silva LI. 2015. Antibacterial mode of action of the hydroethanolic extract of Leonotis nepetifolia (L.) R. Br. involves bacterial membrane perturbations. *Journal of Ethnopharmacology* 172:356–363.

Oliver-Bever B. 1986. *Medicinal Plants in Tropical West Africa*. Cambridge: Cambridge University Press.

Padhan AR, Agrahari AK & Meher A. 2010. A study on antipyretic activity of Capparis zeylanica Linn. plant methanolic extract. *International Journal of Pharma Sciences and Research* 1(3):169–171.

Pahuja M, Mehla J, Reeta KH, Joshi S, Gupta YK. 2012. Root extract of Anacyclus pyrethrum ameliorates seizures, seizure-induced oxidative stress and cognitive impairment in experimental animals. *Epilepsy Research* 98(2–3):157–165.

Pal A, Pawar RS. 2011. A study on Ajuga bracteosa Wall ex. Benth for analgesic activity. *International Journal of Current Biological and Medical Sciences* 1(2):12–14.

Pan Y, Liu H, Zhuang Y, Ding L, Chen L, Qiu F. 2008. Studies on isolation and identification of flavonoids in herbs of Agrimonia Pilosa. *Zhongguo Zhongyao Zazhi* 33(24):2925–2928.

Panda H. 2004. Handbook on Ayurvedic Medicines with Formulae, Processes and Their Uses. New Delhi: National Institute of Industrial Research, p. 438.

Panda P, Das D, Dash P, Ghosh G. 2015. Therapeutic potential of Bauhinia Racemosa—a mini review. *International Journal of Pharmaceutical Sciences Review and Research* 32(2):169–179.

Pande PC, Tiwari L, Pande HC. 2007. Ethnoveterinary plants of Uttaranchal – a review. *Indian Journal of Traditional Knowledge* 6(3):444–458.

Pandey M, Rastogi S, Rawat A. 2007. Saussurea costus: Botanical, chemical and pharmacological review of an ayurvedic medicinal plant. *Journal of Ethnopharmacology* 110:379–390.

Pandey S, Rani Kushwaha G, Singh A, Singh A. 2018. Chemical composition and medicinal uses of anacyclus Pyrethrum. *Pharma Science Monitor* 9(1):551–560.

Pant N, Jain DC, Bhakuni RS. 2000. Phytochemicals from genus Swertia and their biological activities. *Indian Journal of Chemistry* 39B:565–586.

Paranjpe R, Gundala SR, Lakshminarayana N, Sagwal A, Asif G, Pandey A, Aneja R. 2013. Piper betel leaf extract: Anticancer benefits and bio-guided fractionation to identify active principles for prostate cancer management. *Carcinogenesis* 34(7):1558–1566.

Parker A, Goulart Peraza G, Sena J, Sinnott Silva E, Flores Soares MC, Cezar Vaz MR, Badiale Furlong L, Muccillo-Baisch AL. 2007. Antinociceptive effects of the aqueous extract of Brugmansia suaveolens flowers in mice. *Biological Research for Nursing* 8:234–239.

Partap S, Kumar A, Sharma NK, Jha KK. 2012. Luffa cylindrica: An important medicinal plant. *Journal of Natural Product and Plant Resources* 2(01):127–134.

Parthasarathy VA, Kandiannan K. 2007. *Spices and Condiments*. Calicut: Indian Institute of Spices Research.

Parveen Z, Mazhar S, Siddique S, Manzoor A, Ali Z. 2017. Chemical composition and antifungal activity of essential oil from Xanthium strumarium L. leaves. *Indian Journal of Pharmaceutical Sciences* 79(2):316–321.

Patel JJ, Acharya SR, Acharya NS. 2014. Clerodendrum serratum (L.) Moon.—a review on traditional uses, phytochemistry and pharmacological activities. *Journal of Ethnopharmaco* 154(2):268–285. https://doi.org/10.1016/j.jep.2014.03.071

Patil DA. 2008. *Herbal Cures: Traditional Approach*. Aavishkar Publishers.

Patil MV. 2006. *Ethnobotany of Nashik District*. Daya Publishing House, p. 54.

Patil PH, Surana SJ. 2010. Toxicological evaluations of Eranthemum roseum R. Br Linn Root extracts. *International Journal of PharmTech Research* 2(1):753–759.

Patil SM, Ramu R, Shirahatti PS, Shivamallu C, Amachawadi RG. 2021. A systematic review on ethnopharmacology, phytochemistry and pharmacological aspects of Thymus vulgaris linn. *Heliyon* 7(5):e07054. https://doi.org/10.1016/j.heliyon.2021.e07054

Patil UH, Gaikwad DK. 2011. Ethno-pharmacological review of a herbal drug: Anogeissus latifolia. *International Journal of Pharma Sciences and Research* 2(1):41–43.

Pawar HA, D'Mello PM. 2011. Cassia tora Linn.: An overview. *International Journal of Pharmaceutical Sciences and Research* 2(9):2286–2291.

Pawar S, Patil DA. 2008. *Ethnobotany of Jalgoan District*. Daya Publishing house.

Payyappallimana U. 2010. Role of traditional medicine in primary health care: An overview of perspectives and challenging. *Yokohama Journal of Social Sciences* 14(6):57–77.

Peng ZG, Chen HS, Guo ZM, Dong B, Tian GY, Wang GQ. 2008. Anti-HIV activities of Achyranthes bidentata polysaccharide sulfate in vitro and in vivo. *Yaoxue Xuebao* 43:702–706.

Pérez Maté, P. 2002. *Especies aromáticas y medicinales (in Spanish)*. Buenos Aires: INTA.

Peter KV. 2004. *Handbook of Herbs and Spices*. Vol. 2. Cambridge: Woodhead Publishing, pp. 83–85.

Phakeovilay C, Disadee W, Sahakitpichan P, Sitthimonchai S, Kittakoop P, Ruchirawat S, Kanchanapoom T. 2013. Phenylethanoid and flavones glycosides from Ruellia tuberosa L. *Journal of Natural Medicine* 67(1):228–233.

Phan TT, Wang L, See P, Grayer RJ, Chan SY, Lee ST. 2001. Phenolic compounds of Chromolaena odorata protect cultured skin cells from oxidative damage: Implication for cutaneous wound healing. *Biological and Pharmceutical Bulletin* 24:1373–1379.

Phondani PC. 2011. Worth of traditional herbal system of medicine for using ailments prevalent across the mountain region of Uttarakhand India. *Journal of Applied Pharmaceutical Science* 1:81–86.

Pirbalouti AG, Azizi S, Koohpayeh A, Hamedi B. 2010. Wound healing activity of Malva sylvestris and Punica granatum in alloxan-induced diabetic rats. *Acta Poloniae Pharmaceutica* 67(5):511–516.

Pirbalouti AG, Yousefi M, Heshmetollah N, Karimi I, Koohpayeh A. 2009. Evaluation of burn healing properties of Arnehia euchroma and Malva sylvestris. *Electronic Journal of Biology* 5:62–66.

Policegoudra RS, Abiraj K, Channe Gowda D, Aradhya SM. 2007a. Isolation and characterization of antioxidant and antibacterial compound from mango ginger (Curcuma amada Roxb.) rhizome. *Journal of Chromatography B* 854:40–48.

Policegoudra RS, Divakar S, Aradhya SM 2007b. Identification of difurocumenonol, a novel antimicrobial compound from mango ginger (Curcuma amada Roxb.) rhizome. *Journal of Applied Microbiology* 102:1596–1602.

Policegoudra RS, Rehna K, Rao LJ, Aradhya SM. 2010. Antimicrobial, antioxidant, cytotoxicity and platelet aggregation inhibitory activity of a novel molecule isolated and characterized from mango ginger (Curcuma amada Roxb.) rhizome. *Journal of Bioscience* 35:231–240. https://doi.org/10.1007/s12038-010-0027-1

Prajapati MS, Patel JB, Modi K, Shah MB. 2010. Leucas aspera: A review. *Pharmacognosy Reviews* 4(7):85–87. https://doi.org/10.4103/0973-7847.65330

Prajapati ND, Kumar U. 2003. *Agro's Dictionary of Medicinal Plants*. India: Agrobios, p. 22.

Prajapati ND, Purohit SS, Sharma AK, Kumar T. 2007. *A Hand Book of Medicinal Plants: A Complete Source Book*. Jodhpur: Agrobios (India), pp. 266–267.

Prakash A, Khosa RL. 1976. Chemical studies on Bauhiniaracemosa. *Current Science* 45:705–707.

Prashith Kekuda TR, Sudharshan SJ. 2018. Ethnobotanical uses, phytochemistry and biological activities of Clerodendrum paniculatum L. (Lamiaceae): A comprehensive review. *Journal of Drug Delivery & Therapeutics* 8:28–34.

Priya P, Anbarasu M, Christina VS, Abdul Majeed S, Sivamurugan V, Lakshmi Sundaram R. 2022. Isolation of potential compound from the leaves of Elytraria acaulis and evaluating its therapeutic properties using in vitro studies against ovarian cancer. *Applied Biochemistry and Biotechnology* 194(12):5607–5626. https://doi.org/10.1007/s12010-022-04048-7

Prusti AB, Behera KK. 2007. Ethno-medico botanical study of Sundargarh District, Orissa, India. *Ethnobotanical Leafets* 11:148–163.

Pullaiah T. 2006. *Encyclopedia of World Medicinal Plants*. Vol. 1. Regency, p.1217.

Punuri JB, Sharma P, Sibyala S, Tamuli R, Bora U. 2012. Piper betle-mediated green synthesis of biocompatible gold nanoparticles. *International Nano Letters* 2:18.

Pushpangadan P, Rajasekharan S, Ratheshkumar PK, Jawahar CR, Velayudhan Nair V, Lakshmi N & Sarada Amma L. 1988. Arogyappacha' (Trichopus zeylanicus Gaerin), the 'ginseng' of Kani tribes of Agasthyar Hills (Kerala) forever green health and vitality. *Ancient Science of Life* 8(1):13–16.

Qi Y, Chin KL, Malekian F, Berhane M, Gager J. 2005. *Biological Characteristics, Nutritional and Medicinal Value of Roselle, Hibiscus sabdariffa*. Southern University and A&M College System.

Quattrocchi U. 2012. *CRC World Dictionary of Medicinal and Poisonous Plants: Common Names, Scientific Names, Eponyms, Synonyms, and Etymology*. Boca Raton, FL: CRC Press.

Qureshi R, Bhatti GR. 2008. Ethnobotany of plants used by the Thari people of Nara Desert, Pakistan. *Fitoterapia* 79:468–473.

Qureshi R, Bhatti GR. 2009. Folklore uses of Amaranthaceae family from Nara desert, Pakistan. *Pakistan Journal of Botany* 41(4):1565–1572.

Ragasa CY, Galian RF, Arenal M, Tan V, Chien-Chang Shen CC. 2014. Triterpenes and sterols from Samanea saman. *Research Journal of Pharmaceutical, Biological and Chemical Sciences* 5(4):1501–1507.

Raghavan M, Ramjan M. 2018. Burmese grape (Baccaurea ramiflora Lour.): A promising fruit crop for future generations. *Journal of Medicinal Plants Studies* 6(3):50–52.

Raghunathan K, Hariharan V, Rangaswami S. 1974. Chrysophanol-1-β-gentiobioside, a new anthraquinone glycoside from Cassia tora Linn. *Indian Journal of Chemistry* 12:1251–1253.

Raghunathan KK, Mitra R. 2003. *Pharmacognosy of Indigenous Drugs*. New Delhi: Central Council for Research in Ayurveda and Siddha, pp. 1101–1112.

Rahman MM, Polfreman D, MacGeachan J, Gray AI. 2005. Antimicrobial activities of Barringtonia acutangula. *Phytotherapy Research* 19(6):543–545.

Rahman MS, Rahman MZ, Ahad Uddin ABM, Rashid MA. 2007. Steriod and Triterpenoid from Anogeissus latifolia. Dhaka Univ. *Journal of Pharmaceutical Sciences* 6(1):47–50.

Rahul J, Jain MK, Singh SP, Kamal RK, Anuradha, Naz A, Gupta AK, Mrityunjay SK. 2015. Adansonia digitata L. (baobab): A review of traditional information and taxonomic description. *Asian Pacific Journal of Tropical Biomedicine* 5(1):79–84.

Rai A, Mishra T. 2020. Ethnomedicinal and therapeutic values of flacourtia jangomas. *The Journal of Indian Botanical Society* 100(3–4):42–45.

Rai L, Sharma E. 1994. *Medicinal Plants of the Sikkim Himalaya: Status, Uses and Potential*, Govind Ballabh Pant Institute. Bishen Singh & Mahendra Pal Singh, pp. 25–27.

Rai LK, Prasad P, Sharma E. 2000. Conservation threats to some medicinal plants of the Sikkim Himalaya. *Biological Conservation* 93(1):27–33. https://doi.org/10.1016/S0006-3207(99)00116–0

Rai V, Agarwal M, Agnihotri AK, Khatoon S, Rawat AK, Mehrotra S. 2005. Pharmacognostical evaluation of Leucas aspera. *Natural Product Science* 11(2):109–114.

Raina R, Behera MC, Chand R, Sharma Y. 2003. Reproductive biology of Gentiana kurroo Royle. *Current Science* 85(5):667–670.

Raina R, Prawez S, Verma P, Pankaj N. 2008. Medicinal plants and their role in wound healing. *Vet Scan* 3:1–7.

Rane JB, Vahanwala SJ, Golatkar SG, Ambaye RY, Khadse BG. 2001. Chemical examination of the flowers of couroupita guianensis Aubl. *Indian Journal of Pharmaceutical Sciences* 63:72–73.

Ranganatham RM, Nagarajan S. 1980. 6-hydroxyluteolin 7–0-apioside from L. cristata. *Phytochemistry* 19(11):2505–2506.

Rani S, Khan SA, Ali M. 2010. Phytochemical investigation of the seeds of Althea officinalis L. *Natural Product Research* 24(14).

Rao AS, Rajanikanth B, Seshadri R. 1989. Volatile aroma components of Curcuma amada Roxb. *Journal of Agricultural and Food Chemistry* 37:740–743.

Rao BR. 2009. *Astanghrdayam*. 1st ed. Varanasi, India: Chaukhambha Visvabharati Oriental Publishers & Distributors.

Rao PS, Seth N, Jayavver KN. 2012. An important spice, Pimenta dioica (Linn.) Merill: A review. *International Current Pharmaceutical Journal* 1(8):221–225.

Rastogi M. 1970. *Compendium of Indian Medicinal Plants*. 2nd ed. New Delhi: CDRI.

Rastogi RP, Mehrotra BN. 1993. *Compendium of Indian Medicinal Plants*. Vol. I. New Delhi: Publication and Information Directorate, CSIR, p. 220.

Ratih GAM, Imawati MF, Purwanti DI, Nugroho RR, Wongso S, Prajogo B, Indrayanto G. 2019. Metabolite profiling of Justicia gendarussa herbal drug preparations. *Natural Product Communications* 14(6). https://doi.org/10.1177/1934578X19856252

Reddy K, Reddy V. 2009. Antihyperglycaemic activity of ethanol extract of Aerva javanica leaves in alloxan-induced diabetic mice. *Journal of Pharmacy Research* 7:1259–1261.

Reddy KK, Rajadurai S, Nayudamma Y. 1965. Studies on Dhava (Anogeissus latifolia) tannins: Part III—polyphenols of bark, sapwood and heartwood of Dhava. *Indian Journal of Chemistry* 3:308–310.

Reddy MB, Reddy KR, Reddy MN. 1991. Ethnobotany of Cuddapah district, Andhra Pradesh, India. *International Journal of Pharmacognosy* 29:273–280.

Reddy MC, Sri Rama Murthy K. 2013. A review on Decalepis hamiltonii Wight & Arn. *Journal of Medicinal Plant Research* 7(41):3014–3029.

Reis RB, Bragagnolo FS, Gianeti TMR, Rodrigues SA, Funari CS, Gonçalves GG, Ming LC. 2019. Brugmansia suaveolens leaf productivity and alkaloid contents under different doses of organic fertilizer. *The Journal of Agricultural Sciences* 11:341–349.

Ripanda A, Luanda A, Sule KS, Mtabazi GS, Makangara JJ. 2023. Galinsoga parviflora (Cav.): A comprehensive review on ethnomedicinal, phytochemical and pharmacological studies. *Heliyon* 9(2):e13517. https://doi.org/10.1016/j.heliyon.2023.e13517

Roberts MF, Michael W (eds.). 1998. *Alkaloids: Biochemistry, Ecology, and Medicinal Applications*. Springer Science & Business Media. ISBN: 0306454653.

Rodrigues MR, Krause LC, Caramao EB, dos Santos JG, Dariva C, Vladimir de Oliveira J. 2004. Chemical composition and extraction yield of the extract of Origanum vulgare obtained from sub- and supercritical CO2. *Journal of Agricultural and Food Chemistry* 52(10):3042–3047.

Rodrigues TLM, Silva MEP, Gurgel ESC, Oliveira MS, Lucas FCA. 2022. Eryngium foetidum L. (Apiaceae): A literature review of traditional uses, chemical composition, and pharmacological activities. *Evidence-based Complementary and Alternative Medicine* 14:2896895. https://doi.org/10.1155/2022/2896895

Safa O, Soltanipoor MA, Rastegar S, Kazemi M Dehkordi KN, Ghannadi A. 2013. An ethnobotanical survey on Hormozgan province, Iran. *Avicenna Journal of Phytomedicine* 3(1):64–81.

Sahoo HB, Patro SK, Sagar R, Santani DD. 2015. Mutagenic evaluation and spectroscopic characterization of flavonoidal fraction of Apium leptophyllum (Pers.) fruit. *International Journal of Nutrition, Pharmacology, Neurological Diseases* 5:82–88.

Sahoo S, Panda PK, Behera PS, Mishra SR, Ellaiah P. 2008. Antifungal activity of Barringtonia acutangula against selected human pathogenic fungi. *Indian Drugs* 45(1):26–30.

Saitoh T, Shibata S. 1969, April. Chemical studies on the oriental plant drugs. XXII. Some new constituents of licorice root. (2). Glycyrol, 5-O-methylglycyrol and isoglycyrol. *Chemical and Pharmaceutical Bulletin* 17(4):729–734. https://doi.org/10.1248/cpb.17.729. PMID: 5812405.

Salih Y, Harisha CR, Shukla VJ, Acharya R. 2013, January. Pharmacognostical evaluation of Launaea sarmentosa (Willd.) schultz-bip.ex Kuntze root. *Ayu* 34(1):90–94. https://doi.org/10.4103/0974-8520.115439. PMID: 24049412; PMCID: PMC3764888.

Salma Khanam UK, Oba S. 2013. Bioactive substances in leaves of two amaranth species, Amaranthus Tricilor and A. Hypochondriacus. *Canadian Journal of Plant Science* 93:47–58.

Samanta SK, Bhattacharya K, Mandal C, Pal BC. 2010. Identification and quantification of the active component quercetin 3-O-rutinoside from Barringtonia racemosa, targets mitochondrial apoptotic pathway in acute lymphoblastic leukemia. *Journal of Asian Natural Products Research* 12(8):639–648.

Sandeep K, Birendra S, Khajuria R, Suri K, Piush S. 2010. Antimicrobial activity of Crotalaria burhia Buch.-Ham. root. *Indian Journal of Natural Products Resources* 1:481–484.

Sanjita D, Rizvan Mohd, Basu SP, Saumya D. 2012. Therapeutic Potentials of Bambusa bambos Druce. *Indo Global Journal of Pharmaceutical Sciences* 2(1):85–87.

Sankara Subramanian S, Nair AGR. 1971. Angiospermae dicotyledonae anacardiaceae polyphenols of Lannea coromandelica. *Phytochemistry* 10:1939.

Saokaew S, Wilairat P, Raktanyakan P, Dilokthornsakul P, Dhippayom T, Kongkaew C, Sruamsiri R, Chuthaputti A, Chaiyakunapruk N. 2017. Clinical effects of Krachaidum (Kaempferia parviflora): A systematic review. *Journal of Evidence Based Complementary Alternative Medicine* 22(3):413–428. https://doi.org/10.1177/2156587216669628

Sarker SD, Nahar L. 2004. Natural medicine: The genus Angelica. *Current Medicinal Chemistry* 11(11):1479–1500. https://doi.org/10.2174/0929867043365189. PMID: 15180579.

Sarvananda L, Premarathna AD. 2018. Ethnopharmacolological potential and medicinal uses of Hygrophila auriculata. *Journal of Ayurvedic and Herbal Medicine* 4(4):185–188.

Sasi S, Anjum N, Tripathi YC. 2018. Ethnomedicinal, Phytochemical and pharmacological aspects of flacourtia jangomas: A review. *International Journal of Pharmacy and Pharmaceutical Sciences* 10(3):9–15.

Satapathy AK, Gunasekaran G, Sahoo SC, Kumar A, Rodriques PV. 2009. Corrosion inhibition by Justicia gendarussa plant extract in hydrochloric acid solution. *Corrosion Science* 51(12):2848–2856.

Sathasivampillai SV, Pholtan Rajamanoharan RS. 2021. Bioactivities of Desmostachya bipinnata (L.) Stapf. *Plant Biotechnology Persa* 3(1):18–25.

Sathish Babu P, Anand Babu K, Chitra K, Chamundeeshwari D. 2015. A review on phytochemical and pharmacological profile of elytraria acaulis landau. *International Journal of Pharmaceutical Sciences and Research* 6(9):3671–3675.

Sathyavathi R, Janardhanan K. 2014, January 10. Wild edible fruits used by Badagas of Nilgiri District, Western Ghats, Tamilnadu, India. *Journal of Medicinal Plants Research* 8(2):128–132.

Sathyavathi R, Janardhanan KJ. 2007. Folklore medicinal practices of badaga community in Nilgiri Biosphere Reserve, Tamilnadu, India. *IJPRD* 3:50–63.

Senthamizhselvi C. 2013. Traditional knowledge of medicinal plants. *Global Research of Analysis* 2(12):230–233.

Shah N. 1982. Herbal folk medicines in Northern India. *Journal of Ethnopharmacology* 6(3):293–301.

Shakya AK. 2016. Medicinal plants: Future source of new drugs. *International Journal of Herbal Medicine* 4:59–64.

Sharafzadeh S, Alizadeh O. 2001. German and Roman chamomile. *Journal of Applied Pharmaceutical Science* 1(10):1–5.

Sharanabasappa GK, Santosh MK, Shaila D, Seetharam YN, Sanjeevarao I. 2007. Phytochemical studies on Bauhinia racemosa Lam, Bauhinia purpurea Linn and Hardwickia binata Roxb. *E-Journal of Chemistry* 4(1):21–31.

Sharma HK, Chhangte L, Dolui AK. 2001. Traditional medicinal plants in Mizoram, India. *Fitoterapia* 72(2):146–161. https://doi.org/10.1016/S0367-326X(00)00278–1

Sharma NK, Priyanka, Jha KK, Singh HK, Srivastava AK. 2014. Hepatoprotective activity of luffa cylindrical (L.) M.J. Roem. leaf extract in paracetamol intoxicated rats. *Indian Journal of Natural Products and Resources* 5(2):143–148.

Sharma PC, Kalkal M. 2018. Nutraceutical and medicinal importance of seabuckthorn (Hippophae sp.). In Grumezescu AM, Holban AM (eds.) *Therapeutic, Probiotic, and Unconventional Foods*. Academic Press.

Sharma PK, Chauhan NS, Lal B. 2004. Observations on the traditional phytotherapy among the inhabitants of Parvati valley in western Himalaya, India. *Journal of Ethnopharmacology* 92(2–3):167–176.

Sharma PV. 1996. *Dravyaguna Vijnana*. 8th ed. Varanasi, India: Chaukhambha Bharati Academy.

Sharma PV. 2007. *Cakradatta*. 5th ed. Varanasi, India: Chaukhambha Publishers.

Sharma S, Kumar S. 2013. Phyllanthus reticulatus poir.—An important medicinal plant: A review of its phytochemistry, traditional uses and pharmacological properties. *International Journal of Pharmaceutical Sciences and Research* 4(7):2528–2534.

Shendge PN, Belemkar S. 2018, October 22. Therapeutic potential of Luffa acutangula: A review on its traditional uses, phytochemistry, pharmacology and toxicological aspects. *Frontiers in Pharmacology* 9:1177. https://doi.org/10.3389/fphar.2018.01177. PMID: 30459601; PMCID: PMC6232903.

Sheshadri Shekhar D, Ashok Kumar BS, Suresh N, Dhanasekaran S. 2007. Evaluation of antidiarrhoeal activity of Thunbergia fragrans Roxb. *Asian Journal of Chemistry* 19(4):2792–2796.

Shibuya H, Bohgaki T, Matsubara T, Watarai M, Ohashi K, Kitagawa I. 1999. Chemical structures of two new isopimarane-type diterpenes, Orthosiphonones A and B and a new benzochromene, orthochromene a from the leaves of Orthosiphon aristatus (Lamiaceae). *Chemical and Pharmaceutical Bulletin* 47:695–698.

Shilpa K, Harish Chandra R, Krishnaveni B. 2012. Anti-inflammatory activity of Crinum defixum Ker-Gawl. *International Journal of Pharmaceutical Sciences and Drug Research* 4(2):147–149.

Shirazi AM. 1947. Studies on Leucas aspera. *Indian Journal of Pharmacology* 9:116–117.

Shojaii A, Fard MA. 2012. Review of pharmacological properties and chemical constituents of Pimpinella anisum. *ISRN Pharmaceutics* 510795. https://doi.org/10.5402/2012/510795

Shukla P, Singh A, Gawri S, Alexande A, Sonwane S. 2011. In vitro propagation of Barleria prionitis Linn and its antibacterial activity. *International Journal of Pharma Professional's Research* 2:198–200.

Shukla S, Bhargava A, Chatterjee A, Srivastava J, Singh N, Singh SP. 2006. Mineral profile and variability in vegetable amaranth (Amaranthus tricolor). *Plant Foods for Human Nutrition* 61:21–26.

Silva MHL, Zoghbi, MDB, Andrade EHA, Maia JGS. 1999. The essential oils of Peperomia pellucida Kunth and P. circinnata Link var. circinnata. *Flavour Fragrance Journal* 14:312–314.

Singh A, Saharan VA, Veerma R, Bhandari A. 2013. Evaluation of antidiarrhoeal activity of E.acaulis extracts on magnesium sulphate- and castor oilinduced diarrhoea in wistar rats. *Malaysian Journal of Pharmaceutical Sciences* 11(2):31–39.

Singh AK, Chandra KK, Kumar R, Bhardwaj AK. 2022. Phytochemical analysis and medicinal properties of Soymida febrifuga (Roxb.) A. Juss: A review. *Advances in Bioresearch* 13(6):37–43.

Singh G, Kachroo P. 1976. *Forest Flora of Srinagar*. Bishen Singh Mahendra Pal Singh.

Singh MK, Gidwani B, Gupta A, Dhongade H, Kaur CD, Kashyap PP, Tripathi DK. 2015. A review of the medicinal plants of genus Orthosiphon (Lamiaceae). *International Journal of Biological Chemistry* 9(6):318–331.

Singh N, Khanna K, Mudgal V, Dixit R. 2001. *Flora of Madhya Pradesh*. Vol. 3. Calcutt: Botanical Survey of India, p. 587.

Singh R, Mewara DK. 2018. Isolation and characterization of sesquiterpene lactones from the genus amberboa. *Internation Research Journal of Chemistry* 23:1–9.

Singh RB, Saxena VK. 1976. Chemical investigations on Spondias mangifera. *Institution of Chemists* 48.

Singh RP, Singh RP. 1987. Flavanoids of the flowers of Ichnocarpus frutescens. *Journal of Indian Chemical Society* 64(11):715–716.

Singh S, Parma N, Patel B. 2015. A review on Shalparni (Desmodium gangeticum DC.) and Desmodium species (Desmodium triflorum DC. & Desmodium laxiflorum DC.)—Ethnomedicinal perspectives. *Journal of Medicinal Plants Studies* 3(4):38–43.

Sinha V, Malik T. 2015. Chemical constituents from the leaves of Callicarpapedunculata. *International Journal of Pharmaceutical and Chemical Sciences* 4(1):161–163.

Sirinthipaporn A, Jiraungkoorskul W. 2017. Wound healing property review of siam weed, Chromolaena odorata. *Pharmacognosy Review* 11(21):35–38. https://doi.org/10.4103/phrev.phrev_53_16

Soni V, Jha AK, Dwivedi J. 2015. Traditional uses, phytochemistry and pharmacology of Bauhinia racemosa Lam—a review. *Cell Med* 5(4):24.1–24.7. https://doi.org/10.5667/tang.2015.0022

Sowjanya P, Srinivasa BP, Lakshmi NM. 2015. Phytochemical analysis and antibacterial efficacy of Amaranthus tricolor (L) methanolic leaf extract against clinical isolates of urinary tract pathogens. *African Journal of Microbiology Research* 9:1381–1385.

Srivastava B, Sharma VC, Sharma H, Pant P, Jadhav AD. 2016. Comparative physicochemical, phytochemical and high performance thin layer chromatography evaluation of heart wood and small branches of Aquilaria agallocha Roxb. *International Journal of Ayurveda and Pharma Research* 4(1):1–6.

Stefan N, Ngugen T, Valcho Z. 1996. Flavonoids from Achyranthes bidentata BI. *International Symposium on Medicinal and Aromatic Plants, ISHS Acta Horticulturae* 426.

Stuart Rev GA. 1911. *Chinese Materia Medica: Vegetable Kingdom*. Shanghai: American Presbyterian Mission Press, p. 558.

Su BN, Pawlus AD, Jung HA, Keller WJ, McLaughlin JL, Kinghorn AD. 2005. Chemical constituents of the fruits of Morinda citrifolia (Noni) and their antioxidant activity. *Journal of Natural Products* 68(4):592–595. https://doi.org/10.1021/np0495985

Su TP, Wu XZ, Cone EJ, Shukla K, Gund TM, Dodge AL, Parish DW. 1991. Sigma compounds derived from phencyclidine: Identification of PRE-084, a new, selective sigma ligand. *Journal of Pharmacology and Experimental Therapeutics* 259:543–550.

Suh, KS, Lee YS, Choi EM. 2014. The protective effects of Achyranthes bidentata root extract on the antimycin A induced damage of osteoblastic MC3T3-E1 cells. *Cytotechnology* 66(6):925–935.

Suksamrarn A, Chotipong A, Suavansri T, Boongird S, TimsuksaiP, Vimuttipong S, Chuaynugul A. 2004. Antimycobacterial activity and cytotoxicity of flavonoids from the flowers of Chromolaenaodorata. *Archives of Pharmacal Research* 5(27):507–511.

Sun CM, Syu WJ, Don MJ, Lu JJ, Lee GH. 2003. Cytotoxic sesquiterpene lactones from the root of Saussurea lappa. *Journal of Natural Products* 66:1175–1180.

Sun HY, Long LJ, Jun Wu. 2006. Chemical constituents of mangrove plant Barringtonia racemosa. *Journal of Chinese Medicinal Materials* 29(7):671–672.

Surana SJ, Tatiya AU, Jain AS, Desai DG, Shastri KV, Katariya MV. 2008. Pharmacognostical and Physicochemical Standardization of Root of Eranthemum roseum. (Vahl) R.Br. *Phcognosy Magazine* 4(13):74–79.

Surywanshi V, Yadava RN. 2015. New potential allelochemicals from Galinsoga purviflora CAV. *Chem Science Review and Letters* 4(13):405–413.

Sutthanut K, Sripanidkulchai B, Yenjai C, Jay M. 2007. Simultaneous identification and quantitation of 11 flavonoid constituents in Kaempferia parviflora by gas chromatography. *Journal of Chromatography A* 1143:227–233.

Swamy MK, Balasubramanya S, Anuradha M. 2010. In vitro multiplication of Pogostemon cablin Benth through direct regeneration. *African Journal of Biotechnology* 9:2069–2075.

Swamy MK, Sinniah UR. 2015. A comprehensive review on the phytochemical constituents and pharmacological activities of Pogostemon cablin Benth.: An aromatic medicinal plant of industrial importance. *Molecules* 20(5):8521–8547.

Taid T, Rajkhowa T, Kalita J. 2014. A study on the medicinal plants used by the local traditional healers of Dhemaji district, Assam, India for curing reproductive health related disorders. *Advances in Applied Science Research* 5(1):296–301.

Tan SP, Tan ENY, Lim QY, Mohd AN. 2020. Phyllanthus acidus (L.) Skeels. A review of its traditional uses, phytochemistry, and pharmacological properties. *Journal of Ethnopharmacology* 253. https://doi.org/10.1016/j.jep.2020.112610

Tandon S, Rastogi RP. 1976. Studies on the chemical constituents of Spondias pinnata. *Planta Medica* 29(2):190–192. https://doi.org/10.1055/s-0028-1097651

Teklehaymanot T, Giday M. 2007. Ethnobotanical study of medicinal plants used by people in Zegie Peninsula, Northwestern Ethiopia. *Journal of Ethnobiology and Ethnomedicine* 3:12. https://doi.org/10.1186/1746-4269-3-12

The Ayurvedic Pharmacopeia of India. 2004. Part I, Vol. 4. New Delhi: Department of AYUSH, Ministry of Health and Family Welfare, Government of India, pp. 4–5.

Tiwari VJ. 2019. Ethnopharmacology of Leonotis nepetifolia (L.) R. Br., Lamiaceae, used to cure jaundice and liver disorders by Baiga Tribe of Mandla District of Madhya Pradesh. *Research Journal of Pharmacognosy and Phytochemistry* 11(1):1–7.

Tiwary JK, Ballabha R, Tiwari P. 2010. Ethnopaediatrics in Garhwal Himalaya, Uttarakhand, India (Psychomedicine and medicine). *New York Science Journal* 3:123–126.

Torki A, Pirbalouty VK, Lorigooini Z, Kopae MR, Sadeghimanesh A, Rabiei Z. 2018. Anchusa italica extract: Phytochemical and neuroprotective evaluation on global cerebral ischemia and reperfusion. *Brazilian Journal of Pharmaceutical Sciences* 54(1):1–9.

Tripathi HP. 2005. Harita Samhita. 1st ed. Varanasi, India: Chaukhambha Krishnadas Academy.

Tun NL, Hu DB, Xia MY, et al. 2019. Chemical constituents from ethanoic extracts of the aerial parts of Leea aequata L., a traditional folk medicine of Myanmar. *Natural Products and Bioprospecting* 9:243–249. https://doi.org/10.1007/s13659-019-0209-y

Turgumbayeva A, Zhakipbekov K, Shimirova Z, Akhelova S, Amirkhanova A, Koilybayeva M, Seitimova G, Abdambayev D. 2022. Study of phytochemical compounds of Plantago major leaves grown in Kazakhstan. *Pharmacia* 69(4):1019–1026. https://doi.org/10.3897/pharmacia.69.e96526

Uphof JC. 1959. Dictionary of Economic Plants, 2nd edn. Lehre. Hafner Publishing Co., New York, p. 400.

Usha K, Kasturi GM, Hemalatha P. 2007. Hepatoprotective effect of Hygrophila spinosa and Cassia occidentalis on carbon tetrachloride induced liver damage in experimental rats. *Indian Journal of Clinical Biochemistry* 22:132–135.

Usman LA, Ismaeel RO. 2020. Chemical composition of root essential oil of Peperomia pellucida (L.) Kunth. Grown in Nigeria. *Journal of Essential Oil Bearing Plants* 23:628–632.

Usmani A, Khushtar M, Arif M, Siddiqui MA, Sing SP, Mujahid M. 2016. Pharmacognostic and phytopharmacology study of Anacyclus pyrethrum: An insight. *Journal of Applied Pharmaceutical Science* 6(3):144–150.

Vaisakh MN, Pandey A. 2012. The invasive weed with healing properties: A review on Chromolaena odorata. *International Journal of Pharmaceutical Sciences and Research* 3(1):80–83.

Varma SK, Sriwastawa DK, Pandey AK. 1999. *Ethnobotany of Santhal Pargana.* New Delhi: Narendra Publishing House, p. 31.

Vedavathy S. 2004. Decalepis hamiltonii Wight. An endangered source of indigenous health drink. *Natural Product Radiance* 3(1):22–23.

Veenstra JP, Johnson JJ. 2019. Oregano (Origanum vulgare) extract for food preservation and improvement in gastrointestinal health. *International Journal of Nutrition* 3(4):43–52.

Venkata Ratan K, Venkata Raju RR. 2008. Folk remedies for insect bite from Gundlabrahmeswaram wild life sanctuary, Andra Pradesh. *Indian Journal of Traditional Knowledge* 7(3):436–437.

Veshkorova O, Golubenko Z, Pshenichnov E, Avzanov I, Uzbekov V, Sultanova E, Salikhov S, Williams HJ, Reibenspies JH, Puckhaber LS, Stipanovic RD. 2010. Malvone A, a phytoalexin found in Malva sylvestris (Family Malvaceae). *Phytochemistry* 67:2376–2379.

Vijaya Bharathi R, Jerad Suresh A, Thirumal M, Sriram L, Geetha Lakshmi S, Kumudhaveni B. 2010. Antibacterial and antifungal screening on various leaf extracts of Barringtonia acutangula. *International Journal of Research in Pharmaceutical Sciences* 1(4):407–410.

Vinayaka KS, Hegde SV, Banakar S, Prashith Kekuda TR. 2009. Antifungal potency of Gnidia glauca (Fresen) Gilg. and Pothos scandens L. *Natural Products* 5(3):146–148.

Vinod VTP, Sashidhar RB, Suresh KI, Rao BR, Saradhi UV, Rao TP. 2008. Morphological, physico-chemical and structural characterization of gum kondagogu (Cochlospermum gossypium): A tree gum from India. *Food Hydrocolloids* 22(5):899–915.

Vital PG, Windell LR. 2009. Antimicrobial activity and cytotoxicityof Chromolaena odorata (L. f.) King and Robinson and Uncariaperrottetii (A. Rich) Merr. Extracts. *Journal of Medicinal Plants Research* 3(7):511–518.

Wafo P, Kamdem RS, Ali Z, Anjum S, Begum A, Oluyemisi OO, Khanc SN, Ngadjuib BT, Etoae XF, Choudhary MI. 2011. Kaurane-type diterpenoids from Chromoleana odorata their X-ray diffraction studies and potent a-glucosidase inhibition of 16-kauren-19-oic acid. *Fitoterapia* 82:642–646.

Wang L, Ye X, Hua Y, Song Y. 2018. Berberine alleviates adipose tissue fibrosis by inducing AMP-activated kinase signaling in high-fat diet-induced obese mice. *Biomedicine and Pharmacotherapy* 105:121–129. https://doi.org/10.1016/j.biopha.2018.05.110

Wang S, Yu Z, Wang C, et al. 2018. Chemical constituents and pharmacological activity of Agarwood and Aquilaria plants. *Molecules* 23(2):342.

Wang YX, Wang E, Jing S, Hao W. 2012. The chemical composition of caffeoylquinic acid in Vernonia anthelmintica. *China Journal of Chinesemateriamedica* 11:1590–1592.

Warrier PK, Nambiar VP, Ramankutty C (eds.). 1996. *Indian Medicinal Plants. A Compendium of 500 Species.* Vol. 5. Hyderabad: Orient Longman, pp. 263–435.

Warrier PK, Nambiar VPK, Ramankutty C. 1994. *Indian Medicinal Plants—A Compendium of 500 Species.* Vol. 1. Chennai: Orient Longman Pvt. Ltd, p. 106.

Wiart C. 2006a. *Medicinal Plants of Asia and Pacific.* Boca Raton: CRC Press/Taylor and Francis.

Wiart C. 2006b. *Medicinal Plants of the Asia-Pacific: Drugs for the Future?* World Scientific Publishing Co. Pte. Ltd, p. 288.

Wong KC, Tie DY. 1995. Volatile constituents of Couroupita guianensis Aubl. flowers. *Journal of Essential Oil Research* 7:225–227.

Wu TS, Hsu HC, Wu PL, Leu YL, Chan YY, Chern CY, Yeh MY, Tien HJ. 1998. Naphthoquinone esters from the root of Rhinacanthus nasutus. *Chemical and Pharmaceutical Bulletin* 46:413–418. https://doi.org/10.1248/cpb.46.413

Wu TS, Hsu HC, Wu PL, Teng CM, Wu YC. 1998. Rhinacanthin-Q, a naphthoquinone from Rhinacanthus nasutus and its biological activity. *Phytochemistry* 49:2001–2003.

Xue SX, Chen XM, Lu JX. 2009. Protective effect of sulfated Achyranthes bidentata polysaccharides on streptozotocin-induced oxidative stress in rats. *Carbohydrate Polymers* 75:415–419.

Yadav S, Bhadoria BK. 2005. Two dimeric flavonoids from Bauhinia purpurea. *Indian Journal of Chemistry* 44:2604–2607.

Yadong, Q., K. Chin, F. Malekian, M. Behane, and J. Gager. 2005. Biological characteristics, nutritional and medicinal value of Roselle, Hibiscus sabdariffa. *Urban Forestry Natural Resources and Environment* (circular), 604, 2pp.

Yamauchi T, Abe F, Wan ASC. 1987 Studies on Cerbera. IV.1 Polar cardenolide glycosides from the leaves of Cerbera odollam and Cerbera manghas. *Chemical and Pharmaceutical Bulletin* 35:4813–4818.

Yoganarasimhan SN. 1996. *Medicinal Plants of India.* Vol. 1. Bangalore: Interline Publishing Pvt. Ltd., p. 1.

Ysrael MC. 2003. Tonkin herbal drug: A multidisciplinary approach to development. *Clinical Hemorheology and Microcirculation* 29(3–4):247–251. PMID: 14724348.

Yusuf M, Begum J, Hoque MN, Chowdhury JU. 2009. *Medicinal Plants of Bangladesh.* Chittagong: BCSIR, p. 749.

Zaidi SMA, Pathan SA, Singh S, Jamil S, Ahmad FJ, Khar RK. 2013. Anticonvulsant, anxiolytic and neurotoxicity profile of Aqarqarha (Anacyclus pyrethrum) DC (Compositae) root ethanolic extract. *Pharmacology and Pharmacy* 4:535–541. https://doi.org/10.4236/pp.2013.47077

Zakynthinos G, Varzakas T. 2015. Hippophae rhamnoides: Safety and nutrition. *Current Research in Nutrition and Food Science Journal* 3(2). https://doi.org/10.12944/CRNFSJ.3.2.01

Zargar BA, Masoodi MH, Ahmed B, Ganie SA. 2011. Phytoconstituents and therapeutic uses of Rheum emodi wall. ex Meissn. *Food Chemistry* 128(3):585–589.

Zhang HJ, Rumschlag-Booms E, Guan YF, Wang DY, Liu KL, Li WF, Nguyen VH, Cuong NM, Soejarto DD, Fong HHS, Rong L. 2017. Potent inhibitor of drug-resistant HIV-1 strains identified from the medicinal plant Justicia gendarussa. *Journal of Natural Products* 80(6):1798–1807.

Zhang L, Lokeshwar BL. 2012. Medicinal properties of the Jamaican pepper plant Pimenta dioica and Allspice. *Current Drug Targets* 13(14):1900–1906.

Zhang X, Pei Y, Liu M, Kang S, Zhang J. 2010. Chemical constituents from the leaves of Cerbera manghas. *Asian Pacific Journal of Tropical Medicine* 3:109–111.

Zheng Y, Liu B, Chen M, Chen T. 2008. Supercritical fluid extraction of ecdysterone from the roots of Achyranthes bidentata BL. *Journal of Separation Science* 31:1393–1398.

Zhou NJ, Geng CA, Huang XY, Ma YB, Zhang XM, Wang JL, Chen JJ. 2015. Anti-hepatitis B virus active constituents from Swertia chirayita. *Fitoterapia* 100:27–34. https://doi.org/10.1016/j.fitote.2014.11.011

Zich FA, Hyland BPM, Whiffen, TR, Kerrigan A. 2020. Callicarpa pedunculata. In *Australian Tropical Rainforest Plants Edition 8 (RFK8).* Australia: Centre for Australian National Biodiversity Research (CANBR).

WEB SOURCES

https://www.worldfloraonline.org/taxon/wfo-0000165003#:~:text=
http://www.flowersofindia.net/catalog/slides/Mexican%20Tea.html
https://en.wikipedia.org/wiki/Cissampelos_pareira
https://www.planetayurveda.com/library/patha-cissampelos-pareira
https://www.easyayurveda.com/2017/06/01/cissempelus-pareira-remedies/
https://www.planetayurveda.com/library/patha-cissampelos-pareira
https://www.efloraofgandhinagar.in/shrub/clerodendrum-phlomidis
http://www.flowers of India.net/catolog/slide/sage%20Glory%20Bower.html
https://www.efloraofgandhinagar.in/shrub/clerodendrum-phlomidis
http://medplants.blogspot.com/search/label/Cocculus%20hirsutus
https://www.gbif.org/species/2928390
https://pfaf.org/User/Plant.aspx?LatinName=Crescentia+cujete#:~:text
https://indiabiodiversity.org/species/show/229336
https://www.planetayurveda.com/library/sudarshana-crinum-latifolium/
https://en.wikipedia.org/wiki/Desmostachya_bipinnata
https://indiabiodiversity.org/species/show/253016
https://www.easyayurveda.com/2017/10/05/kusha-desmostachya-bipinnata/
https://en.wikipedia.org/wiki/Desmostachya_bipinnata
https://www.easyayurveda.com/2017/10/05/kusha-desmostachya-bipinnata/
https://www.indianetzone.com/38/jingini_plant.htm
http://en.wikipedia.org/wiki/Plasmodium_falciparum
http://www.worldagroforestrycentre.org/sea/products/afdbases/af/asp/SpeciesInfo.asp?SpID=690
https://indiabiodiversity.org/species/show/229620
https://en.wikipedia.org/wiki/Elephantopus_scaber#:~:text
https://ayurwiki.org/Ayurwiki/Elephantopus_scaber;
https://www.flowersofindia.net/catalog/slides/Elephant%20Foot.html#
https://indiabiodiversity.org/species/sh
https://ayurwiki.org/Ayurwiki/Flacourtia_montana
https://en.wikipedia.org/wiki/Flacourtia_montana
https://ayurwiki.org/Ayurwiki/Flacourtia_montana
en.wikipedia.org/wiki/Galium_aparine
https://indiabiodiversity.org/species/show/31369
https://www.webmd.com/vitamins/ai/ingredientmono-1488/gardenia
http://proseanet.org/
http://www.stuartxchange.org/CompleteList.html
https://identify.plantnet.org/ml/the-plant-list/species/Gnidia%20glauca%20(Fresen.)%20Gilg/data
https://www.1mg.com/otc/sbl-helleborus-niger-dilution-200-ch-otc347855?wpsrc=Google+Organic+Search
https://indiabiodiversity.org/species/show/229952
https://www.flowersofindia.net/catalog/slides/Madhavi%20Lata.html
https://indiabiodiversity.org/species/show/32219#diagnostic
https://en.wikipedia.org/wiki/Humulus_lupulus#:
https://efloraindia.bsi.gov.in/eFlora/taxonList.action?id
https://pfaf.org/user/Plant.aspx?LatinName=Hydnocarpus+kurzii
https://en.wikipedia.org/wiki/*Hyssopus officinalis*
https://www.ncbi.nlm.nih.gov/books/NBK548391/#:~:text=Hyssop
https://en.wikipedia.org/wiki/Iberis_amara
https://www.bimbima.com/ayurveda/medicinal-herb-vidarialligator-yam-ipomoea-digitata/363/#
https://plants.jstor.org/stable/10.5555/al.ap.upwta.1_1103
http://tropical.theferns.info/viewtropical.php?id=Jasminum+grandiflorum
http://www.flowersofindia.net/catalo
https://www.indianetzone.com/38/parpatakah.htm
https://www.sciencedirect.com/topics/agricultural-and-biological-sciences/kigelia-africana
https://www.zambiaflora.com/speciesdata/species.php?species_id=152450
https://www.gbif.org/species/7403263
https://en.wikipedia.org/wiki/Lawsonia_inermis
https://indiabiodiversity.org/species/show/263498

https://en.wikipedia.org/wiki/Leonotis nepetifolia
https://en.wikipedia.org/wiki/Leonurus_cardiaca
https://www.efloraofgandhinagar.in/herb/phyla-nodiflora
https://pharmaveda.com/kb/Devdangri.html#:~:text=Luffa%20echinata
https://indiabiodiversity.org/species/show/31616
https://www.rxlist.com/supplements/marjoram.htm
https://www.easyayurveda.com/2017/02/01/gojihva-onosma-bracteatum/
https://herbfinder.himalayawellness.in/onosma-bracteatum.htm; https://www.easyayurveda.com/2017/02/01/gojihva-onosma-bracteatum/
https://www.rxlist.com/supplements/marjoram.htm
https://indiabiodiversity.org/species/show/230522
https://indiabiodiversity.org/species/show/266545
https://en.wikipedia.org/wiki/Anise)
https://en.wikipedia.org/wiki/Plantago_asiatica#cite_note-yoshitaka-8
https://en.wikipedia.org/wiki/Premna_serratifolia
https://wcc.edu.in/premna-serratifolia/
http://loriorganics.com/benefits-of-rhinacanthus-nasutus-l-kurz/
https://en.wikipedia.org/wiki/Rubia_cordifolia; https://www.indianmedicinalplants.info/herbs/index.php/home/983-rubia-cordifolia; https://www.indianmedicinalplants.info/herbs/index.php/sanskrit-names-of-medicinal-plants/44-2012-02-24-07-34-36/429-rubia-cordifolia;
https://pharmeasy.in/blog/ayurveda-uses-benefits-side-effects-of-manjistha/https://en.wikipedia.org/wiki/Rubia_cordifolia; https://www.indianmedicinalplants.info/herbs/index.php/home/983-rubia-cordifolia
https://www.indianmedicinalplants.info/herbs/index.php/sanskrit-names-of-medicinal-plants/44-2012-02-24-07-34-36/429-rubia-cordifolia
https://pharmeasy.in/blog/ayurveda-uses-benefits-side-effects-of-manjistha/
https://en.wikipedia.org/wiki/Ruellia_tuberosa#cite_note-8
https://www.easyayurveda.com/2017/08/13/ushava-smilax-zeylanica/
https://medplants.blogspot.com/2015/04/thunbergia-fragrans-indrapushpam-chimin.html
https://medplants.blogspot.com/2015/04/thunbergia-fragrans-indrapushpam-chimin.html
Pushpangadan et al; 1988; https://en.wikipedia.org/wiki/Trichopus zeylanicus
https://www.nparks.gov.sg/florafaunaweb/flora/2/5/2537
http://proseanet.org/
https://indiabiodiversity.org/species/show/262963
https://en.wikipedia.org/wiki/Wikstroemia_indica
https://wssa.net/wp-content/uploads/Wikstroemia-indica.pdf
https://apps.worldagroforestry.org/treedb/AFTPDFS/Acacia_mellifera.PDF
https://tropical.theferns.info/viewtropical.php?id=Senegalia+mellifera&redir=Acacia+mellifera
https://indiabiodiversity.org/species/show/228556#habitat-and-distribution
https://indiabiodiversity.org/species/show/228556#habitat-and-distribution.
http://allpedia.dkart.in/herbal-plants/3418-acalypha-ciliata-forsk.html
https://tropical.theferns.info/viewtropical.php?id=Acalypha+ciliata
https://www.flowersofindia.net/catalog/slides/Ox%20Knee.html
https://www.google.com/search?q=Actaea+spicata+description&oq=Actaea+spicata+description+
https://indiabiodiversity.org/species/show/6924
https://indiabiodiversity.org/species/show/225386
https://indiabiodiversity.org/species/show/32885
https://indiabiodiversity.org/observation/show/272589
https://upload.wikimedia.org/wikipedia/commons/9/93/Aganosma_dichotoma_%281094439019%29.jpg
https://en.wikipedia.org/wiki/Aglaia_roxburghiana#/media/File:Aglaia_roxburghiana_04.JPG
https://ayurwiki.org/Ayurwiki/Agrimonia_eupatoria_-_Agrimony#/media/File:Odermennig.jpg
https://thenaturopathicherbalist.com/herbs/a-2/agropyron-repens/